PLASTICS & PLASTICS TECHNOLOGY

R.S. PARMAR
B.E. (Mech.), M.E. Hons. (Prod. Engg.), Ph.D. (Welding)
FIIW, FIE (India), MISME, MISTE
Ex Professor Deptt. of Mech. Engineering, IIT Delhi

KHANNA BOOK PUBLISHING CO. (P) LTD.
Publisher of Engineering and Computer Books
4C/4344, Ansari Road, Darya Ganj, New Delhi-110002
Phone : 011-23244447-48 **Mobile:** +91-9910909320
E-mail : contact@khannabooks.com
Website : www.khannabooks.com

Price: ₹ 325.00

PLASTICS AND PLASTICS TECHNOLOGY

R.S. PARMAR

ISBN: 978-93-86173-74-4

Edition: 2020

Published by:

Khanna Book Publishing Co. (P) Ltd.
4C/4344, Ansari Road, Darya Ganj, New Delhi-110 002
Phone: 011-23244447-48 **Mobile:** +91-9910909320
E-mail: contact@khannabooks.com

Printed in India by:
Vaibhav Products

DEDICATED
TO
DEAR MUKUL SETH
M.D., K.B.P.

for his persistent though gentle pestering over 3 years
to get this book squeezed out of me.

PREFACE

Plastics in the form of synthetic plastic were invented by Leo Baekeland in the from of bakelite in New York (USA) in the year 1907. They have increased in general use to such an extent that they have become a major world problem because of their non-degradable properties. Most of the underdeveloped and developing countries are full of plastics as waste making life miserable for general public. But plastics which are used in manufacturing Engineering Products are serving the world in many useful ways and have become indispensible. In fact one cannot imagine the world without plastics. They are used in every sphere of life, right from clothing, personal effects (Bags, shoes, tooth brush, tooth paste tubes, etc.) to packaging and plumbing, automobiles; boats, yachts and ships; aeroplanes, satellites, engineering components, electrical insulation, computers, calculators, T.V., cameras, fire arms, filters, surgical equipment, eye lens, body implants and what not. But carry bags which have become indispensible in everyday life have become a real source of concern by creating millions of tons of waste annually are harming very seriously animals on the land as well as in Sea.

In spite of all their harmful properties they are going to stay and serve the humanity in almost every field of life. Keeping that in mind this book has been written to give sufficient knowledge to the readers right from basics of plastics to their properties and processing for producing engineering and general purpose goods. A case study on Automobile Body has been included to make the reader know the extent of plastic parts used in the common cars. Identification and recycling of plastics has been included at the end to help identify plastics for their optimal selection and applications and to determine their reuse and recyclability as per International Norms.

Hope this book will come up to the expectations of students, working engineers and industrialists. It is expected that it will fulfill the void due to the absence of books on the subject in our country.

R.S. Parmar

New Delhi

15.10.2019

CONTENTS

1 POLYMERS, PLASTICS AND ELASTOMERS (RUBBERS)

Natural polymeric materials have always been around but synthetic or man-made polymers (Plastics, rubbers, and composites) have had significant impact on our lives only since about the Second World War. The subsequent growth has, however, been very rapid and synthetic polymers are now widely used for common household use and as an engineering material.

The term *polymer* is derived by shortening the term *polymonomer* (*poly* = many, *monomer* = basic building block or molecule of a particular substance). Thus, a polymer is a material which is composed of a very large number of comparatively simple recurring unit or molecule or chemical compound called monomer.

Well known examples of polymers that occur in nature are the polysacchrides, e.g., *Starch* and *Cellulose,* which are made up of very large number of glucose molecules. Another well-known polymer is natural rubber, the repeating unit or monomer in which is *isoprene,*

$$CH_2 = \underset{\displaystyle CH_3}{\underset{|}{C}} - CH = CH_2$$

Monomers are, thus, capable of being converted into different types of polymers viz; plastics, synthetic resins, or elastomers by combining them with themselves or other simple substances.

Plastics are a wide group of solid composite materials which are largely organic, usually produced synthetically or by modifying polymers of natural origin, and *possess appreciable mechanical strength.*

Plastics, in the finished product form, seldom comprise exclusively of polymers, but also include other ingredients such as fillers, reinforcements, pigments, stabilizers, and processing aids like lubricants. However, designation of the plastic material or moulding compound is always taken from the polymer designation.

Resin is a term which is used synonymously with the terms plastic and polymer e.g., synthetically produced polymer or plastic may be referred to as synthetic resin. This is more often used in the case of thermosetting plastics.

Elastomer is the term used to designate both natural rubber and synthetic rubber-like material that can be stretched under load, at room temperature, to at least twice its length and snaps back to its original length, if the load is removed.

Rubber is defined as a natural or synthetic material (Compound of base elastomer in which other ingredients have been added to modify its properties) that can be or is already vulcanized to a state which it has high extensibility and forcible quick retraction.

Vulcanization is the process of heating rubber with sulphur to remove not only its tackiness (stickiness) but also to change it from plastic condition to an elastic state that is even superior to that of the original product.

From the above defined polymeric products it may be said that plastics and rubbers (elastomers) are the final engineering materials used very widely in the industry for making such items as friction bearings, gears, worms, brake components, vehicle bodies; automobile, aircraft and missile components; key elements of pumps and turbines; parts for conveyors, jigs and fixtures, cases for electrical instruments and electronic devices; refrigerator liners, insulating materials, protective coatings; piping and tubing, and structural members.

Plastic are used not only in the manufacture of engineering goods, but also as spare parts for the human body – bones, joints, aortas, and other major blood vessels. Plastics are increasingly being used for making household goods like dinner sets, clothing, artificial furs, footwear and fabrics. Also, in buildings the plastics are extensively used as tiles, flooring, plumbing, as well as furniture.

Rubber is widely used for automotive tyres and tubes, shock absorbing components, etc. The list of items made of plastics and rubbers can be very long.

Although plastics and rubbers or, in general, the polymers can be made from oil, gas, coal, farm products, and other generally available materials but the major feedstock is oil and a new industry called *petrochemicals* has developed to provide the monomers for manufacturing plastics, and the wide range of additives to modify their behaviour as per requirement in service.

1.1 Monomers and Polymerisation

The starting element for the formation of a polymer is a monomer. A monomer (M) is a small, low molecular weight molecule with an inherent capability of forming chemical bonds (B) with the similar or other monomers in such a manner that long chains called *polymeric chains* or *macromolecules* [P] are formed, as shown in Fig. 1.1. A typical polymeric chain often involves hundreds of monomeric molecules. The process of forming polymeric chains is called *polymerisation* and material consisting of polymeric chains is known as a *polymer*. Although chemical formation of polymeric chains is all the time carried out by nature to produce biological macromolecules which are essential parts of living cells but the polymers produced for industrial and general use are man-made and are referred to as *synthetic polymers*.

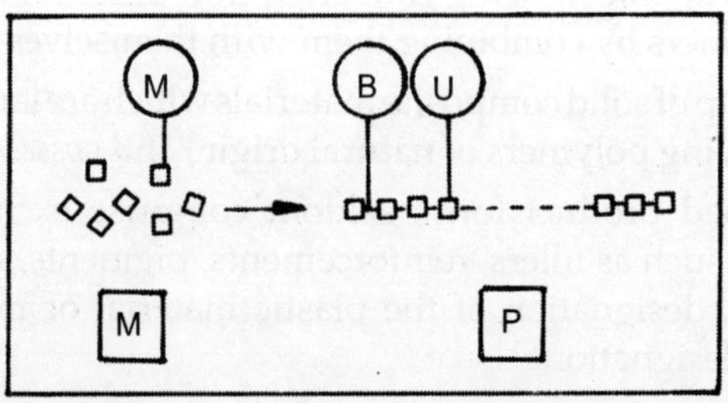

Fig 1.1: Polymerization mechanism; [M] Monomers; [P] Polymeric chain; (M) Monomers; (B) Backbone; (U) Repeat unit.

The commercial manufacture of polymers involves two steps, viz.,

(i) the production of the monomer from existing raw materials by chemical or other processes, and

(ii) the polymerisation of the so produced monomers.

Polymerisation processes are in many respects distinct from regular chemical processes.

In the simplest and most general case, individual macromolecules or polymeric chain in a polymer, consists of a single characteristic group of atoms, closely associated with the employed monomers, which is regularly repeated all along the chain or back bone; such a polymer is referred to as *homopolymer*, as shown in Fig 1.2.

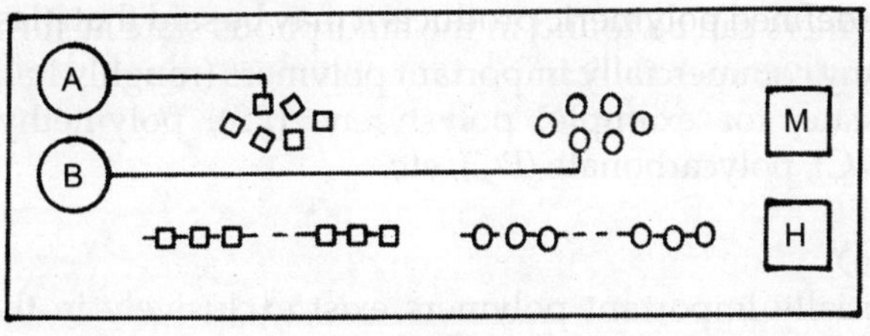

Fig 1.2: Monomers and their Homopolymeric chains;
(A) Monomers A; (B) Monomers B; [M] Monomers; [H] Homopolymeric chains.

The characteristic group is referred to here as the *repeat unit* (U) which is also called *momomeric unit* or *structural unit* or *building block*. The repeat unit is the most essential characteristic of a homopolymer, and its features are primarily responsible for the physical and chemical properties of the corresponding polymer.

1.2 Amorphous and Crystalline States

In terms of the shape of the polymeric chains, the polymers can exist in two states viz., amorphous and crystalline.

1.2.1 Amorphous State

The term amorphous generally indicates the absence of shape that implies the absence of the characteristic regular arrangement of polymeric chain.

At high temperature or in the presence of small solvent molecules, the chemical bonds between the building blocks of the polymeric chain allow significant motion, primarily rotational, that can be quite free except at the kinks. The result of this relative freedom is the characteristic random-like conformation (U), shown in Fig 1.3 and is sometimes referred to as *random coil conformation*.

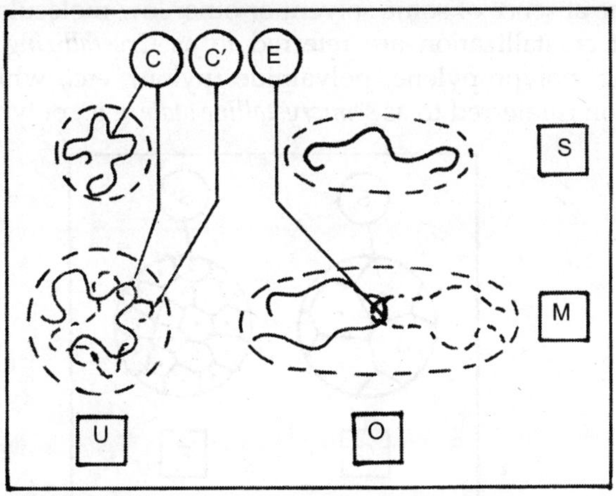

Fig 1.3: Molecular conformations. [U] Unoriented (random);
[O] Partially oriented; [S] Single chain; [M] Multiple chains; (C), (C′) I dividual chains; (E) Entanglement.

Above a characteristic temperature zone for each polymer, molecular or Brownian motion is very significant and allows a long range displacement of portions of polymeric chains, and in particular their inter-penetration. Inter-molecular interactions and physical entanglements shown at [E] in Fig. 1.3 assure a rather strong cohesion of the polymeric chains which is responsible for the very special properties of polymers in the fluidic state in particular.

Practically all polymers can be found in the amorphous state at sufficiently high temperatures or in solution, and many commercially important polymers (roughly half of them) exist exclusively in the amosphous state, for example, polystyrene (PS), polymethyl methacrylate (PMMA), polyvinylchloride (PVC), polycarbonate (PC), etc.

1.2.2 Crystallinity

While many commercially important polymers exist exclusively in the amorphous state, many others are capable of forming regions where the random conformation of polymeric chains of amorphous state is replaced by a regular ordered conformation resembling that found in regular crystals of low-molecular weight chemicals.

In a linear homopolymer, the regular succession of identical repeat units along the backbone of each polymeric chain affords the possibility of at least some portions of the same or different chains piling up regularly, as shown in Fig 1.4 with all atoms and group of atoms arranged in a manner characteristic of ordinary crystals with a specific unit cell for each polymer. Such an arrangement, however, is possible and stable only if sufficiently strong intermolecular forces are developed and maintained to overcome the tendency of polymeric chains to adopt a random conformation, but it is primarily the structure of the repeat unit that determines whether a polymer will crystallise, for example, polyethylene (PE), polytetrafluoroethylene (PTFE), etc.

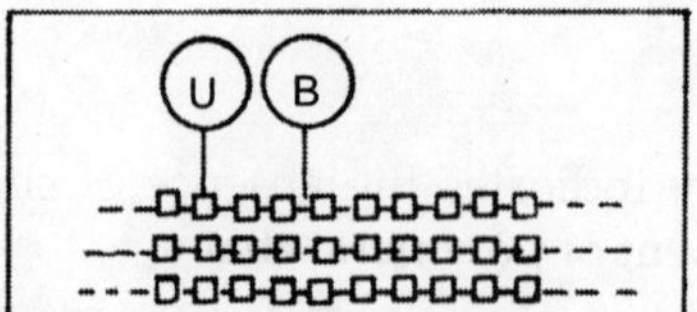

Fig 1.4: Crystalline order in polymers, (U) Repeat unit; (B) Chemical bond.

If may well be realized that it is practically impossible to achieve perfect crystallization of polymers in bulk in the absence of some solvent or other low molecular weight diluent. Polymers capable of significant crystallization are referred to as *Crystallizing or crystalline polymers*, for example, polyethylene, polypropylene, polyamide (nylon), etc., while their state of partial or imperfect crystallization is referred to as *semicrystalline state*, e.g., polybutylene (PBI), etc.

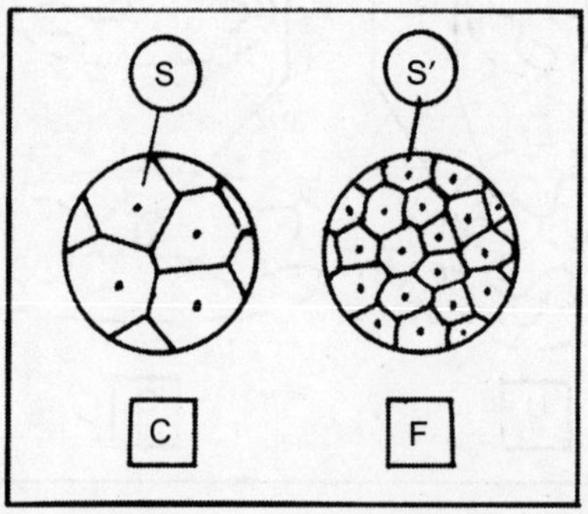

Fig 1.5: Spherulitic morphologies. [C] Coarse morphology; [F] Fine morphology; (S) Large spherulite; (S′) Small spherulite.

Crystallization is a time-dependent process and several factors affect not only the speed at which it takes place but also the resulting morphology. Fig. 1.5 shows two different morphologies that may be found in the same polymer. A coarse morphology [C] is normally associated with slow cooling and a fine morphology [F] is associated with fast cooling, as is the case with metals. Generally, the coarse morphology gives a stiffer but more brittle and less transparent material.

Such situations are typical of polyethylene for example, for which the lower degree of crystallinity in low density polyethylene (LDPE) is attributed to a high level of short branching [S] which is often combined with long branching [M]. Linear low density polyethylene (LLDPE) appears to have a molecular structure similar to case [S], although in this case the equivalents of short branches are associated with copolymerization mechanism.

Because of the limited flexibility of the backbone of a polymeric chain, interactions cannot take place within the repeat unit or even between close neighbours. They involve pairs of repeat units belonging to different chains or perhaps to portions of chains that are widely separated along the backbone. These intermolecular interactions, which range from very weak to relatively strong, play an important role in the properties of the polymers. These are generally classified as *hydrogen bonding, polar bonding, Vander Waals forces,* and *ionic bonding*.

Apart from the described physical-chemical interactions, the nature of polymeric chains, particularly their lengths and relatively high flexibility results in a purely physical type of interaction that leads to molecular entangling resulting in chain entanglements.

Contrary to chemical bonds, which are basically stable below a certain decomposition temperature, physical-chemical interactions and physical entanglements are rather strongly affected by temperature in a somewhat progressive manner.

1.3 Homopolymers and Polymeric chains

The fact that all polymeric chains in a homopolymer feature a single repeat unit does not generally mean that all chains are identical. Differences concern primarily the variation in length of the chains and whether those are linear or have branches.

The polymeric chains, assumed linear for present discussion, resulting from most polymerization processes vary widely in the number of repeat units involved. The number of repeat units, often called the *degree of polymerization* (DP), is the most logical quantitative way of characterizing a polymeric chain although in practice the molecular weight (MW) is more commonly used. The molecular weight of polymers may run into tens, hundreds or even millions of units.

When the chemical structure of the repeat unit is known, it is easy to calculate its molecular weight if its mole weight M_1 is known. For example, for polystyrene (PS), $M_1 = 104$ g per gram-mole, that is for 6.023×10^{23} actual repeat units, just 104 in short. The molecular weight of the polymeric chain is the product $M_1 \times$ DP. For a representative value MW = 100,000 for a polystyrene chain, the degree of polymerization is, therefore, 100,000/104 = 961.54 i.e., about 1000.

A polymer consists of a huge number of polymeric chains, typically about 10^{19} chains per cubic centimeter which differs widely in length. The length of each chain is thousands of times its width. The repeat units at the chain ends differ from the inner links.

In general, for a given polymer, higher molecular weights tend to improve properties, but by rapidly increasing the resistance to flow (viscosity) in the fluidic state, they make processing more difficult and costly. A specific range of molecular weights is generally found most appropriate for each polymer. Molecular weights (MW) and degrees of polymerization (DP) for a variety of common polymers are given in Table 1.1.

Table 1.1: Molecular Weights and Degrees of Polymerization for some common Polymers

S.No.	Polymer	Abbreviation	Molecular Weight (approx) (MW)	Degree of Polymerization (approx) (DP)
1.	Polyethylene	PE	300,000	10,000
2.	Ultrahigh MW Polythylene	UHMWPE	5×10^6	200,000
3.	Polystyrene	PS	300,000	3,000
4.	Polyvinylchloride	PVC	100,000	1,500
5.	Polyamides (Nylons)	PA	15,000	120
6.	Polycarbonate	PC	40,000	200
7.	Polyethylene terephthalate	PET	20,000	100
8.	Acetals (Polymethyleneoxides)	PMO	40,000	1,000

The values of molecular weight and degree of polymerization in Table 1.1 show that some polymers achieve good properties for very different degrees of polymerization than others.

The second important aspect of molecular structure concerns whether there is deviation from the *linear model* of homopolymer wherein repeat units are strung along a single line i.e. the chain backbone marked by (B) in Fig 1.6. During the polymerization, secondary reactions lead to the growth of side chains or branches indicated by (R) and can take different forms as shown in figure. One extreme case (L) features a few long branches while another extreme case (S) corresponds to many short branches. Any combination or intermediate case such as (M) may, of course, be encountered.

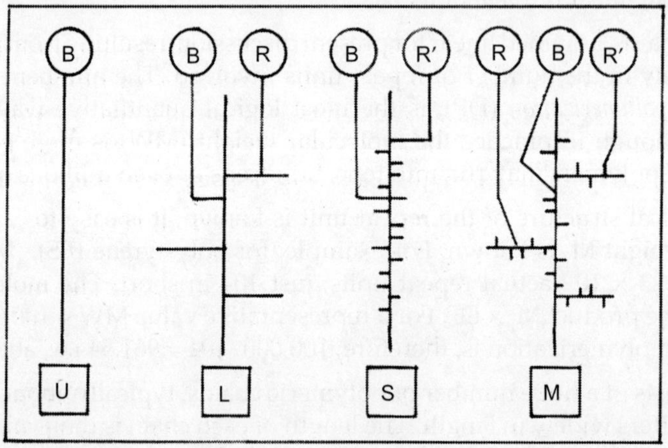

Fig 1.6: Molecular architectures. [U] Unbranched; [L] Long Branching; [S] Short Branching (B) Backbone; (R) Long Branch; (R′) Short branch; [M] Mixed Branching.

The effect of branching on the properties of a given polymer can be illustrated by comparing the two extreme cases of branching viz., Long (L) and Short (S) with that of linear chain (U) in the case of crystallizing polymers. For comparable molecular weights, linear chains (U) are likely to behave differently in the fluidic state than the bulkier, less slender chains (L); hence the meltviscosity or the processing characteristics of corresponding polymers could be markedly different. Short branches in chains of type (S) are small but so numerous and so irregularly spaced that they may reduce considerably the ability of portions of the linear chains to line up closely to form crystal-like domains; the corresponding polymer would display or lower the degree of crystallinity and, thus, a lower stiffness.

1.4 Liquid Crystal Polymers

A comparatively newer variety of polymers with outstanding mechanical and thermal properties which possess molecular order of a kind different from that found in conventional crystallizing polymers is referred to as Liquid Crystal Polymers (LCP).

Polymeric liquid crystals are associated with polymeric chains that are basically stiff, most commonly as a result of the presence of aromatic (phenyl, etc.) rings and include aromatic polyamides and polyesters.

Two very notable features of liquid crystal polymers are a very low viscosity in the temperature region (Tf-Tc), i.e. fluidic temperature and the crystallization temperature and a very small contraction on solidification. These characteristics, and their resistance to high temperature and chemicals, make them very suitable for production of small precision parts for demanding applications.

The mechanical properties of highly oriented LCPs in the direction of molecular orientation are outstanding, ranging from a modulus of about 15 GPa and a tensile strength of about 200 MPa for a moulded sample to a modulus of about 100 GPa and a tensile strength of about 350 MPa for a fibre.

Some of the commercial polyester-based LCPs include Exxcel, Xydar, Vectra, Ultrax, Granlar while Kevlar is an aromatic polyamide.

1.5 Cross-Linking and Molecular Networks

So far the discussion has been about those polymers wherein a polymer has been viewed as a collection of individual, independent polymeric chains with chemical bonds found only within each chain, i.e. intermolecular chemical bonds. The basically non-permanent physical-chemical interactions between chains are relatively weak and diminish or vanish with increasing temperature to the point where individual chains can move fairly freely relative to each other over long distances — a characteristic of the fluidic state. This type of polymer is referred to as a *thermoplastic* or sometimes as a *linear polymer*.

There exists another class of polymeric materials where, instead of individual polymeric chains, the whole material consists of polymeric sections that are fully interconnected through chemical bonds to form *three-dimensional* molecular network as shown in Fig. 1.7.

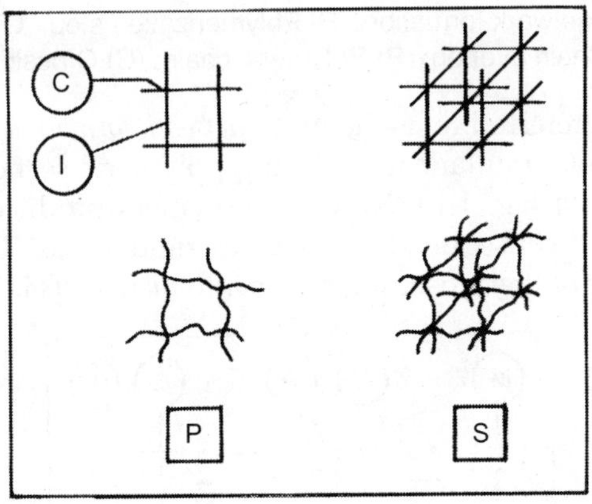

Fig 1.7: Polymeric molecular networks. [P] Plane representation; [S] 3-D representations; (C) Interchain link (crosslink); (I) Interlink chain.

When relatively few repeat units are involved in the polymeric section (I) between interconnections (C), the network is termed *tight* or *dense* [T]; in the opposite case the network is referred to as *loose* or *light* [L], as shown in Fig 1.8. The rigidity of a network polymer depends on the nature of the repeat units involved and is affected by temperature, but generally tight networks correspond to stiffer materials, while loose networks are more likely to be associated with rubbery materials. However, in no case, network polymers can be found in a fluidic state.

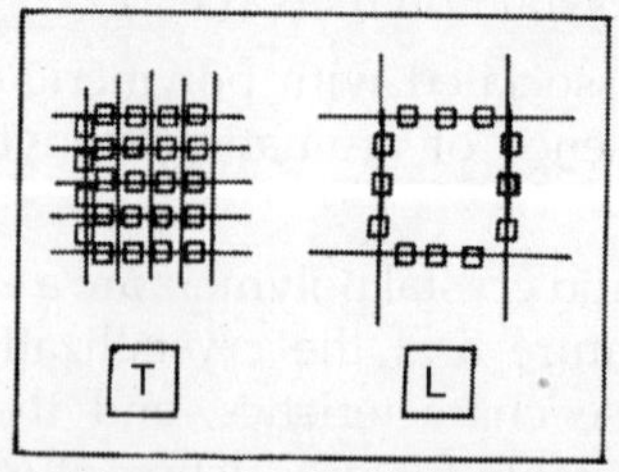

Fig 1.8: Network density [T] Tight; [L] Loose.

Two basic mechanisms are involved in the formation of network polymers. The first mechanism, called the *two-step network formation* involves first the polymerization [P] of monomers [M] to individual, independent fully polymeric chains [P], followed by the introduction of chemical bonds points of the initial polymeric chains as shown in Fig. 1.9. These chemical interchain links (C) are called *Chemical cross-links*, and the second step as cross-linking (C). The networks formed by this mechanism are generally *loose networks* with *rubber-like* properties. The conventional rubber industry has been based on the use of this mechanism where terms such as *Vulcanization* or *curing* are used for the reactions which correspond to cross-linking and are achieved by the use of suitable chemicals and/or energy sources.

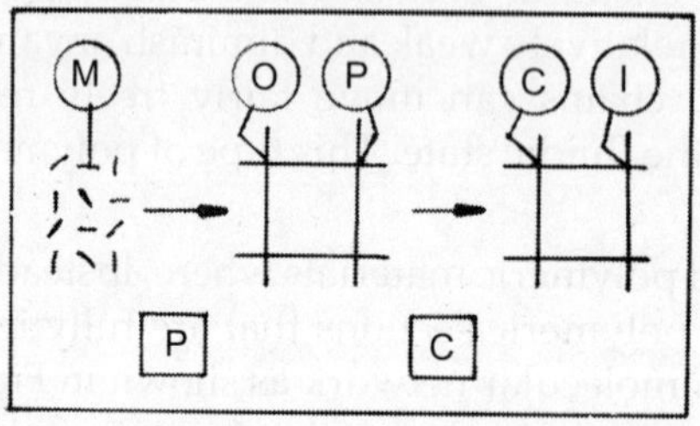

Fig 1.9: Two-step network formation. [P] Polymerization step; [C] Crosslinking step; (M) Monomer; (O) Chain overlap; (P) Polymeric chain; (C) Crosslink; (I) Interlink chain.

The second mechanism, referred to as the *direct network formation* generally implies monomer molecules that are capable of simultaneous forming polymeric sections and connecting them in network fashion, as shown *in* Fig. 1.10. Network polymers produced by this mechanism tend to have tighter networks and consequently, behave as rigid solids. This mechanism is primarily employed for the processing of thermosetting resins into *thermosets* i.e., thermosetting plastics.

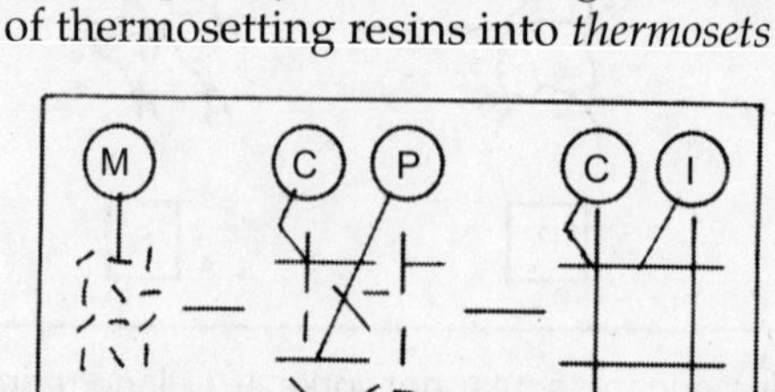

Fig 1.10: Direct network formation. [N] Early stage; (N′) Later stage; (M) Monomer; (C) Crosslink; (P) Polymeric chain; (I) Interlink chain.

All network polymers are the result of a physically irreversible process and no physical change can return the processed material back to the initial state or even to the fluidic state. Chemical processes that could destroy the cross-links or interchain links would probably cut the interlink chains as well; thus would have no real practical value.

1.6 Copolymers

When monomers are used in a polymerization process correspond to a single repeat unit in the polymer chain, the resulting polymer as already discussed, is called a *homopolymer*.

It is, however, sometimes possible to polymerize two or more monomers corresponding to different repeat units and to have these different repeat units incorporated in the same polymeric chain. The material produced of such mixed structure is called a *copolymer*.

Different possible structures that may be produced by the copolymerization of monomers of two different units M_A and M_B (squres and circles respectively) are shown in Fig. 1.11 and are deseribed as follows. Corresponding homopolymers, denoted by $P(M_A - M_B)$. In all cases discussed hereunder, it has been assumed that comparable numbers of M_A and M_B repeat units are involved in each polymer.

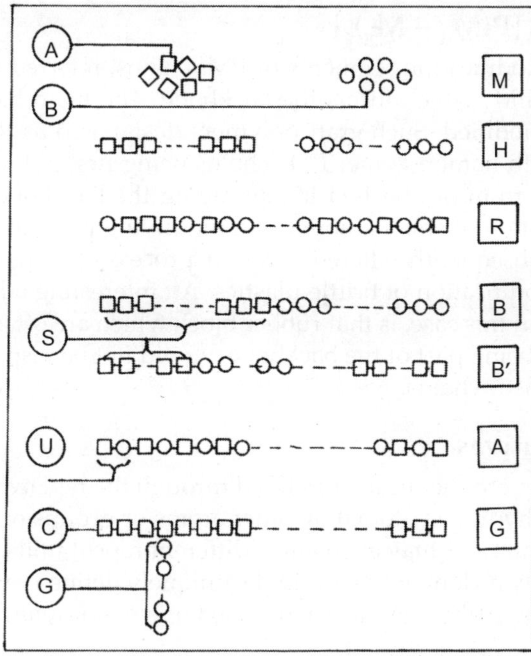

Fig 1.11: Copolymer structures; [M] Monomers; [H] Homopolymers; [R] Random copolymer; [B] Di-block copolymer; [B′] Multi-block copolymer; [A] Alternating Copolymer; [G] Grafted polymer; (A) Type-A monomer; (B) Type-B monomer; (S) Segment (block); (U) Repeat unit; (C) A-backbone; (G) B-graft.

1.6.1 Random Copolymer (R)

This type of copolymer features repeat units M_A and M_B randomly distributed along the chain as a result of a random (statistically) incorporation of M_A and M_B to a growing chain. The resulting structure is called a *random copolymer* and denoted as $P(M_A - M_B)_r$. The lack of a regular repetition of M_A and M_B repeat units may have important consequences, e.g., characteristic properties of PM_A and PM_B may not be found in $P(M_A - M_B)_r$. A well known example is that of ethylene and propylene, where both homopolymers $P(M_A - M_B)_r$ are of the crystallizing type, giving solid plastics at ambient temperature, while a 50/50 random copolymer does not crystallize and behaves as a rubber (ethylene-propylene rubber).

1.6.2 Block Polymers (B_1 and B_2)

Copolymers B_1 and B_2 feature relatively long segments (blocks) of repeat units M_A and M_B that are attached to form the chain. Such a polymer is called *block or segmented copolymer* and is denoted by $P(M_A - M_B)_b$. Each block is sufficiently long to impart to the copolymer the characteristics of the corresponding homopolymers, and the overall behaviour is that of a mixture, provided the components are extremely well dispersed in each other down to the molecular level. Type B_1 or B_2 could, in principle, result from a sequential (i.e. one at a time) addition of the monomers to the reactor.

1.6.3 Alternating Copolymers [$P(M_A - M_B)_a$]

This type of copolymer involves regularly alternating M_A and M_B repeat units in its polymeric chain. Such a copolymer can be viewed as a homopolymer whose repeat unit is $M_A - M_B$ and the properties are likely to be characteristic of a homopolymer.

1.6.4 Graft Polymers [$P(M_A - M_B)_g$]

In this mode of polymer modification segments of PM_B chains, referred to as grafts, are attached to the polymeric chain of PM_A which forms the backbone. The main backbone chain consisting of M_A repeat units is not modified. Such graft polymers designated as $P(M_A - M_B)_g$ are normally produced in two stages with homopolymer PM_A chains being first polymerized and grafts being subsequently polymerized from or attached to sites along the backbone. This technique allows, in particular, the incorporation of small amount of a modifying polymer in a perfectly dispersed manner that cannot be subsequently altered through processing stages. If has been used, for instance, for the rubber-modification of brittle plastics. An interesting advantage of grafting over regular copolymerization, in this case, is that rubber block which are often susceptible to chemical attack and rupture, by not being part of the backbone, would not be responsible for a reduction of molecular weight of the plastic chains.

1.6.5 Polymer Derivatives

A number of homopolymers are chemically modified through the replacement of a relatively small number of atoms (usually hydrogen atoms) by other atoms or groups of atoms. This substitution does not affect all repeat units and may even affect different repeat units to a different extent. The modified polymer, thus, may no longer have a single, uniquely defined repeat unit (homopolymer) or two distinct repeat units (copolymer). Such a modified polymer is referred to as *polymer derivative* or *modified polymer*.

It is evident from above description that controlled polymerization provides almost endless possibilities of slightly or drastically modifying the properties of conventional polymers. A majority of commercial polymers feature at least small amounts of a so called *comonomer* that can be considered as a modifier rather than full fledged component. Many commercial polymers contain comparable amounts of two or more comonomers. The term *terpolymer* is sometimes used when three repeat units are involved in copolymerization to obtain a specific product.

1.7 Polyalloys

Polyalloys are mixtures of two homopolymers or copolymers to achieve desired properties in the material to serve satisfactorily in the intended service. Cost, properties, and processibility are major considerations in the assessment of potential benefits. Mixtures are defined here as combinations of materials not involving chemical reactions, i.e. no formation or rupture of chemical bonds. There are three possibilities when mixing polymers, as follows.

Many pairs of polymers are almost impossible to mix or disperse into one another, and an attempt to do so yield, at best, rather useless coarse aggregates with little or no cohesion between phases. Polystyrene and polypropylene are a well known example of this behaviour. Such pairs of polymers are referred to as *incompatible polymers*.

The second case is that of pairs of polymers that are *miscible*, i.e., they are capable of forming potentially useful homogeneous, one phase, mixtures in which molecules of the two types are thoroughly mixed down to the molecular level. This implies a high level intermolecular interaction related to the chemical structure of the repeat units involved. Such mixtures are largely analogous to solutions of simple miscible liquids, and factors such as temperature or concentration can play a role in the solubility. Typical examples of such polyalloys include polyvinyal chloride (PVC) and nitrile rubber (NBR); the homogeneous mixture of which may be denoted as A (PVC/NBR), or polypherylene oxide (PPO) and polystyrene (PS) that may be represented as A (PPO/PS).

In the third case a pair of polymers behave in a somewhat intermediate manner between incompatible and miscible pairs of polymers. Under certain conditions, a dispersion of small particles of one of the polymers into a matrix of the other polymer, and sometimes an interpenetrating arrangement of two phases, can be achieved where the combination of reasonable interaction between phases and a uniform dispersion yield a potentially useful material. The polymers involved in such pairs are said to be *compatible*.

Miscible and compatible pairs are usually referred to as *homogeneous* (one-phase) and heterogeneous (two-phase) polyalloys and may be denoted as A(PM_A/PM_B). In general such polymer alloys are called *polyalloys* or polyblends. The development of methods of increasing the compatibility of pairs has lead to the production of many newer polyalloys.

Copolymerization or grafting of one or both components, with segments of the other component, or coblending with a polymer compatible with both the major components (compatibiliser), are among the methods of increasing the compatibility of pairs of polymeric materials.

There is a certain parallel between random copolymers and homogeneous polyalloys, on one hand, and block copolymers and heterogeneous polyalloys, on the other hand.

The structure of polyalloys is determined partly by the nature of the component polymers and partly by the blending procedure and conditions, as well as the final processing.

1.8 Plasticizers

The primary role of a plasticizer is to reduce the rigidity of a plastic, that is, to render it more flexible. Polyvinylchloride (PVC), for which plasticizers are widely used, can thus be obtained in wide range of stiffeners from rigid and somewhat brittle types, to very flexible rubber-like types.

One of the earliest commercial applications of synthetic plastics (celluloid) was based on the use of camphor as a plasticizer for cellulose nitrate in the fabrication of billiard balls.

Plasticizers are normally relatively small molecules that are sufficiently compatible with the polymer molecules to penetrate them like a solvent. They form non-chemical bonds with polymeric molecules of varying strength or permanance. Highly compatible plasticizers remain locked-in while less compatible ones may gradually migrate towards the surface and form a powdry layer sometimes referred to as *blooming*.

Plasticizers often play a so-called lubricating role in the processing of thermoplastics, largely through reduction in viscosity. However, plasticizers are used primarily with amosphous polymers as their use in crystallizing polymers tends to interfere with the crystallization process. A direct effect of their use is the lowering of the *Vitreous-or the glass transition temperature zone* of the polymer.

The number of commercial plasticizers is very large and esters form the largest group. Examples include di-2-ethyl hexyl (octyl) phthalate (DOP), dibutyl phthalate (DBP), butyl benzl phthalate (BBP), and diphenyl phthalate (DPP).

Phosphates are esters of trifunctional phosphoric acid and generally also play a flame retardant role. They include tributoxy ethyl phosphate (TEP), tricresyl phosphate (TCP) and triphenyl phosphate (TPP).

Adipets are esters of adipic acid and are generally suitable for food contact; for example di-2-ethyl hexyl (octyl) adipet (DOA).

Other esters include azelates (Z) citrates, glutarates, oleates and stearates.

Amid-based plasticizers, such as sulfonamides, lauramides, acetoacetamilide, include n-ethyl-toluene and p-toluene sulfonamide.

Some regular polymers can play a strong plasticizing role; the mechanism is really similar to that concerning polymer alloys. This appears to be the case for ethylene-vinylacetate (EVA) copolymer and polyisobutylene (PIB).

1.9 Fillers and Reinforcements

The intrinsic stiffness of bulk polymers in the vitreous or glass state, even with high degree of crystallinity, is relatively low compared to that of most inorganic metals. Typically the modulus of elasticity, E, is below about 3.5 GPa (≈ 0.5 M psi) for polymers, compared to a range of about 70 – 200 GPa (≈ 10-30 M psi) for inorganic metals.

Stiff inorganic materials can often the combined with polymeric materials in the form of composite systems, where the polymeric material forms the matrix into which the inorganic material is dispersed as a discrete phase. The discrete inorganic components of such composite systems can be classified, according to their shape, into particles (P) (spheres, cubes, etc.), short fibres (F) and disks (D) (Platelets and flakes) as shown in Fig. 1.12, and for very long elements as continuous fibres (F) and tapes (T), as illustrated in Fig. 1.13.

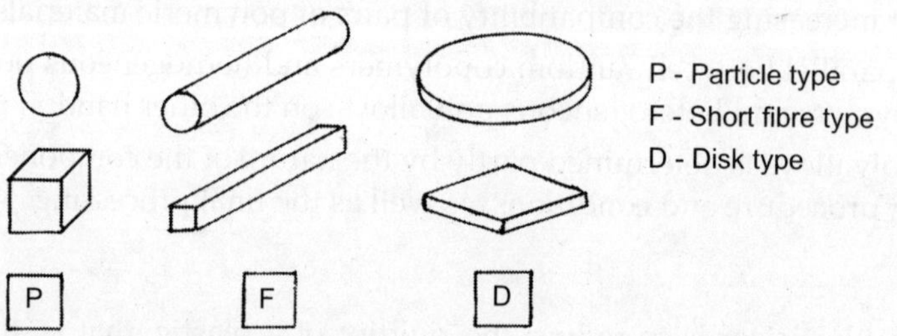

Fig 1.12: Discontinuous filler/reinforcement shapes.

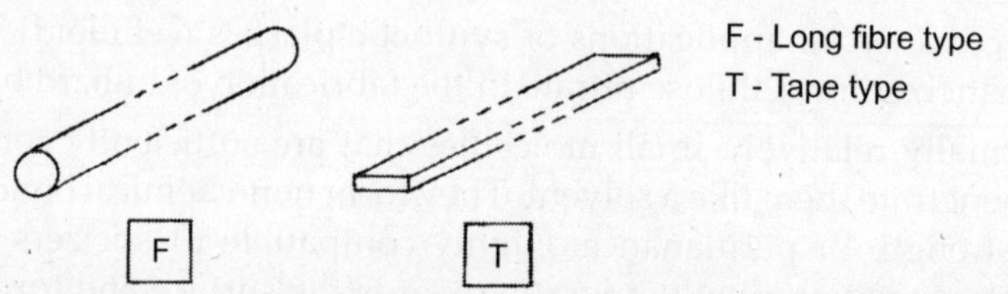

Fig 1.13: Continuous reinforcement shapes.

Inorganic additives in polymers are referred to as fillers, if their role is primarily to increase the stiffness of the polymeric material, though it may lead to cost advantage also. *In case where they have a favourable effect on* some aspects of the *mechanical strength they are called reinforcements.*

Fillers and reinforcements generally possess such chemical and heat resistance that they are unaffected by processing with either reactive or thermoplastic polymeric systems.

The degree of adhesion between the polymeric matrix and the additive component often determines the difference between mere filling or true reinforcing. Discrete components are often surface treated with chemical coupling agents to enhance adhesion. Organosilanes, titanates and zirconates are often used as coupling agents.

1.9.1 Particle Class

Fillers and reinforcements primarily in the particle class include calcium carbonate, silica, kaolin, talc, alumina trihydrate, feldspar, calcium sulfate solid and hollow spheres of glass or other ceramic, and carbon black. However, a disadvantage of most inorganic fillers is their high density compared to that of polymeric materials.

Ceramic-type materials such as fly-ash, a by-product of combustion of coal, and temperature-resistant polymers, such as phenolics, have been made into solid or hollow spheres.

Carbon black is widely used as a reinforcing filler in rubbers and as a pigment, electrically conductive filler or shield against ultraviolet radiation in plastics. The various reinforcing grades like thermal, furnace or channel result from the incomplete combustion of hydrocarbons. The small particles used are in the range of 0.03 to 0.30 microns and have porous structure and form good bonds with rubbers.

1.9.2 Fibre Class

Fillers and reinforcements in the fibre class include glass fibres, carbon and graphite fibres, asbestos, whiskers, and polymeric fibres.

Glass-fibres have been extensively used in plastics, most often in continuous form in catalyst-activated thermosetting resins and, in short form in thermoplastics.

Much like glass fibres, carbon and graphite fibres are supplied in a variety of forms with monofilament diameters in the range of 5-10 microns. These fibres are particularly attractive for their high stiffness and low density, and in certain applications their thermal or electrical conductivity, x-ray transparency, and their low abrasive characteristics.

Even though carbon and graphite fibres were developed in the late 1950s, their high cost long restricted their use to so-called advanced composites primarily for defence and aerospace applications. However, the price is now getting more affordable and their market is rapidly growing.

A number of polymeric fibres are used for reinforcement. They include conventional polyamide and polyester fibres and high performance fibres, like aromatic polyamides, polyetheretherketone (PEEK) fibres, and even speciality-oriented (extended chain) polyethylene.

1.9.3 Disk Class

Fillers and reinforcements primarily in the disk class include, mica, glass flakes, and aluminium flakes; the last ones are intended primarily as conductive filler for electromagnetic interference (EMI) shielding.

1.10 Miscellaneous Additives

Apart from the already described, a large number of additives are available to improve specific properties of polymers. However, they tend to have an effect, often negative, on the other properties. The cost and efficiency of these additives can vary widely.

Some of the important commercial additives include the following:

1. UV Absorbers and Light Stabilisers

2. Antioxidants
3. Heat Stabilisers
4. Flame Retardants
5. Antimicrobials
6. Colorants
7. Odorants
8. Surface-Modifying Additives

Brief description of these additives follows.

1.10.1 UV Absorbers and Light Stabilizers

Light in itself may not directly do much harm to polymers but the radiations, in particular ultraviolet (UV) radiations, tend to initiate or catalyse chemical degradation such as oxidation wich is often referred to as *photo-oxidation*.

The additives employed to reduce the ill effects of oxidation are divided into two classes viz., UV absorbers (UA), which reduce the essential factors of the degradation, and light stabilizers (LS) which control the progress of degradation; the difference between the two may be very subtle.

There are many classes of UV absorbers and light stabilizers and they are generally added in the range of 0.1 to 0.6%. Some of the well known amongst them include benzophenones (primarily UA), for example : 2-hydroxy-4-dodecycloxy benzophenone (DOBP); acrylic or aryl esters (UA or LS), like resorcinol monobenzoate (RMB); metal salts or organic metal compounds such as, zinc or nickel dibutyl dithiocarbamate; alkylated phenol phosphonate nickel complex. Carbon black (CB) acts as a barrier to penetration of UV radiations, provide good protection when its colour can be tolerated.

Among the plastics susceptible to photoxidation are polyethylene (PE), polypropylene (PP), polystyrene (PS), unsaturated polyesters, polyvinylchloride (PVC), cellulose esters (CA, CAB), acrylics, and polyethylene tetraphthalate (PET).

1.10.2 Antioxidants

Quite a few polymers are susceptible to oxidation which causes either a reduction in molecular chain length and thus molecular weight, for example, in polyethylene, leading to a loss of stiffness and strength in the first instance followed by embrittlement later. Oxidation can be slow at moderate processing temperatures and its effects are then associated with ageing, or quite fast at high processing temperatures, and its effects are then related to thermal stability. Radiations or impurities such as metal traces can also initiate or catalyse oxidation (photo-oxidation, metal activation).

Primary antioxidants which can be used alone are subdivided into two classes. *Hindered phenolics* of low or higher molecular weight operate through reactive OH groups and include butylated hydroxy toulene (BHT) and tetrakis methane. *Hindered Secondary arylamines* operate through reactive NH groups and include substituated paraphenylene diamines and diphenyl amines.

Secondary antioxidants engage in mutually reinforcing (synergistic) action with primary oxidants but would not react strongly alone. Phosphorus containing types include organophosphites and sulphur containing types include thioesters such as thiopropionate.

1.10.3 Heat Stabilizers

A form of chemical degradation involves some decomposition at processing temperatures with the evolution of corrosive or toxic chemicals. PVC is a well known and perhaps the most commercially important example, and the resulting HCI is a particularly undesirable chemical. Although other polymers require or benefit from suitable heat stabilizers, the term is generally associated with PVC heat stabilization.

Most of the primary heat stabilizers are metal based. Single metal types include organo-metal compounds, such as butyl tin mercaptides; antimony mercaptide; and lead in the form of phathalate, sulphate, and carbonate. Mixed metal types may correspond to salts and soaps such as stearates including those of calcium and zinc; barium and cadmium; barium, cadmium and zinc; magnesium; and aluminium.

Secondary or auxiliary stabilizers, such as those involving phosphorus as phosphites or biphenylene diphosphonite, are also available for use with primary stabilizers.

1.10.4 Flame Retardants

Practically all polymers are combustible but there are great variations in their fire-or flame resistance. For many applications, additives must be compounded with the most fire-prone polymers to reduce their flammability, referred to as *flame retardants* (FR) or to reduce the production of smoke, called smoke suppressants (SS) or smoke reducers or smoke inhibitors.

There is a large number of flame retardants and smoke suppresants but most of these are such that they are effective only with specific polymers. They are divided into two types viz., mineral based and organic related.

Mineral-based types include alumina trihydrate (ATH), which dehydrates at high temperatures, liberating water and is widely used in unsaturated polyesters for its smoke suppressant role in particular; hydrated magnesium sulphate (Epsom salt); boron compounds; such as zinc or aluminium borates; magnesium hydroxide; and antimonytrioxide (SbO_3).

Organic-related types feature halogens and phosphorus which appear to play a crucial role and include brominated or chlorinated aliphatic or cycloaliphatic compounds that tend to produce toxic fumes, such as decabromo dipheynl oxide; and phosphorus-based compounds, such as aromatic phosphate esters with or without halogenation.

1.10.5 Antimicrobials

Micro-organism like bacteria, fungi, and algae being carbon scavengers, can sometime grow on the surface of polymers when temperature and humidity are favourable particularly when the polymer compound offers appropriate nutrients, which are often low molecular weight additives. Such a situation is important in applications such as humidifiers, air-conditiong filters, or foam mattresses. Microbial growth like bactrial, fungal, algal, spores, or mildew can be countered with a number of antimicrobial additives available as liquid solutions or in powder form.

There are three types of antimicrobials that are found suitable for the said purpose. The first type consists of organo-metallics which include tin-based types such as trialkyl tin compounds, arsenic-based types, and copper-and antimony-based types. The second type consists of organic sulphur compounds of the captan or phaltan types. The third type consists of quaternary ammonium compounds Chloroxylenol has also been reported to act as effective antimicrobial.

1.10.6 Colourants

One major advantage of polymers over metals is that most often they can be produced with attractive built-in colours thus eliminating secondary finishing operations. The additives used to impart colour to polymers are basically divided into two types viz., pigments and dyes.

The *pigments* are finely divided powders of 0.01 – 1 micron particle size that are basically insoluble in the polymers and must be well dispersed and distributed to give maximum colouring effect with the smallest possible concentration, usually less than 10%. They tend to act as fillers when used in large proportions and thus cause deterioration of certain mechanical properties.

The pigments are either inorganic or organic. *Inorganic pigments* consist of two basic types. Carbon black is widely used for black colouration, but also for its radiation screening and reinforcing effects in certain grades. Metal-based pigments, which yield colours generally characteristics of the metal involved, mainly include oxides of Ti, Fe, Cr, of which white titanium dioxide (TiO_2) is an important example; sulphides of Cd, Hg, Zn; selenides of Cd; sulphates of Ba and Pb; chromates of Pb; molybdates of Pb, and titnates of Ni and Cr. Strontium Sr, Ca, Ba, and Al are also involved in compounds.

Organic pigments are available in two major classes, viz., azo type and phthalo type, for example copper phthalocyanine. Other classes include carbazole, perylene, and aniline types.

Dyes consist of relatively small molecules that are compatible with the polymer molecules and they are therefore basically soluble in polymers, getting dispensed at a molecular level.

Dyes are normally used to impart colour to transparent polymers, such as styrenics or acrylics. They are generally supplied in liquid form. The active ingredient is often in solution in a solvent or 0.1 carrier, to facilitate its mixing with the polymer.

Dyes are available in three types, viz., anthraquinone, types typically used in automotive tail-lights; nigrosine types, alternative for carbon black, and oil-or water-soluble azo types.

Besides regular colourants whose purpose is just to impart colour, other additives are used to produce special appearance-related effects. *Fluorescent additives* usually organic dyes, impart the well-known shining effect in relative darkness when exposed to light.

Phosphorscent additives, usually inorganic pigments store light when exposed to a light source and can subsequently shine giving an after glow for a while in darkness, and are usefully employed to indicate roadside limits. *Pearlescent additives* can impart pearl-like sheen mainly to transparent polymers. *Metallic powders,* mainly that of aluminium and bronze, are used as an alternative to plating for metal-like appearance or for other special effects.

1.10.7 Odorants

Additives are also sometimes employed to give a specific odour to a polymer and are referred to as *odorants* or *fragrances* while when used to control or mask an objectionable odour, they are called *dedorants* or *sanitizers*. A sanitizer may also serve as an antimicrobial agent.

1.10.8 Surface Modifying Additives

Some additives are used to modify the surface properties of polymers to achieve the desired aim. Three such types of additives are discussed in this section.

The first type relates to cases involving an interface between molten polymer and hot metal to avoid sticking of polymer to the metal surface as in rolling and extrusion sizing. Additives employed for the purpose are referred to as *external lubricants*, while *internal lubricants* are employed to affect the viscous flow of the polymer.

External lubricants are of particular importance to the PVC industry. Some of the additives employed for the purpose include fatty acids such as stearic acid, glycerol monostearate, fatty acid amides such as stearamide, oleamide, and crucamide; and fatty acid metallic soaps, e.g., lead stearate, calcium stearate, paraffin waxes, etc.

These lubricants also serve well for use as internal release agents for detachment of solidified polymeric products from metal mould surface, and are then often referred to as *parting agents*.

The second type concerns cases involving solid polymer to solid polymer interface. Some solidified polymeric parts like *films* often tend to stick to each other in an undesirable manner. This sticking tendency is often attributed to static electrical changes and intimate contact of smooth flexible surfaces. Additives used to reduce this sticking tendency are referred to as *slip agents* or *antiblocking agents*. Inorganic powders such as silica or silicates are reportedly useful in this respect.

The third type concerns cases involving static electricity build-up on a solid polymer surface. Such a situation is attributed to friction, particularly at high speeds, and the low volume and surface conductivity of polymers which prevent the dissipation of accumulated electric charge. Static charges can be the cause of sparks and shocks particularly in textile and chemical industries, tend to favour the collection of dust and dirt on records, and contribute to the sticking together of polymeric films and fibres. Additives used to reduce static electric build-up are called *antistatic agents* or *antistats*.

Large concentrations (10 - 30%) of carbon black can give acceptable results which can also be achieved with lower loadings of metallic powders. However, real antistatic additives added to polymers in small amounts of 0.1 to 0.4% exude or bloom to the surface. They include ethoxylated fatty amines and ethoxylated glycerol esters of fatty acids as well as quaternary ammonium salts of fatty acids.

Most of the above described additives also help in giving the polymer an *antifogging property* whereby condensation of water on transparent containers or on lenses are avoided.

2 POLYMERIC MATERIALS

2.0 Classification of Polymeric Materials

Polymers or polymeric systems, which are fairly widely used, have good though not outstanding properties and are priced at moderate levels. They are referred to as *General Purpose* or *Commodity Plastics or Rubbers*; well known amongst them are polyethylene (PE), polyvinyl chloride (PVC), styrene butadiene rubber (SBR), etc. The term *Engineering Plastic or Rubber* is often used for materials that have a good set of excellent properties like mechanical resistance, temperature limit, etc.; and are also sometimes called *High Performance Plastics or Rubbers*. They are usually significantly more expensive; examples include polyamide (PA), polycarbonate (PC), polysulfon (PSU), etc. Then there are specialty polymers which are normally associated with one or more outstanding properties like low friction, electrical conductivity, etc.; though other properties may be mediocre, for example, polytetra fluoro ethylene (PTFE).

Because it is intended to consider mainly the Engineering Plastics so in the rest of the chapter the term plastics will be more often employed rather than the generic term polymers.

All the plastics and rubbers can be classified into five main groups, viz., thermoplastics, thermosets (thermosetting resins), Elastomers or rubbers, specialty plastics, and reinforced plastics and composites, as shown in Fig. 2.1.

It may be noted that some materials are associated with more than one group or some groups may include materials of different classes. Urethane systems, for example, include thermoplastic and elastomeric materials in addition to thermosetting resins. Imide polymers and silicone polymers are other examples. Brief description of important members of different groups follows in this section.

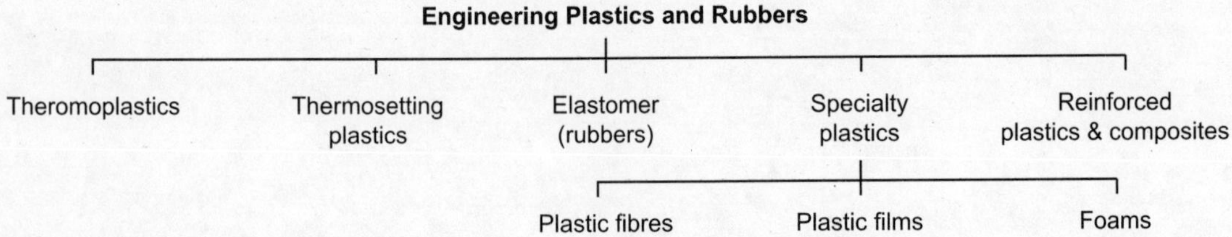

Fig 2.1: Classification of Engineering Plastics & Rubbers.

2.1 Thermoplastics

Thermoplastics are linear chain polymers which soften on heating and solidify again on cooling. When reheated, thermoplastics can repeatedly pass into a highly elastic state and then into a viscous fludic or plastic state without losing their original properties on cooling again. These plastic materials are worked by the application of heat and pressure.

Because of their high molecular weight, these plastics never become thin fluids. There are two subgroups of thermoplastics viz., *non-crystallizing or amorphous thermoplastics* and the other is *crystallizing thermoplastics*.

2.1.1 General Behaviour of Thermoplastics

Plastics drastically differ from metals in their behaviour when deformed at room temperature, as also at elevated temperatures. At room temperature, the stress-strain relationship is very peculiar. This difference in behaviour under loading between metals, thermoplastics, and rubber is shown in Fig. 2.2 in which point B marks the limit of elasticity.

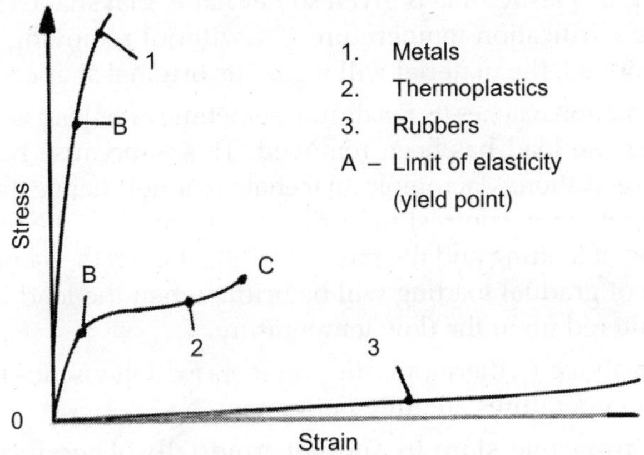

Fig 2.2: Stress-strain diagram for metals, thermoplastics and rubbers.

In the range OB both plastics and metals obey Hook's law (i.e. strain is proportional to stress). Beyond the limit of elasticity, that is within the range BC, the law of proportionality is no longer obeyed, and the plastic specimen experiences a marked deformation even under a relatively small increase in load and then the plastic is said to flow. Past point B, the deformation becomes permanent. In contrast to metals, the region BC is of a considerable extent for plastics. In the case of most of the metals, the deformation is limited and rupture occurs rapidly beyond the elastic limit.

Highly elastic materials like rubber, as shown, reach point B under a relatively small amount of loading, exhibit marked deformation, and regain the initial condition after the load is removed.

Another distinction of plastics from metals is that they have a relatively low hardness and also that their hardness does not show a well defined dependence on tensile strength, unlike steels.

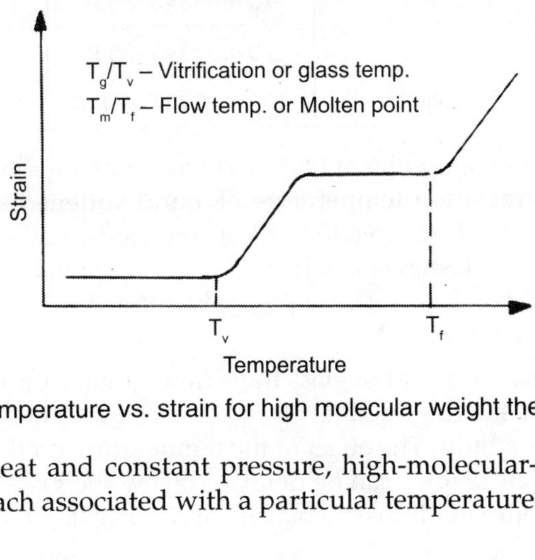

Fig 2.3: Temperature vs. strain for high molecular weight thermoplastics.

Under the action of heat and constant pressure, high-molecular-weight polymers reside in three consecutive states, each associated with a particular temperature range, as shown in Fig. 2.3, namely vitreous or glass-like, highly elastic, and plastic or viscous fluidic. As the temperature is raised to the vitrification point, T_v (or glass transition temperature T_g), strain under constant load does not change practically, as is evident from the figure (Fig. 2.3). Further rise in temperature brings about a sudden change in strain up to a certain limit beyond which it remains almost constant until the temperature reaches the flow point, T_f. Above this point the material becomes fluidic. In the temperature in interval $T_v - T_f$, the deforation is reversible for thermoplastic polymers. Below the vitrification point, T_v, plastics behave as brittle materials like glass. In the interval $T_v - T_f$, they act as highly elastic rubber-like substances. Above T_f, they behave like a highly viscous fluid.

If a plastic in the highly elastic state is given some shape, that shape can be retained by cooling the article to below the vitrification temperature, T_v, without removing the pressure. If reheated after the pressure is removed, the material will regain its original shape.

Highly elastic deformation reaches its maximum sometime after load is applied, and it disappears likewise sometime after the load has been removed. This is because highly elastic deformation exhibits the property of relaxation. Macromolecular chains cannot change their shape the same instant as load is changed. Therefore, in contrast to brittle transformation, the highly elastic deformation depends on the duration of loading and the rate of loading. Given the same temperature, a material which is elastic in case of gradual loading will be brittle when the load is applied suddenly. This behaviour remains unaltered up to the flow temperature, T_f.

As discussed earlier, above T_f, thermoplastic polymers exhibit viscous fluidity. The strain arising in the material at such temperatures are fully permanent.

A plastic changes from one state to another gradually. Therefore, both the vitrification temperature and the flow temperature ought to be visualized as some temperature intervals rather than some definite temperatures as is evident for some of the well known thermoplastic polymers listed in *Table 2.1*. A long stay at elevated temperature or further increase in temperature causes the plastic to break up, i.e. get thermally destroyed. Within the safe temperature range of $T_v - T_f$ the thermoplastic polymers may be reheated many times without much deterioration in properties.

Table 2.1: Vitrification and Flow Temperatures of some Common Thermoplastics

Polymer	Abbreviation	T_v, ºC (K)	T_f, ºC (K)
Polyethylene	PE	– 21 (252)	120 (323)
Polyamide	PA	55–65 (328–338)	200–240 (473–513)
Polystyrene	PS	70–80 (343–353)	220 (493)
Polyvinylchloride	PVC	70–80 (343–353)	190 (463)
Polymethylacrylate	PMA	100–110 (373–383)	220 (493)

From the above description it is evident that thermoplastics are rigid solids below the vitreous temperature (T_v) or glass transition temperature (T_g) and softened elastic material above that temperature. This change in physical state is often referred to as *glass transition* because the structure and properties of the plastics below it are in good resemblance to those of ordinary glass. It is important to note that T_v or T_g is not very precisely defined or measured, but it is very useful characteristics of every homopolymer.

In the following, Charrier's concept of glass transition for amorphous thermoplastics is shown through schematic graph. Considering a quantity R as *resistance to deformation*, it corresponds to stiffness for a solid, and viscosity for a fluid. The effect of the temperature on the quantity R is represented schematically in Fig. 2.4. Four regions can be defined. Below the glass transition temperature the material is solid and rigid with a modulus of elasticity of about 3500 KPa (500,000 psi); it is called the

Vitreous or glass state. In the glass transition region, the material is still a solid though considerably softened. Just above the T_v or T_g, polymers of a sufficient molecular weight retain a good level of elasticity but with a low modulus of elasticity of 0.7–7KPa (100–1000 psi) which is a characteristic of rubbers. The range of temperature, over which the value of R does not change, strongly depends primarily on the molecular weight of the thermoplastic, as shown by curves L and H. This region is known as the *rubbery plateau*. This rubbery behaviour is attributed to molecular entanglements. As the temperature is increased, the visco-ielastic material becomes increasingly viscous-fluidic.

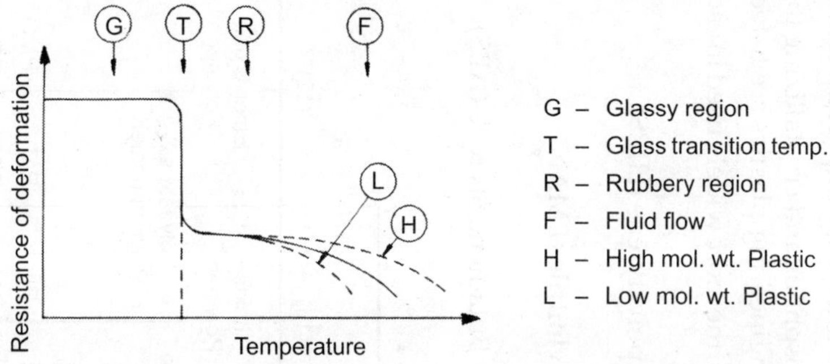

Fig 2.4: Effect of temperature on an amorphous thermoplastics.

Amorphous thermoplastics are transparent and have lower resistance to chemicals and solvents, than the crystallising thermoplastics.

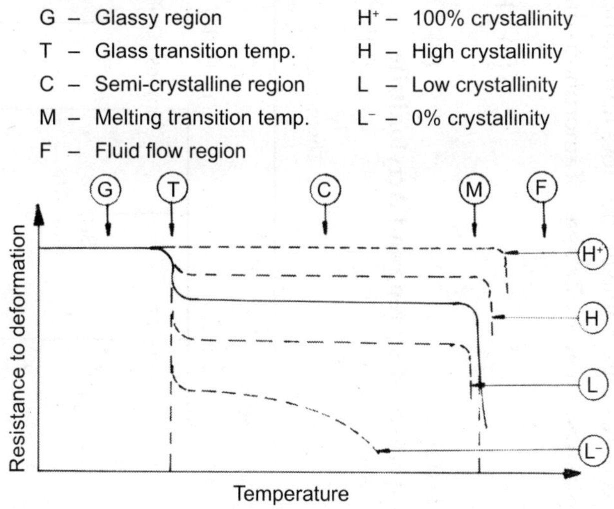

Fig 2.5: Effect of termperature on Crystallizing Thermoplastics.

In the case of crystallising thermoplastics, the resistance to deformation, R, of the solid at low temperatures is high, as shown in Fig. 2.5, but comparable to that of glassy/vitreous amorphous thermoplastics. As the temperature reaches T_g or T_v of the polymer, a drop in stiffness occurs that basically corresponds to the softening of the amorphous zones of the semi-crystalline material. The crystalline regions are, in principle, not affected. Further, heating below the melting point, T_m, hardly causes any further softening of the material and the material behaviour is, therefore, represented by a horizontal or a slightly drooping line. On reaching the temperature, T_m, the crystals, that were holding the material together as a solid, melt and the material becomes visco-fluidic with hardly any rubber-like elasticity unless the molecular weight is very high. The level of the semi-crystalline plateau between T_g (or T_v) and T_m depends upon the degree of crystallinity of the polymer as is shown by the four cases, from 100% to 0% crystallinity, in the decreasing order by H_1, H_2, L_1 and $_2$ respectively.

Crystallizing thermoplastics are generally opaque or transluscent and have better chemical and impact resistance than amorphous thermoplastics.

2.2 Classification of Thermoplastics

Detailed classification of thermoplastics is given in Fig. 2.6 and the description of commercially important amongst them follows.

As is evident from Fig. 2.6, there are two main groups of thermoplastics, viz., amosphous and crystallizing thermoplastics, and then there are their copolymers and polyalloys. Some of the more important industrial engineering plastics are listed as follows:

(A) Amorphous Thermoplastics: There are 7 classes of amorphous thermoplastics; the more widely used in each class are listed below.

1. **Styrenics**
   (i) Polystyrene (PS) (ii) High impact polystyrene (HIPS)
2. **Acrylonitrilics**
   (i) Polyacrylonitrile (PAN) (ii) Polymethacrylnitrile (PMAN)
3. **Copolymers and Polyalloys of Styrene and Acrylonitrile**
   (i) Styrene-acrylonitrile (SAN) (ii) Acrylonitrile Butadiene Styrene (ABS)

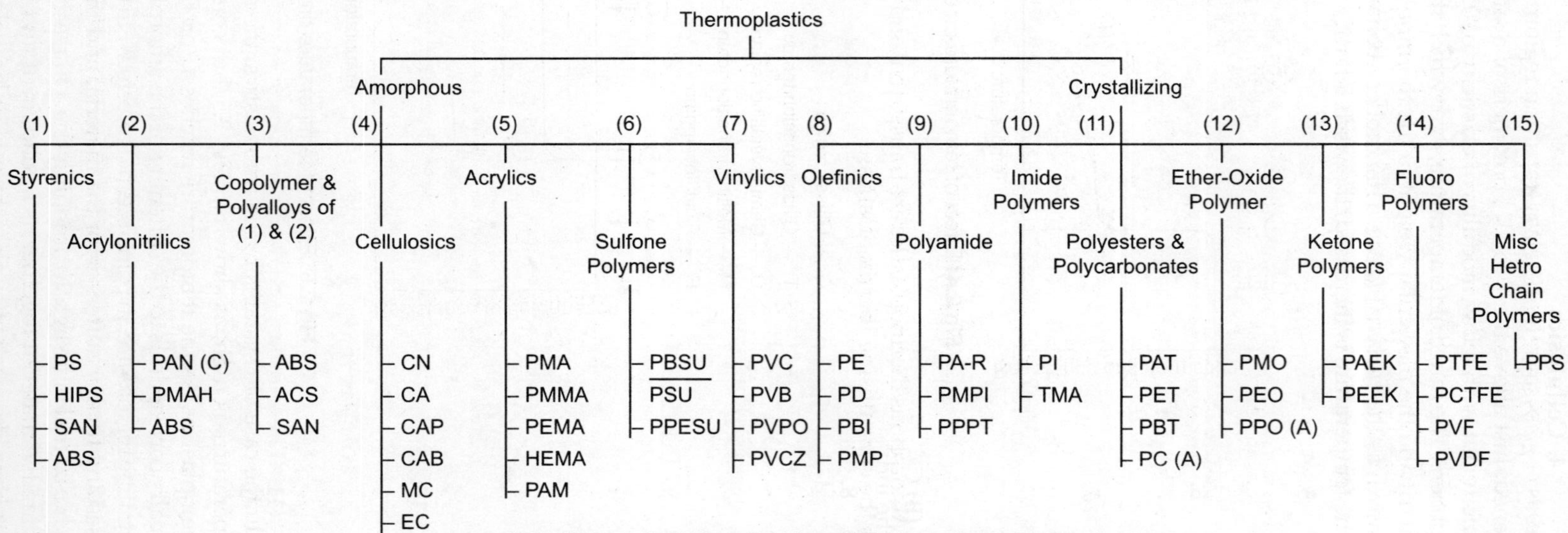

Fig 2.6: Detailed Classification of Thermoplastic Polymers.

4. Cellulosics

(i) Cellulosenitrates (CN)
(ii) Celluloseacetates (CA)
(iii) Cellulose acetate-propionate (CAP)
(iv) Cellulose acetate-butyrate (CAB)
(v) Methylcellulose (MC)
(vi) Ethyl Cellulose (EC)

5. Acrylics

(i) Polymethyl methacrylate (PMMA)
(ii) Hydroxyethyl methacrylate (HEMA)

6. Sulfone Polymers

(i) Polysulfone (PSU)/Polybisphenol sulfone (PBSU)
(ii) Polyether sulfone (PESU)

7. Vinylics

(i) Polyvinylchloride (PVC)
(ii) Polyvinyl butyral (PVB)
(iii) Polyvinyl pyrrolidone (PVPO)
(iv) Polyvinyl carbazole (PVCZ)

(B) Crystallizing Thermoplastics : There are eight classes of crystallizing thermoplastics. Some of the industrially more important amongst them are listed as follows:

8. Olefincs

(i) Polyethylene (PE)
(ii) Polypropylene (PP)
(iii) Polybutylene (PBI)
(iv) Polymethylpentene (PMP)

9. Polyamides (PA)/Nylons

(i) PA–R Aliphatic Polyamides (General Amides)
(ii) Aromatic Polyamides
 (a) Poly 'm' phenylene isophthal amide (PMPI)
 (b) Poly 'p' phenylene terephthal amide (PPPT)
(iii) Transparent Polyamide
 Polyamide (PA–(6, T))

10. Amide Polymers

(i) General Polyimide (PI–R)
(ii) Thermosetting and Thermoplastic, Polyimide
(a) Trimetallic acid anhydride (TMA)

11. Thermoplastic Polyesters & Polycarbonates

(i) Polyalkylene Terephthalates (PAT)
(ii) Polyethylene Terephthalates (PET)
(iii) Polybutylene Terephthalates (PBT)
(iv) Polycarbonates (PC) [Amorphous]

12. Ether-Oxide Polymers

(i) Polymethylene Oxides (PMO)/Acetals
(ii) Polyphenylene Oxides (PPO)–[Amorphous]

13. **Ketone Polymers**
    (i) Polyether ether Ketone (PEEK)
14. **Fluoropolymers**
    (i) Polytetra fluoroethylene (PTFE)
    (ii) Polychloro trifluroethylene (PCTFE)
    (iii) Polyvinyl fluoride (PVF)
    (iv) Polyvinylidene fluoride (PVDF)
15. **Miscellaneous Heterochain Polymers**
    (i) Polyphenylene sulfide (PPS)

Brief description of all the above listed thermoplastics follows:

(A) Amorphous Thermoplastics

2.3 Styrenics

The two widely used members of this class are:

(i) Polystyrene (PS) (ii) High impact polystyrene (HIPS)

2.3.1 Polystyrene (PS)

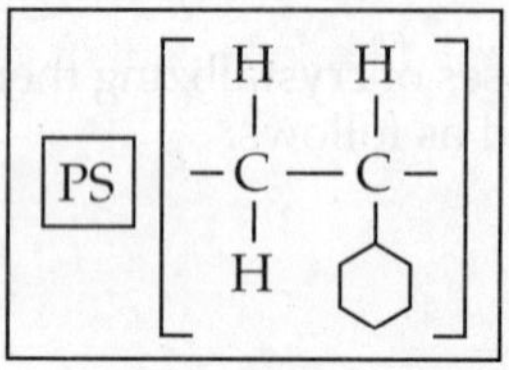

PS

T_g: 100° C (70 – 115)
212° F (158 – 239)

Polystyrene is one of the four (PE, PP, PS, PVC) major thermoplastics; and is available in a wide variety of grades. The homopolymer is of the amorphous type and below its glass transition temperature, T_g, it is very stiff and highly transparent (clear) glassy material. Its high gloss and sparkling appearance are associated with a high index of refraction. It can take up colour and be in opaque and transluscent form, all of which exhibit a brilliant surface gloss and appearance. Hence, low cost polystyrene finds numerous applications in decorative and commodity packaging; examples are cosmetics for the former and food stuffs for the latter.

The specific gravity of polystyrene is 1.07 and the plastic is hard and brittle. When polystyrene is in a stressed state, it has a tendency to craze (i.e. form network of fine cracks) at a stress lower than the failure stress. Polystyrene has a low resistance to weathering and ultraviolet (UV) radiation that causes yellowing and embrittlement; it is therefore not normally used in outdoor applications.

Many organic fluids, such as aromatic and chlorinated hydrocarbons as well as food, oil and fats can cause undesirable changes to polystyrene though properties and dimensions are not affected by moisture and water.

Polystyren may be fabricated by injection moulding, extrusion, and blow moulding; secondary operations like thermoforming, heat staking, ultrasonic welding, and hot stamping are also commonly used. Polystyrene has remarkable dielectric strength and is therefore widely used in electronics as insulator and capacitor material.

General-purpose polystyrene is used for novelities, toys, cigarette packs, low-cost disposable serviceware, cabinets as well as for pipes and sheets which are subsequently thermoformed into small or large objects.

Polystyrene is also widely used as a foam for insulation or packaging applications.

2.2.1.1 *High Impact Polystyrene (HIPS)*

Polystyrene can be modified for improved impact strength by the addition of rubber polymer to the styrene. Rubbers such as polybutadiene (PB) and styrene butadiene (SBR) are added to the polystyrene by blending or graft polymerization. These rubbers/elastomers may be combined with polystyrene in percentages varying from 10 to 40%, depending on the degree of ductility desired in the finished product. Rubber modified styrene called high impact polystyrene exhibits lower strength, hardness, clarity, and chemical resistance compared with general-purpose polystyrene.

HIPS finds use in making refrigerator door liners, appliance housings, luggage, furniture components, closures for bottles, shoe heels, toys, and sports equipment. Heat-resistant version is selected when warpage of other grades may occur. Typical uses of this form of HIPS include radio and television housings; telephones, and automotive interior trim parts.

2.3.2 Acrylonitrilics

There are several polymers that have repeat units that feature a nitrile (C ≡ N) side group, however only the following two members deserve discussion from the engineering point of view:

(i) Polyacrylonitrile (PAN) ———— Crystallizing

(ii) Polymethacrylonitrile (PMAN)

2.3.2.1 *Polyacrylonitrile (PAN)*

PAN

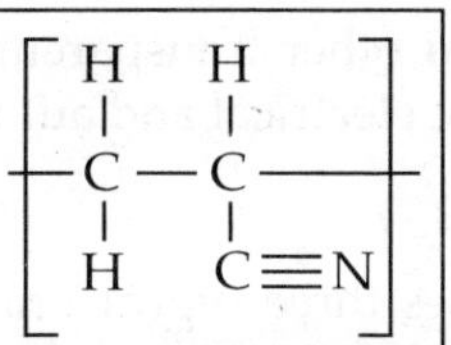

PAN

T_g: 100° C (80 – 130)
212° F (176 – 266)

T_m: 317° C
603° F

Polyacrylonitrile (or polyvinylcynide) homopolymer is polar crystallizing polymer that undergoes chemical decomposition before crystalline melting at temperatures above 300° C (572° F). It is, therefore, not a true thermoplastic material. The almost pure homopolymer, however, can be processed into fibres (acrylic fibres) through its spinning from solutions.

Commercial acrylic textile fibres contain at least 85% acrylonitrile, combined, for example, with 5–15% of methylmethacrylate repeat units.

PAN has a very good resistance to most chemicals, as well as to heat and weathering. It is also a superior barrier material. Acrylonitrile repeat units help confer these properities to thermoplastic copolymers in which they are involved.

2.3.2.2 *Polymethacrylonitrile (PMAN)*

PMAN

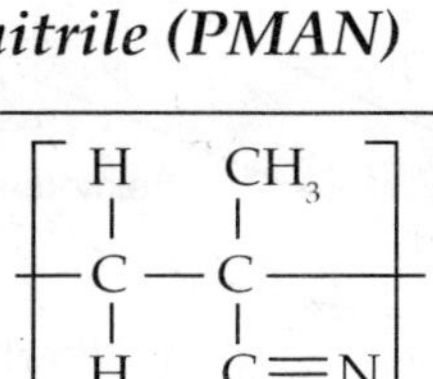

T_g: 120° C
248° F

Polymethacrylonitrile homopolymer appears to be an amorphous thermoplastic with its glass transition temperature at 120° C (248° F). Methacrylonitrile repeat units are dominant (90%) in copolymers, involving about 5% of styrene repeat units and also 5% of alphamethyl styrene. Such a copolymer is transparent, is an excellent barrier to carbondioxide and possesses good creep resistance making it a suitable material for carbonated beverage containers.

2.3.3 Copolymers and Polyalloys of Styrene and Acrylonitrile

Commercial polymer systems can become very complex when their chemical and physical structures are tailored to fulfil the desired needs of the final product. Manufacturers, obviously, do not generally publicise the structure of their specific resins and grade. Two such plastics of considerable importance, in this class are:

1. Styrene-acrylonitrile (SAN)
2. Acrylonitrile-Butadiene-Styrene (ABS)

2.3.3.1 *Styrene-Acrylonitrile (SAN)*

SAN is obtained by random copolymerization of 20–30% polyacrylonitrile repeat units with polystyrene. These plastics have many of the useful properties of polystyrene like surface appearance and ease of processing as well as develop improved temperature, and chemical resistance.

SAN has a softening point about 10° C higher than that of polystyrene, retains transparency, although with a tendency to yellowing, particularly on exposure, and is somewhat tougher and stronger than polystyrene. The characteristic yellow tint of SAN is, however, usually masked by adding dye to produce an effect similar to smoked glass.

The better resistance to food and body oils makes SAN plastics prime material for transparent houseware personal care items. They are also widely used for transparent medical products and for meter or light lenses.

SAN is also used for drinking tumblers jugs and other transparent domestic wares and for transparent and reasonably tough covers for domestic electrical and automotive equipment.

2.3.3.2 *Acrylonitrile Butadiene Styrene (ABS)*

ABS as a family of thermoplastic polymers comprises three repeat units that vary according to the grade with representative ranges, such as acrylonitrile 20–30%, butadiene 20–30% and styrene 40–60%. In a simplified way it can be said that acrylonitrile provides chemical and temperature resistance, butadiene accounts for improved strength, and styrene takes care of cost and processibility.

By varying the content of both acrylonitrile and butadiene in styrene, a wide range of specific materials with concurrent wide property ranges becomes available. ABS in all versions possesses greater impact strength and ductility than does polystyrene but with comparatively low rigidity and strength for structural use.

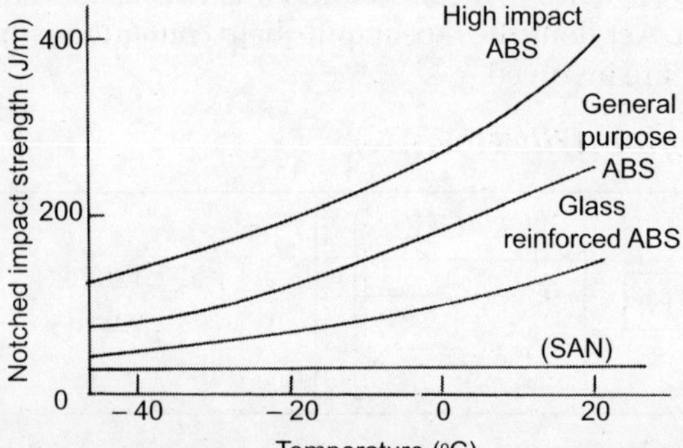

Fig 2.7: Impact properties of different grades of ABS and that of SAN.

The rigidity of ABS can be improved by reinforcing fillers; glass fibre composites are quite popular, although the excellent surface finish and impact strength are reduced. Fig. 2.7 shows that impact properties as affected by temperature, for SAN and various grades of ABS. Fatigue properties of ABS are superior to those of HIPS, as is evident from Fig. 2.8.

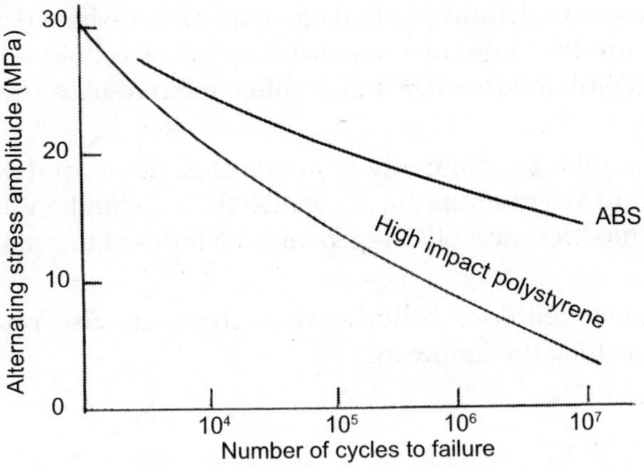

Fig 2.8: Fatigue behaviour of ABS and HIPS.

Normally classified as slow burning, ABS can be blended with PVC to make it flame resistant for applications demanding this characteristic.

Chemical resistance of ABS to alkalies and some acids is good, although ABS is subjected to stress cracking when exposed to certain acids and organic solvent.

Applications of ABS plastics reflect the processing behaviour and the high impact strength; vacuum cleaner covers, carpet sweeper housings, lawn mower housings, telephone handsets and hair dryer housings of higher softening points than HIPS.

High gloss and unlimited colourability make ABS plastics well suited in decorative applications, such as automotive interior trim parts, instrument panels, and television and radio cabinets. The first plastic to be successfully chrome-plated in large volume parts, ABS has found its way into demanding applications like automobile grills. High impact resistance and moderate cost enable ABS to be used in safety helmets, sports car underbodies, and some furniture parts.

2.3.4 Cellulosics

Cellulosics are the oldest of the family of plastics and were discovered in 1833. They found commercialization in celluloid collars that were popular by the late ninteenth century. Unlike other plastics that are produced by polymerization, cellulose occurs abundantly in nature. Cotton linters and wood pulp are the usual raw materials for cellulose polymers.

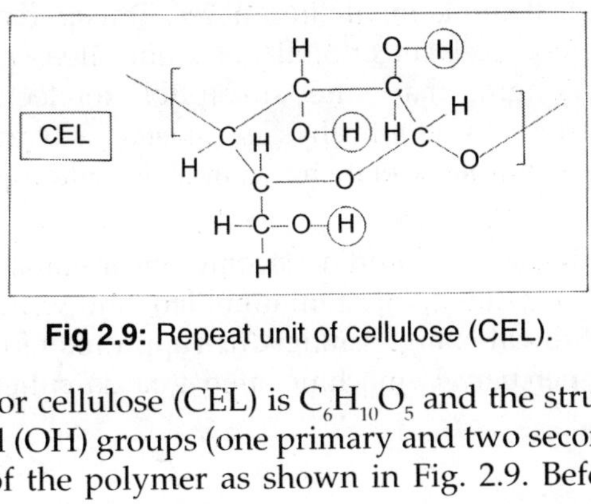

Fig 2.9: Repeat unit of cellulose (CEL).

The empirical formula for cellulose (CEL) is $C_6H_{10}O_5$ and the structure of repeat unit features three characteristic hydroxyl (OH) groups (one primary and two secondary) with hydrogen bonds and the high crystallinity of the polymer as shown in Fig. 2.9. Before cellulose crystals can be melted, chemical degradation occurs, and thus cellulose does not behave as thermoplastic nor has a measurable melting temperature, T_m.

Cellulose can be given a chemical treatment that can lead to the destruction of crystalline structure, and a cellulose-based fluid is obtained that can be shaped into films and fibres. A subsequent treatment returns the molecules back to their original cellulose structure and the material to crystalline state. The typical intermediate fluid is often referred to as *viscose* and the final product as regenerated cellulose.

Regenerated cellulose film is commonly known as *Cellophane* and was the first transparent packaging or wrapping film to be manufactured commercially. Cellophane is rather strongly affected by moisture, but not by food fats and oils. Regenerated cellulose fibres are known as *Viscose rayon* or just *rayon*.

Apart from regenerated cellulose, cellulose derivatives are also produced; the industrially important among them include the following:

1. Cellulose nitrates (CN)
2. Cellulose acetates (CA)
3. Cellulose acetate-propionate (CAP)
4. Cellulose-acetate-butyrate (CAB)
5. Ethyl Cellulose (EC)

Brief descriptions of these follows.

2.3.4.1 *Cellulose nitrates (CN)*

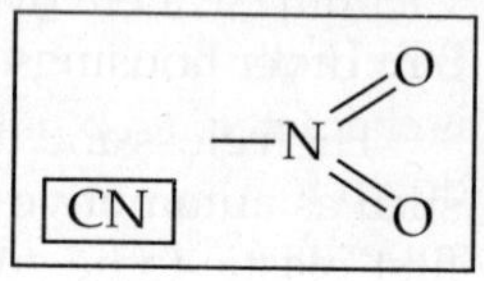

The substituent in cellulose-nitrates, for the encircled hydrogen atoms of the cellulose structure (Fig. 2.9) is NO_2. CN is an organic ester and was the first man-made plastic introduced in mid-ninteenth century. The properties of CN depend upon the degree of nitration (Nitrogen content varies between 11 and 14%) and the type and amount of plasticizer, such as camphor. Advantages claimed for this plastic include toughness, water replency, and good surface appearance. But it is highly inflammable and difficult to process by the established methods.

The most significant use of CN or celluloid is in the manufacture of table-tennis balls and spectacles frames.

2.3.4.2 *Cellulose acetates (CA)*

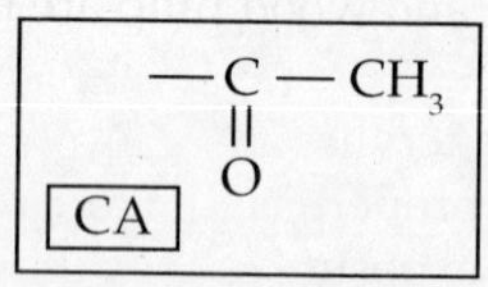

The substituent in cellulose acetate is the acetyl group $COCH_3$. The fully substituted derivative (44.8% substitution) like cellulose, is not thermoplastic. As other crystalline polymers, it can be made into fibres. Transparent films are also made, which have good resistance to oils, fats, and greases. CA is also used for transparent packaging that is not susceptible to electrostatic charge and hence does not attract dust. Optically clear sheets of CA are used for audiovisual purposes, for example, acetate transparencies, photographic films, etc.

Less substituted derivatives or secondary acetate, with substitution up to 40%, are true thermoplastics and are easily extruded, injection moulded, and vacuum formed. Unlike cellulose nitrates they are slow burning to self-extinguishing. Chief applications include tool handles, furniture wear strips, sunglass frames, pen barrels, machine safety guards, cutlery handles, blister packaging, and toys.

2.3.4.3 *Cellulose-acetate-propionate (CAP)*

It is a mixed cellulosic ester involving propionyl substituent (CO C_2H_5) (39–47%) along with acetyl substituents (2–9%).

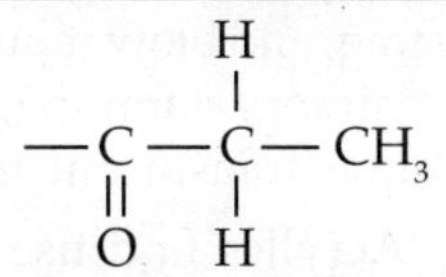

CP Cellulose propionate

Propionate materials combine good toughness, transparency, dimensional stability, and weatherability. All conventional thermoplastic processing secondary operations like hot stamping can be performed with propionate. Solvent welding may be used to join propionate parts.

Major applications of CAP include automobile steering wheels, toothbrush handles, pen and pencil barrels, flashbulb shields, tool handles requiring impact, television screen shields, outdoor signs, and lettering. Control buttons, telephone dials, toys, tabulator keys, spectacle and sunglass frames, decorative trims for cars and domestic appliances are some other common applications of CAP.

2.3.4.4 *Cellulose Acetate Butyrate (CAB)*

CAB is also mixed cellulosic organic acid with butyryl (CO C_3H_7) substituent (26–39%) alongwith acetyl substituents (12–15%).

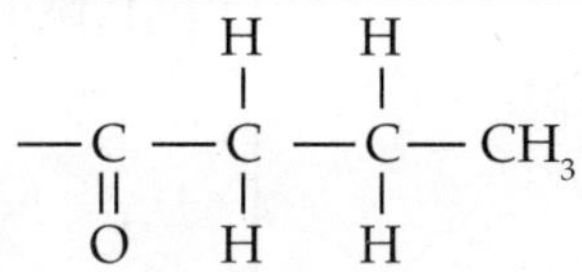

CB Cellulose butyrate

CAB competes closely in price and properties with CAP. It has high impact strength and toughness, combined with higher tensile strength than CAP. Applications include type writer and calculator machine keys, skylights, windshields for boats and snowmobiles, lenses for boat and vehicle lights, street lamps, blister packaging, glass-frames, and outdoor signs and lettering.

2.3.4.5 *Ethyl Cellulose (EC)*

Ethyl cellulose involves 45–48% substitution and is a genuine thermoplastic that is characterised by its ability to withstand shock over a wide range of temperature down to – 40°C. While tensile strength is moderate, its modulus of elasticity, and superior resistance to alkalies relative to that of other cellulosics enable ethyl cellulose to be used in coatings and adhesives, automobile arm rests, helmets, radio cabinets, and rollers. EC has a unique property of resistance to nitroglycerine and black powder which leads to its use in defence applications.

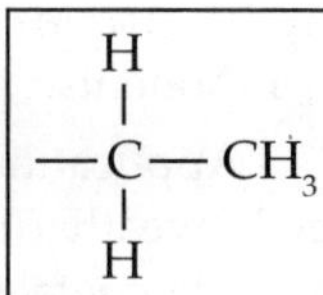

EC

While designing parts to be made of any of the cellulosics, it should however be kept in mind that dimensions of such parts will vary with change in humidity and temperature. Also, cellulosics have a tendency to creep or cold flow when under any load so their use should be avoided in load bearing components.

2.3.5 Acrylics

Acrylics are amorphous thermoplastics which are renowned for their clarity and light transmission properties combined with excellent retention of their physical and optical properties even after long exposure to the outdoor conditions.

Although the strength and stiffness of acrylics are similar to those of nylon and acetal but their use is generally limited to temperature below about 95°C. Electrical properties of acrylics are good at low frequencies. These materials burn slowly and a low degree of smoke is emitted that makes acrylics useful in many building applications like decorative panels and light diffusers.

Acrylics may be modified and copolymerized to improve certain properties like impact strength and heat resistance. However, such modifications generally impair optical properties which are the main attractive feature of acrylics. Other properties affected badly by such modifications are the dimensional stability and colour retention when acrylics are exposed to ultravidet light.

Manufacturing processes employed for the fabrication of acrylics include moulding, extrusion, casting, and blow moulding. Secondary methods employed for shaping acrylics are thermoforming, machining, stamping, printing, and metallizing. Colours are possible in all types of acrylics viz., opaque, translucent and transparent acrylics.

Acrylics find use in airplane canopies, window glazing, tail lamp assemblies, medallions, reflectors, outdoor signs and lighting, building panels, watch crystals, and fibre optic light pipes.

Two well-known commercial acrylics are:

1. PMMA — Polymethyl Methacrylate
2. HEMA — Hydroxyethyl Methacrylate

Brief description of these thermoplastics follows.

2.3.5.2 *Polymethyl methacrylate (PMMA)*

PMMA

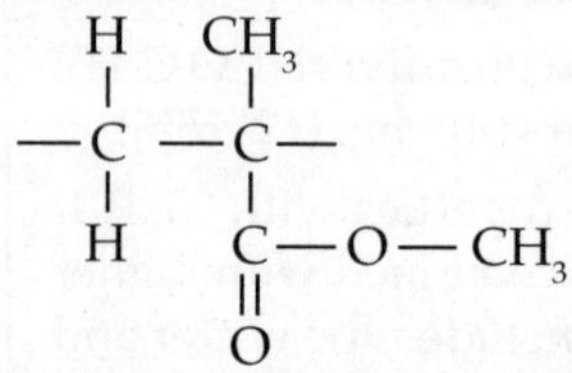

T_g: 105°C (70–120)
221°F (158–248)

PMMA is best known for its remarkable property of excellent transparency. Its optical properties are supported by good rigidity and acceptable impact strength and abrasion resistance, outstanding weatherability, and good general chamical resistance to salts, alkalies and dilute acids, detergent solutions, greases and oils. However, some solvents including alcohols, ketones, and chlorinated hydrocarbons are partial or complete solvents and promote crazing particularly in stressed components.

Applications of PMMA are principally concerned with transparency and light transfer, with good weathering resistance, high surface hardness and durability. They include such diverse objects as hard contact lenses, aircraft and other transport glazing, safety and security glazing, skylights, transparent guards and covers, rear light assemblies for road transport, bath tubs, wash basins and sanitary wares, PMMA cast sheet has become established as the most suitable material for bath tubs as it is rigid and hard, so that the surface does not scratch easily; moreover it is not affected by the different chemicals used in baths. It is also used for coatings, such as floor waxes.

2.3.5.2 *Hydroxy Ethyl Methacrylate (HEMA)*

HEMA

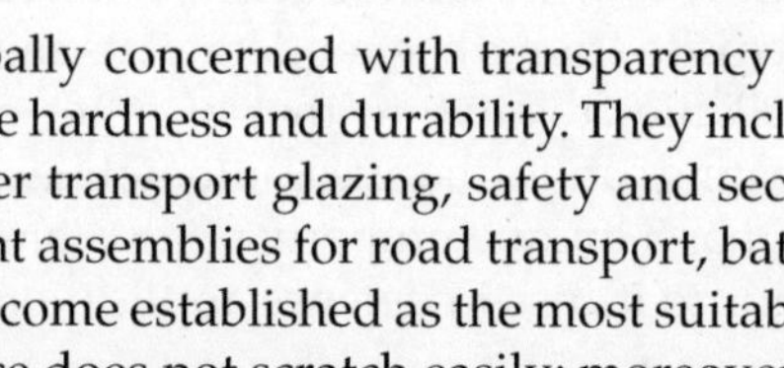

HEMA homopolymer is water soluble, but when it is polymerized in the presence of ethylene glycol dimethacrylate (EGDMA), which acts as a cross-linking agent through its four available links, a network polymer is formed. The network polymer so formed is not soluble in water but can be swollen to an extent, depending upon the degree of cross-linking, to form what is referred to as a *hydrogel*. The soft hydrogel possesses the combination of properties, including good oxygen permeability, required to produce soft contact lenses.

2.3.6 Sulfone Polymers

This is comparatively a new group of plastics that emerged in mid-1960s. Sulfone polymers are so named because of the characteristic - SO_2- (Sulfone) group which is always flanked along the chain backbone by two phenyl groups that contribute rigidity, thermal-, oxidative-, and hydrolysis for resistance to a polymer chain. The chemplate or structural formula for DPSU (diphenyl sulfone) is as shown.

DPSU

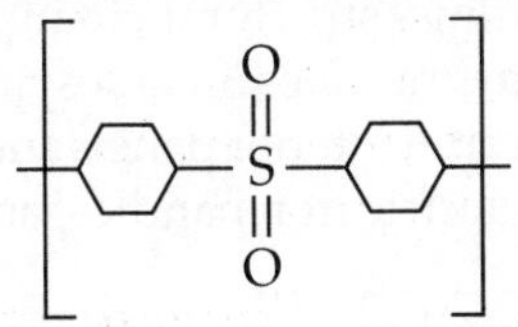

The common feature of sulfone polymers is that although the structures appear formally to be linear, they in fact all contain a pronounced *kink* or *dog-leg* which leads to a reluctance to crystallize. Indeed one member of the family which does not possess this kinked feature is crystalline with a melting point exceeding 500°C.

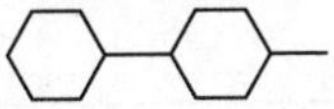 — SO_2— melting point > 500°C with decomposition.

Although possessing an apparently regular structure, most polysulfones available commercially are amorphous, and consequently transparent, after melt processing. Two of the well known thermoplastics of this group are:

1. Polysulfone (PSU)
2. Polyethersulfone (PESU)

Brief description of these plastics follows:

2.2.6.1 *Polysulfone (PSU)*

PSU

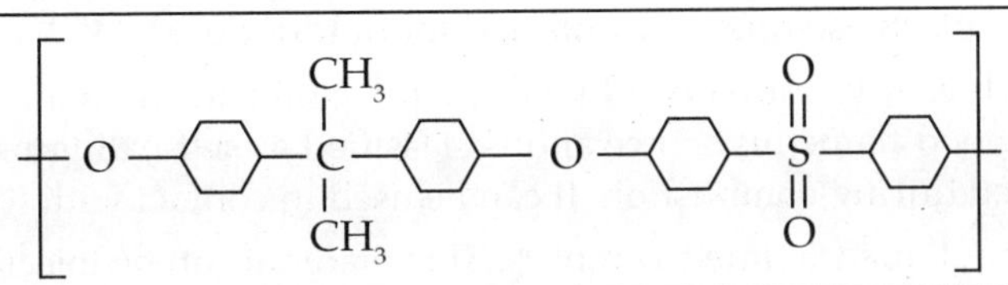

T_g: 190°C (175–190)
374°F (350–374)

Polysulfone which is also referred to as polybisphenolsulfone (PBSU) is a polymer with chain that is generally stiff, giving a high glass transition temperature. The material is amorphous and consequently transparent and is available in light amber-tinted grades.

PSU has good mechanical properties particularly rigidity, creep resistance and toughness, which are little affected by large temperature changes. Electrical properties are also good and stable.

The hydrolysis resistance is particularly effective and PSU can undergo repeated steam sterlization. The resistance to chemicals including detergents, oils, gasoline, and alcohols is generally good. However, strong polar solvents like ketones, aromatic hydrocarbon solvents, and halogenated solvents attack this plastic adversely. Like other amorphous polymers, PSU can be subject to environmental stress-cracking (ESC).

Resistance to ultraviolet (UV) radiation or sunlight is not very good, so adequate protection is required for outdoor exposure. PSU is rated as self-extinguishing with low smoke and low toxic fume generation so is found effective for fire-sensitive applications.

Injection moulding is well suited for PSU as the melt is very stable at high processing temperature up to 370°C (700°F). The viscosity remains quite high and the mould temperatures are around 95°C (203°F). Extrusion is used to produce thin films and sheets which can be thermoformed.

Low mould shrinkage, dimensional stability, self-extinguishing nature, ability to be electroplated and withstand steam sterlization enable PSU to be used in varied industry, for example, medical field for sterlization, food industry, and for appliances (piping and fittings, coffee makers, microwave wares); in the electrical-electronic industry for connectors, fuse boxes, coil bobbins, capicitor films, structural circuit boards; in the chemical industry electroplating tanks, pumps and piping, membranes; in aerospace for astronaut helmets and face shields; and in transportation for aircraft cabin interior panels and trim. PSU is also used for kitchen ware, alkaline battery cases, wire coating, electric iron handles, and shower heads.

2.2.6.2 *Polyethersulfone (PESU)*

PESU 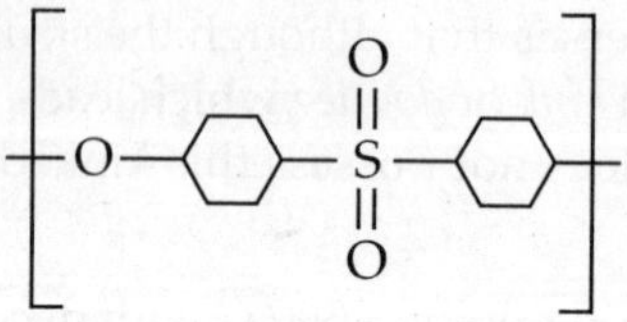	T_g: 221°C (210–230) 430°F (410–446)

The glass transition temperature, T_g, of PESU is high but manageable for thermoplastic processing. Since the materials are amorphous they are likely to be transparent.

PESU is rigid, strong and tough with very low creep over a wide range of temperature reaching up to 150°C (302°F).

PESU is an aromatic highly polar polymer; thus, as expected, it is resistant to aqueous acids, bases and most salts and to hydrocarbons including petrol, oils and greases even at high temperatures. However, it is attacked by esters, ketones and some aromatic or halogenated solvents. As an amorphous polymer it can be sensitive to environmental stress cracking (ESC). It is not good weathering or UV-resistance plastic so must be protected for outdoor use.

PESU is resistant to high energy radiations like β, γ, and x-rays in the temperature range of 20–200°C (68–392°F). It has good flame resistance and is classified as self-extinguishing, with little smoke or toxic fumes emitted during combustion. It can be used in contact with foods.

PESU has an ability to be shaped to fine tolerances. This material can be injection moulded on conventional equipment.

PESU can be used for tens of thousands of hours at 200°C (392°F) without significant loss in strength. This makes this material suitable for a number of applications including internally illuminated push-buttons, switches, and indicator lights for cookers. Most other applications also normally involve high temperature situations encountered in appliances such as oven, dryers, irons, heaters, high intensity lamp sockets, reflectors and lenses, and steam-sterilizable medical equipment. It also finds use in aerospace applications, bases for integrated circuits. A typical application is the inlet manifold between carburettor and the cylinder in a petrol engine power chain saw.

2.3.7 Vinylics

A few commercially important polymers are discussed here under as a group, referred to as *Vinylics*, because of certain similarities in their chemical structures and the presence of the word *vinyl* in their common name. The major polymer in this group is polyvinylchloride (PVC); however a few other commercially significant polymers are also briefly described, and in all include the following:

1. Polyvinylchloride (PVC)
2. Polyvinylbutyral (PVB)
3. Polyvinyl pyrrolidone (PVPO)
4. Polyvinyl carbazole (PVCZ).

2.3.7.1 *Polyvinyl Chloride (PVC)*

PVC

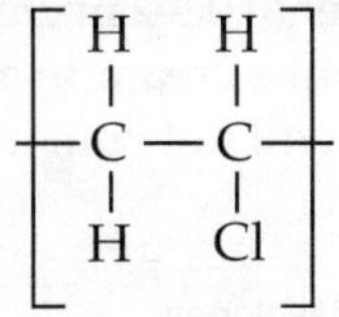

T_g: 70°C (70–90)
158°F (158–194)

PVC is one of the four major, highest volume plastics although it is one of the least thermally stable polymers. The monomer is now almost exclusively obtained from ethylene.

PVC has good general chemical resistance though some polar organic solvents allow easy solvent welding. Weatherability can be excellent, especially with the incorporation of suitable additives that has led to its wide spread exterior use in construction.

PVC is flame-resistant and is classed as a self-extinguishing polymer. PVC has high molecular weight, often denoted by *K-value**, that tends to improve many properties, but create processing difficulties which are overcome by the addition of heat stabilizers. PVC has good insulation properties for low frequency electrical systems. At high frequencies dielectric losses are high and in fact, the resulting heat generation is utilized usefully in dielectric heating or sealing.

In general, it can be said that all vinyl materials are used below 93°C (200°F). They are resistant to most reagents but are soluble in ketones, and resist water permeation.

One attractive feature of PVC is its ability to accept large amounts of *modifiers* that can change its properties considerably and lead to a wide range of applications. While unmodified/unplasticized PVC (rigid PVC) homopolymer is stiff and relatively brittle; the plasticized (flexible) PVC can show rubber like behaviour.

Rigid PVC is a stiff material and is tough under ambient conditions provided there is no stress concentration. The softening point is about 85°C; but it is not reliably load-bearing at 60°C or above. It is resistant to water and hydrocarbons, but is soluble in polar solvents.

Successful processing of rigid PVC requires the use of lubricants. Mixtures may be employed, for example, lead stearate, calcium stearate, and wax or stearic acid. Rigid PVC is frequently extruded and lubricants having melting point of 100–120°C are most effective for extrusion at 165°C. Rigid PVC is also widely used in sheet form, obtained by compression moulding multiple piles of calendered foil.

Rigid PVC is self-extinguishing and resist creep within a specific temperature range, thus making it suitable for chemical, agriculture, and domestic piping.

Typical applications of rigid PVC include rainwater goods, sheet used in hygenic cladding, clear bottles, curtain rails, PVC storm windows, electrical connectors, rain gutters, down spouts, conduits, and containers, etc.

Plasticized/Flexible PVC: Plasticizers are high boiling point liquids which are solvent for PVC; their addition reduces glass transition temperature, T_g, for the PVC compound until the material is rubbery. Most of the plasticizers are derivatives of phthalic acid, for example, dioctyl phthalate. Other plasticizers used include phosphate plasticizers like trixylyl phosphate, and aliphatic acid esters with alcohols.

Flexible PVC has replaced rubber and other materials in areas like gasketing, automobile wire connectors and vacuum tubing, packaging, toys, floor mats, crash pad covers, garden hoses, baby pants, shower curtains, electric wire insulation, leathercloth, footwear, flexible films, shoe soles, etc.

---

* K-Value: Molecular weight is recorded as K-value, which is derived from solution viscosity measurements. K-values from 45 to 80 (cyclohexanone) cover number-average molecular weight from 25,000 to 75,000.

PVC is also available in liquid-like form as water-based emulsions (latex) or as suspensions of suitable particles in plasticizer (plastisols, organosols) that are used for special coating techniques.

Vinyl Plastisols: When a finely divided PVC powder (0.02–2 microns) is dispersed in a plasticizer, it is called a *plastisol*. Plastisols vary in consistency from a cream to a heavy paste. If the plastisol is thickened with gelling or thickening agents so that it will not run, it is called a *plastigel*.

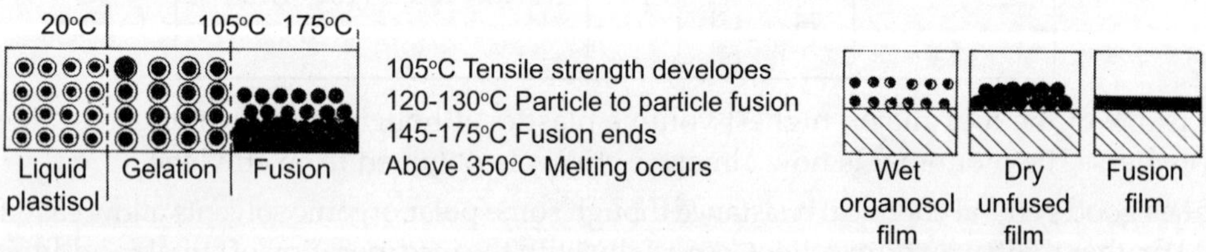

(a) Plastisol (b) Organosol

Fig 2.10: Effect of temperature on the behaviour of the two PVC dispersions; vinyl plastisol and vinyl organosol.

When heat is applied to a plastisol, the resin concentration increases until a final fusion temperature of about 177°C (350°F) is reached, as is illustrated in Fig. 2.10 (a). Plastisols are used in slush and rotational moulding, dipping, spraying, film casting, and coating.

Vinyl Organosol: An organosol is a suspension of a finely divided resin (generally PVC) in a plasticizer with a volatile organic liquid. If a plastisol is modified with a volatile solvent or diluent, and the solvent or diluent evaporates upon heating, the plastisol is considered an *organosol*. Thus, an organosol can also be prepared from a plastisol, by merely adding a volatile diluent or solvent that serves to lower the viscosity and evaporates when the compound is heated, as is illustrated in Fig. 2.10 (b).

Organosols are used in coating substrates such as sheet metal. The substrate is cleaned, primed, coated and baked. The solvents are first heated off from the organosol, and final fusion takes in a fashion similar to that in a plastisol.

There is also an important copolymer of PVC called Polyvinylchloride-vinylacetate [P (VC–VAC)]; which is briefly described as follows:

Polyvinyl Chloride-Vinylacetate [P(VC–VAC)]: This copolymer is obtained by internally plasticizing PVC with 3-30% of vinylacetate repeat units. Impact properties, in particular and processing ease are improved. Moulded, extruded or calendered parts often contain 10-15% vinylacetate. P (VC – VAC) is used for phonograph records and floor coverings. Fibres are also made from this copolymers.

2.3.7.2 *Polyvinyl Butyral (PVB)*

Polyvinyl butyral corresponds to butyraldehyde, and in a partially converted form that may be considered as a random copolymer P (VB–VAL), i.e. monomers of vinylbutyral and vinylalcohol containing 10-15% of vinylalcohol repeat units. It has long been used in plasticized form for the adhesive layer of laminated (triplex) safety glass, which requires toughness, transparency, weatherability, and excellent adhesion to glass.

PVB

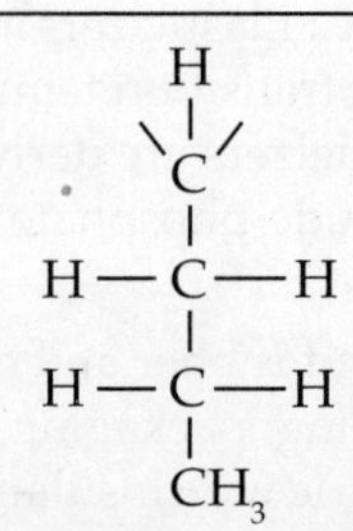

T_g: 50°C (49–60)
122°F (120–140)

2.3.7.3 *Polyvinyl Pyrrolidone (PVPO)*

PVPO 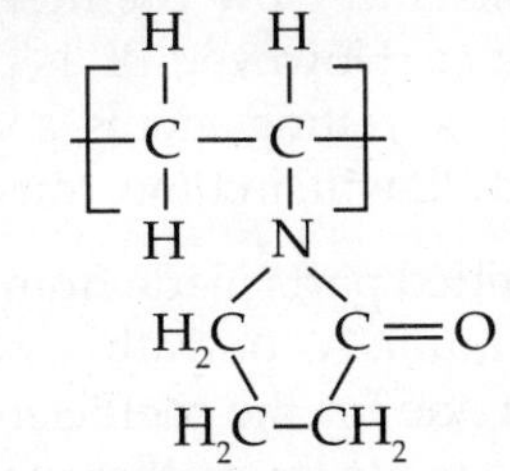	T_g: 175°C 347°F

PVPO is not considered a true plastic. It is highly polar and water soluble, and finds use in adhesives and as a water thickner. Water solutions of PVPO can be used as blood plasma substitute referred to as *artificial blood*.

2.3.7.4 *Polyvinyl Carbazole (PVCZ)*

PVCZ 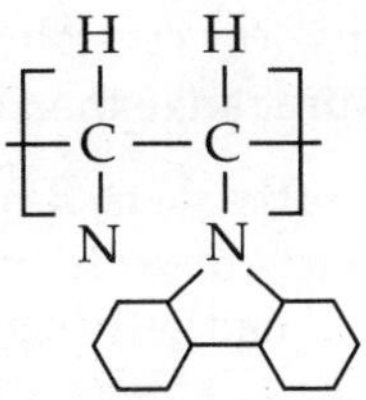	T_g: 190°C (180–210) 347°F (356–410)

PVCZ is a true plastic with a high softening point. Reported uses of PVCZ include high frequency dielectrics and photoconductive polymer for xerography.

(B) Crystallizing Thermoplastics

2.3.8 Olifincs

This group of crystallizing thermoplastics consist of important commercial plastics that feature exclusively repeat units with carbon (C) and hydrogen (H) atoms and no ring structures. Four important plastics and a copolymer in this group are:

1. Polyethylene (PE)
2. Polypropylene (PP)
3. Polybutylene (PB1)
4. Polymethyl pentene (PMP)
5. Inomers — copolymer

2.3.8.1 *Polyethylene (PE)*

PE 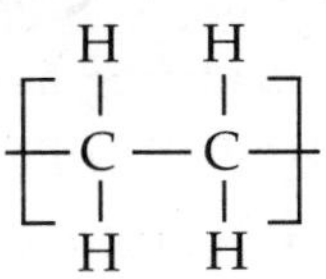	T_g: – 100°C (– 20 to – 180) – 148°F (– 4 to – 292) T_m: 120°C (150–138) 248°F (221–280)

Polythylene is the oldest of the olifincs and is one of the most widely used low-cost plastics available. It is, infact, one of the four high volume commodity plastics; the other three being PP, PS, and PVC. Its discovery in the 1930s was one of the landmarks of the industry.

Processable by all manufacturing processes applicable to thermoplastics, polyethylene is noted for its flexability, low temperature impact resistance, low coefficient of friction, good dielectric strength, and generally good chemical resistance. However, PE is prone to environmental stress cracking (ESC). Thus, ordinary chemicals such as wetting agents, soaps, detergents, silicones, and certain hydrocarbons when contacting stressed PE will, in time, cause cracking.

Close tolerances on the dimensions of moulded parts made from PE are difficult to hold. Close tolerances should be avoided or kept to aminimum. As polyethylene expands about twice that of most common plastics, allowance should be made for the coefficient of thermal expansion when the polythylene part is to be assembled to parts made from other materials.

Not a single material, but a whole family of materials with wide range of properties, polyethylenes are classed into low density polyethylene (LDPE), high density polyethylene (HDPE), and ultrahigh molecular weight polyethylene.

Low Density Polyethylene (LDPE): Low density PEs exhibit branching of the molecular chain that minimises the degree of crystallinity possible and hence the spaces between the molecules cause a low density in the range of 0.910–0.925 g/cm^3. Thus, low crystalline LDPE is flexible, transparent to translucent, and has a lower maximum temperature range than does HDPE.

LDPE finds use in covers for refrigerator dishes, soft gaskets, flexible film for packaging, squeeze bottles, general domestic wares like bowls, buckets, etc., domestic and agricultural plumbing items, wire insulation, cold water tanks, paper coating, and road safty signs, etc.

High-Density Polyethylene (HDPE): High density polyethylene is comparatible free of branching and is, therefore, called linear polyethylene. Linear PE is crystalline, therefore, compared with low density materials these polyethylenes are stiffer, stronger, more opaque, less resistant to impact, but more resistant to higher temperatures.

Applications for HD polyethylenes include dishpans, garbage pails, automobile cowl and trim panels, rigid detergent bottles, beverage cases, toys, refrigerator dishes, wire insulation, mudguards for transport vehicles, and milk bottle crates, etc. A unique application of very thin film of HDPE is as tissue paper, point-of-sale wrapping, particularly of wet food-stuffs, such as meat and fish.

Ultrahigh Molecular Weight Polyethylene (UHMWPE): Ultrahigh molecular weight polyethylenes have very high molecular weight in the range of 3×10^6 – 6×10^6. This leads to high density polyethylenes with the densities ranging between 0.93 - 0.94 g/cm^3. Above the fluidic temperature (T_m) of abcut 130°C, the material behaves in a rubber-like rather than fluid-like manner, causing considerable processing difficulties. That is why methods have been developed to produce parts directly from powdered resin.

The chemical inertness of UHMWPE is almost unmatched and includes environmental stress-cracking (ESC) resistance and resistance to foods and physiological fluids. Its most outstanding property is wear or abrasion resistance, a very low coefficient of friction, excellent impact resistance or toughness, and fatigue resistance. These properties and a moderate cost make its use popular in large scale materials handling equipment, e.g. for chemical and mining, as well as in many specialized applications such as pen tips, wear surfaces of artificial human body parts, gears, etc.

The wide range of properties and processing methods combined with unlimited colourability enable PE to retain its position as one of the most demanded plastics in use. In addition to being used in itself, PE may be blended or copolymerized with other materials to obtain an even wider range of properties that may be required for some specific applications.

2.3.8.2 *Polypropylene (PP)*

PP 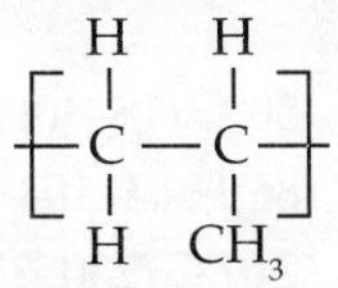	T_g: – 20°C (– 5 to – 24) – 4°F (– 23 to – 11.2) T_m: 175°C (165 to 180) 347°F (329 to 356)

Polypropylene is one of the four very high volume plastics; others being PE, PS and PVC. In many respects PP is similar to medium- or high-density PE.

Mechanical properties generally correspond to moderately rigid and tough material. Creep resistance is particularly good in comparison with corresponding polyethylene. PP is a crystalline thermoplastic that is stiffer than PE and has one of the lowest densities of all plastics. It is translucent and has a milky-white natural colour. It has excellent colourability and can be pigmented to unlimited array of colours.

The excellent physical and electrical insulating properties of PP are due to its crystalline structure. It is immune to environmental stress cracking (ESC); however it is prone to low-temperature brittleness which can be overcome in its copolymers. PP can be modified to obtain improved properties by adding fillers such as asbestos or glass fibres.

Polypropylene is easy to process by injection moulding, extrusion, and blow moulding. A special feature of PP is its ability to be moulded with an integral hinge that has outstanding flex life. PP can be decorated by the coventional methods such as spray painting, hot stamping, vacuum metallizing, etc.

Current applications for PP include housings with moulded integral hinges joining two halves, appliance parts, housewares, toys, cable insulation, automobile interior panels, and party picnicware. It is also widely used as a packaging film, electrical capacitors, film for carpet backing, sacks, twines, ropes, textiles and outdoor furniture. The film of the PP can be laminated to paper or cloth.

2.3.8.3 *Polybutylene (PB1)*

PB1 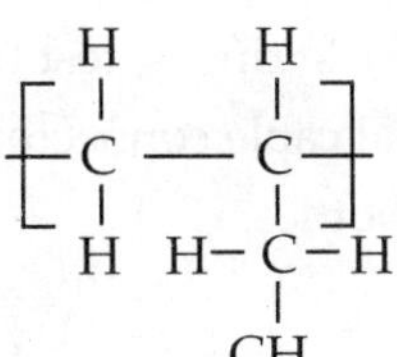	T_g: – 25°C 13°F T_m: 125°C (98–135) 257°F (208–275)

Polybutylene is based on butene-1 monomer and is also known as *Polybutene-1*. It closely resembles polyethylene in properties, appearance and process ability; but offers some specific property advantage over it. Its glass transition temperature, T_g, is closer to that of PP, while its metting point is comparable to that of polyethylene.

Retaining similar flexability to that of PE, PB1 exhibits higher tensile strength and retains a greater percentage of tensile strength at elevated temperatures. Still more significant for certain applications is PB1 resistance to environmental stress cracking and a creep rate that is lower than that of PE. For these reasons, PB1 finds applications in the field of pressure piping particularly for cold water, gas distribution and irrigation uses, as well as for hot water installations. PB1 is also abrasion resistant, in respect of which it finds its use for transportation of aqueous mineral slurries by pipeline, e.g. gypsum.

2.3.8.4 *Polymethyl pentene (PMP)*

PMP

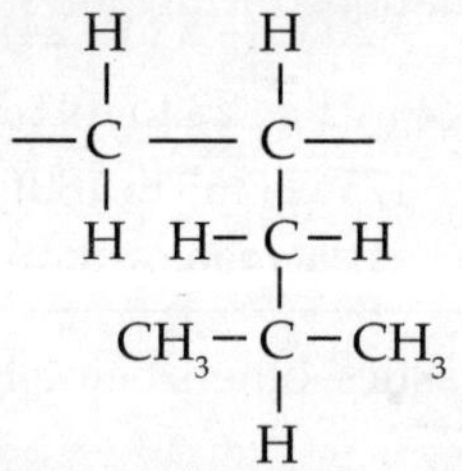

T_g: 30°C (18–40)
86°F (64–104)

T_m: 240°C (230–250)
464°F (446–482)

PMP or, more specifically, poly 4-methyl pentene-1 (PUMP) is the lightest of the olifins with a density of 0.83.

It is the only bulk transparent polyolifin in commercial production but is very susceptible to UV (sunlight), and high energy radiations attack and is quite flammable.

Chemical resistance of PMP is generally good and is, in particular, not affected by cold or hot water. PMP is, however, attacked by oxidising agents and affected by light hydrocarbon solvents. It is sensitive to environmental stress cracking.

Generally, properties of PMP are midway between those of PE and PP at temperatures below 93°C (200°F). However, at elevated temperatures, PMP far outshines the other olifins.

PMP may be obtained crystal clear or opaque and may be compounded with colourants. Clarity is retained even after sterilization temperatures, and PMP resists staining by dyes. Its electrical properties are comparable to those of PTFE.

PMP is primarily processed by injection moulding, extrusion, and blow moulding.

Applications generally utilize the high transparency of PMP and include lighting elements like diffusers, lenses, reflectors; liquid level and flow indicators, spirits despensers, car interior light covers.

Because of its inertness, PMP is very suitable for use in food packaging, processing and service, as cook-in (bake-in, boil-in) containers, trays and bags. It is compatible with hot-air or microwave ovens. It is also used for automatic milking equipment and parts of coffee makers.

PMP is suitable for laboratory and medical wares, like syringes, test tubes, connectors, etc.

Electrical applications include wire insulation, coaxial cable connectors, etc.

PMP also finds use in making closures for cosmetic bottles.

2.3.8.5 *Ionomers (IO)*

Ionomers are copolymers of ethylene and methacrylic acid. They are prepared by copolymerization of ethylene with 1-10% methacrylic acid using a high pressure process. The copolymer is then treated with a derivative of a metal, generally sodium or zinc.

Although the presence of carboxyl groups reduces crystallinity in the polymer, the ionic cross-links give enhanced stiffness and toughness; as a result the ionomers have many physical properties similar to those of LDPE. They are, however, much clearer in thick sections and have better abrasion resistance, higher impact strength and higher elongation. However, ionomers have low softening point of about 71°C (160°F) and are some what moisture sensitive. The ionic cross-links are stable at normal ambient temperatures, but reversably breakdown on heating so that ionomers can be processed by the normal techniques used for thermoplastics. They retain flex, puncture, impact, and abrasion properties at low temperatures.

Notable applications of ionomers include golf balls, bowling pin covers, ski boot outer shell, wire coatings, safety shields, and food packaging film.

2.3.9 Polyamides (PA)/Nylons

A relatively broad class of polymers involves the formation, during polymerization, of characteristic amide groups – CO – NH –. These polymers are referred to as polyamides (PA) or more popularly as *nylons*.

The most common polyamides are the so called *aliphatic polyamides* in which the repeat groups consist of CH_2 units. *Aromatic polyamides,* also known as *aramids* involve aromatic rings in repeat groups. There are also commercial polyamides of a mixed (aliphatic-aromatic) type called *transparent polyamides*.

The most popular ones used worldwide are: type 6, type 66, type 610, type 11 and type 12 from amongst the aliphatic polyamides, PPPT (Poly p phenylene terephthalamide) of the aromatic group and PA– (6, T) of the mixed type. The aliphatic polyamides are designated by the number of carbon atoms in the diamine, followed by the number of carbon atoms in the diacid. All these polyamides are listed below and discussed briefly in that order in this section.

(A) Aliphatic Polyamides

1. Nylon 6
2. Nylon 11
3. Nylon 12
4. Nylon 66
5. Nylon 610

(B) Aromatic Polyamides

1. Poly p phenylene terephthalamide (PPPT)

(C) Transparent Polyamide

1. Polyamide PA–(6, T)

2.3.9.1 *Aliphatic Polyamides*

PA-n

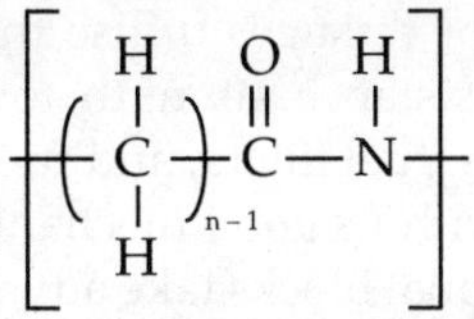

PA–6 (Nylon-6)

T_g: 50°C (20–60)
122°F (68–140)

T_m: 220°C (215–230)
428°F (419–446)

PA–12 (Nylon-12)

T_g: 40°C (104°F)

T_m: 179°C (175–180)
354°F (347–356)

PA–11 (Nylon-11)

T_g: 47°C (117°F)

T_m: 187°C (179–194)
369°F (354–381)

PA (n_1, n_2)

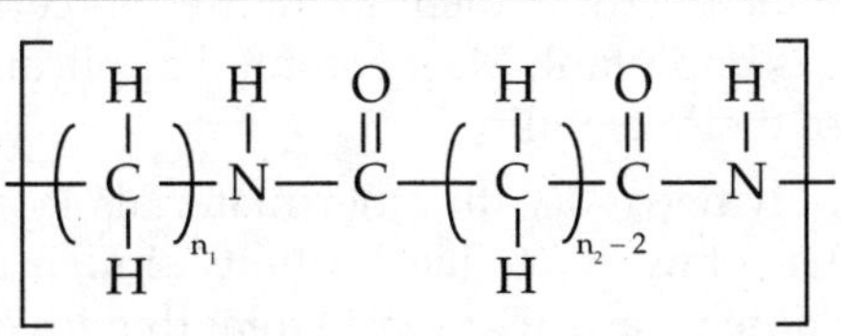

PA–66 (Nylon-66)	
T_g:	50°C (35–90)
	122°F (95–194)
T_m:	260°C (215–265)
	500°F (419–509)

PA–610 (Nylon-610)	
T_g:	40°C (104°F)
T_m:	220°C (210–227)
	428°F (410–441)

All nylons share some property similarity. The common characteristics include good toughness, fatigue resistance, impact resistance, inertness to aromatic hydrocarbons, and low friction. While nylons 610, 11 and 12 exhibit lower total moisture absorption than do types 6 and 66, all nylons absorb some moisture. The absorbed moisture acts as a plasticizer, thus increasing impact strength and decreasing stiffness. Moisture absorption is a reversible phenomenon, thus moisture will be rejected if parts are dried and that affects the change in properties accordingly. Pigmented formulations are available in a wide variety of colours.

The chemical resistance of nylon is generally good. Strong acids and oxidising agents can have deteremental effect. Resistance to hydrocarbons, fungi, bacteria, and food chemicals is particularly good. Very few solvents are known. Nylons have a fair resistance to UV light (weathering). Flame resistance is generally good. Nylons are well known for their self-lubricating property as well as good wear and abrasion resistance. Nylons are readily machinable for single or small lot production.

All types of nylons are fabricated by conventional melt-process equipment, except calendering. Secondary processes include hot stamping, ultrasonic welding, friction welding, heat sealing and staking.

Nylon 6 and 66 are the most popular, because of their favourable cost and useful property ranges. Although mainly used in textiles, moulded parts of these materials also take advantage of many good properties of nylons. Door latch hardware, automobile speedometer gears, windshield-wiper gears, seat sliders, and carburettor linkages utilise the combination of good properties of nylon including low friction, abrasion resistance, strength, toughness, fatigue resistance, and temperature resistance. Timing-chain sprockets, fuel filters, and fuel lines use the heat resistance and resistance to hydrocarbons of nylon, alongwith its good mechanical properties. Automobile trim clips, wire-harness retaining straps, and terminal blocks take advantage of high elongation, toughness, impact strength, and self-extinguishing nature of nylons. Other uses of nylon 6 and 66 include oil filler caps for motor vehicles, teeth in plastic zip fasteners, castors for light furniture, e.g. television sets; radiator tanks for cars, hose connectors for vacuum cleaners; gears in food processing equipment, and switch components in telephone exchanges.

Nylon 610 is more flexible than either Nylon 6 or 66, but exhibits lower moisture absorption than either of these. Thus nylon 610 is used in Vending machine and home laundry water mixing valves, because it retains properties and dimensions when in contact with water better than do the other nylons. Other uses of nylon 610 include tubing and food packaging.

Nylon 11 and nylon 12 are the most flexible of the group and enjoy the unique property of resisting zinc chloride, which permits them to be used in certain types of tubing on automobiles where this characteristic is important. Most common applications of nylon 11 and nylon 12 are in film or tubing, because of their flexibility.

All nylons may be polymerized with other materials to achieve specific property levels. For example, some copolymers of nylon 6 offer flexibility similar to that of nylon 11 and at lower cost. Other copolymers contribute greatly to impact strength with only slight loss in stiffness.

Fillers can be used to improve stiffness of nylons, particularly at elevated temperatures. Glass and asbestos have been used to improve stiffness, and molybdenum disulfide has been used to increase lubricity.

Apart from other applications listed above, the use of nylons as fibres is well known. Monofilaments are used for ropes, tyre cords, fishing lines, tennis racket strings, surgical sutures (threads for stitching wounds and cuts, etc.), brush bristles, and artificial hair and fur. Chopped and spun yarn is used in conventional textiles.

2.3.9.2 *Aromatic Polyamides Poly p Phenylene Terephthalamide (PPPT)*

PPPT

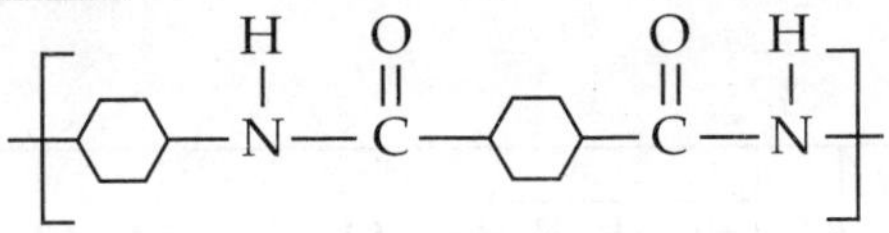

T_g: > 300°C
> 572°F
T_m: 500°C
932°F

In these polyamides, aromatic rings in the backbone corresponds to high chain rigidity and high glass transition temperatures.

Aromatic polyamides are mainly produced as fibres by the unconventional spinning processes which are related to *liquid crystal polymer* (LCD) technology. As all polymeric fibres, aramid fibres are highly crystalline but the crystalline structure differs significantly from that of conventional polymeric fibres and involves extended straight chain crystals.

Kevlar 29, developed from PPPT, as a textile fibre, is one of the stiffest and strongest organic fibres, and is used in products like tyre cord, ropes, cables, protective fabrics, and coated fabrics for inflatable structures.

Aramid fibres have the typical low density of polymeric materials, but remarkably high modulus of elasticity and strength, comparable with those of glass. Carbon fibres or metals give them outstanding specific properties such as modulus/density or strength/density ratio.

Typical applications of aromatic polyamides in woven fabric or composite form include radial tyre belts, ballistic protection items such as vests, jackets, helmets and armour plates, and sports equipment such as skis, tennis rackets, and fishing rods. They are used extensively in high-performance marine or aerospace applications. Their heat resistance has encouraged their use in brake lining materials, as well as for certain gaskets and packings.

Nomex (m-phenylene diamine isophthalic acid) finds applications in re-entry parachutes for space vehicles and in textiles for fire fighters apparel. Aramids are also available as stock shapes, i.e. rods, bars, sections, etc.

2.3.9.3 *Transparent Polyamides*

PA-(6, T)

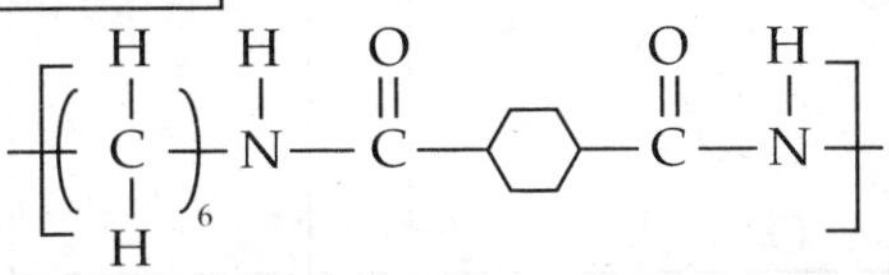

T_g: 120°C
248°F
T_m: 370°C (210 – 370)
698°F (410 – 698)

Conventional amorphous engineering plastics that are intrinsically transparent often lack certain aspects of chemical resistance, which polyamide can offer, in particular the resistance to hydrocarbon chemicals often found in engineering environment. A number of transparent polyamides have been developed which are now commercially available. One such transparent polyamide involving copolymer based on PA-6 is PA-(6, T). Chemplate and temperature data for which is shown above. This is achieved by incorporating aromatic groups in the repeat unit to stiffen the chain.

Because of the involvement of a variety of chemical structures, the properties and, in particular, the chemical resistance of transparent polyamides can vary signifacantly. Though resistance to hydrocarbons is generally good but some types may be adversely affected, for example, by alcohols and hot water. Prolonged expouser to high temperatures may also trigger crystallization and the consequential clouding of parts.

Some of the typical applications of transparent polyamides include filter bowls, headlamp reflectors, and rotameter parts.

2.3.10 Imide Polymers

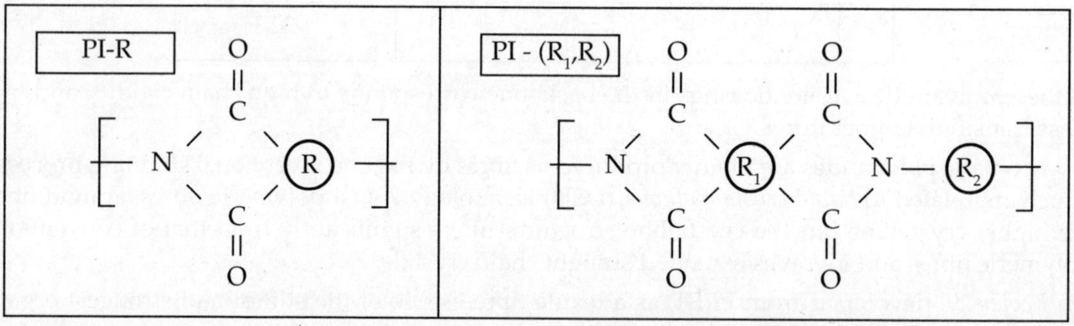

A family of temperature resistant polymers involves imide rings in the repeat unit which contain an atom of nitrogen N other than the usual carbon C atoms. They correspond to the reaction of certain acid anhydrides with amines just as amide groups correspond to the reaction of certain acids with amines. Polyimides of types PI-R or PI-(R_1, R_2) can be formed where the Rs are characteristic groups.

Polyimides are of both thermoplastic and thermosetting types. However, in this section discussion will be confined to only thermoplastic type while thermosetting polyimides are detailed in the latter section on thermosetting plastics. The two commercially used thermoplastic imide polymers are:

1. Polyamideimides (PAI)
2. Polyetherimides (PEI)

2.3.10.1 *Polyamideimide (PAI)*

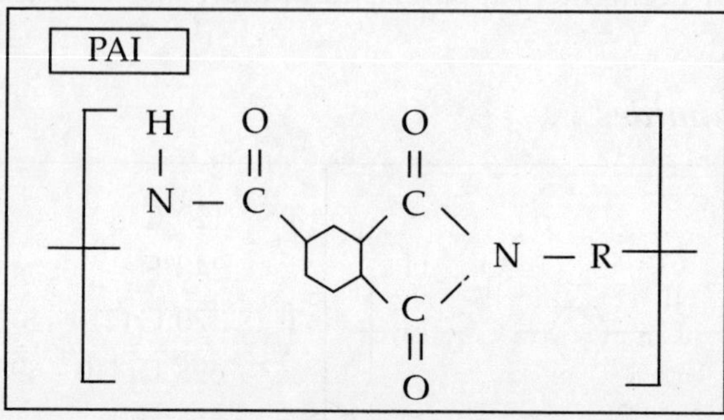

T_g: 275ºC
527ºF

PAI is formed by the reaction of trimellitic acid anhydride (TMA) with aromatic diamines and features both amide and imide groups. Temperature resistance is primarily associated with the aromatic and imide groups, while the degree of flexibility is the result of presence of amide group.

PAI retains good mechanical properties, particularly creep and impact resistance, at temperatures as high as 275°C (527°F). It is resistant to hydrocarbons, solvents, and solutions of acids and bases but can be attacked at elevated temperatures by alkalis (bases), some acids, and steam. The amide groups also allow a significant water absorption thus necessitates predrying prior to processing.

PAI is processed mainly by injection moulding at high temperatures (343°C/650°F) in hot moulds (232°C/450°F). Moulded parts achieve higher property levels if gradually postcured to 260°C (500°F), immediately after moulding.

Applications include electrical-electronic and mechanical high-performance parts in automotive, aircraft, and aerospace where they strongly compete with metal parts. PAI are also used as high-temperature electrical insulants.

2.3.10.2 *Polyetherimide (PEI)*

PEI

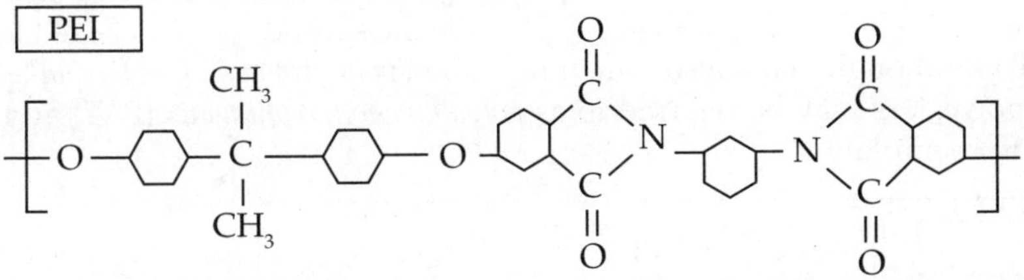

Tg: 217°C (210 — 220)
423°F (410 — 428)

The repeat unit of PEI features two imide rings separated by an m-phenylene diamine residue, as well as a bisphenol-A residue, also found in several other engineering thermoplastics, which comprises two other linkages – O – and contributes enough flexibility to the chain to make it a true thermoplastic. The amrophous character of the polymer gives it good dimensional stability and inherent transparency.

PEI has good mechanical properties with a tensile strength for the unreinforced resin highest among thermoplastics. These properties are little affected over a wide range of temperature allowing, for example, hand or wave soldering on printed circuit boards.

PEI has good general chemical resistance, including resistance to most corrosive automotive fluids, and is comparable to that of good crystalline polymers. Resistance to weathering (UV radiation) is also good. The resin is less affected by water and steam than PAI.

PEI has an exceptionally high inherent flame resistance with very low smoke evolution, making it well suited for use in construction and transportation systems.

Electrical properties of PEI are good with high dielectric strength and suitability for microwave applications.

PEI is used in the electrical-electronic area, e.g. for transparent parts like filter bowls, fibre optic connectors, in aircraft engine components and in high intensity lighting systems in plated reflectors. PEI can also be made into thin films which have very good thermal stability. Typical applications include sterilizable home ventelator manifold and styling comb/curator brush for high heat personal care electrical appliances.

Note: It may be noted that Imide Polymers are rather unique among plastics in the sense that they are available not only as thermoplastic and thermosetting polymers but also both amorphous and crystallizing types amongst thermosetting resins. Thermosetting PI is discussed latter in the text in section dealing with thermosetting plastics.

2.3.11 Thermoplastic Polyesters, Polycarbonates

TP–R

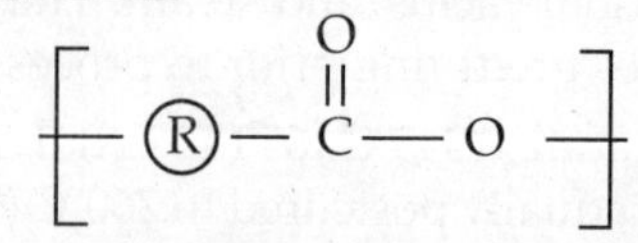

TP–R_1,R_2

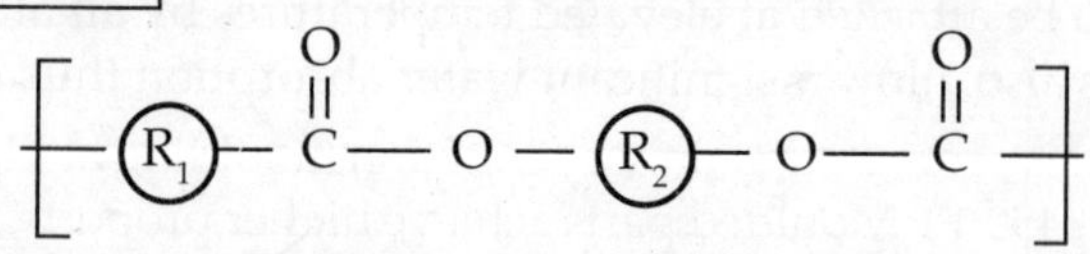

Thermoplastic polyesters (TP) is a class of thermoplastic polymers that involves the formation, during polymerization, of characteristic ester linkages or groups –CO–O. Groups of atoms *R* or R_1 and R_2 interspereced between ester linkages determine differences between types of thermoplastic polyesters. Most common thermoplastic polyesters involve both aliphatic (alkyl) and aromatic (aryl) groups, while some polyesters are fully aromatic.

Two high volume polyesters involve terephthalate groups and either ethylene or butylene groups. The first type is called *polyethylene terephthalate* (PET) while the second type is referred to as *polybutylene terephthalate* (PBT), although that is also sometimes called *polytetramethylene terephthalate* (PTMT). In general, these polyesters may be referred to as *polyalkylene terephthalate* (PAT) and be represented by the following chemplate.

PAT

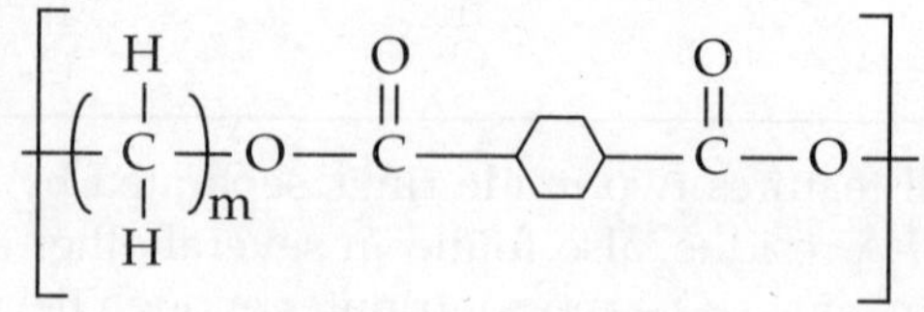

2.3.11.1 *Polyethylene terephthalate (PET)*

PET

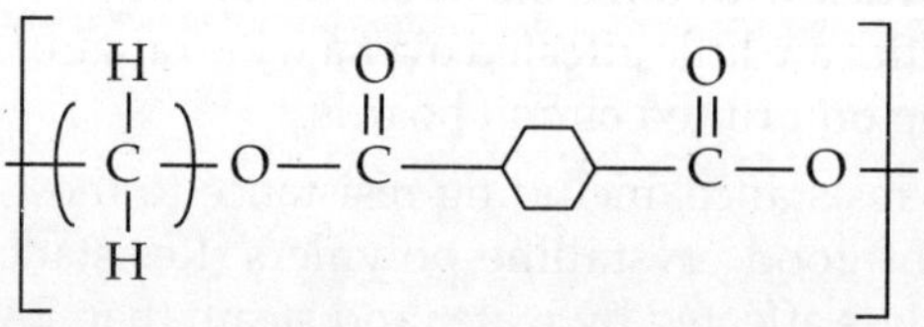

Tg: 70ºC (60 — 81)
159ºF (140 — 178)
T_m: 265ºC (245 — 271)
509ºC (473 — 520)

PET is particularly noted for its very slow crystallization outside a relatively narrow temperature range centered around 190ºC (374ºF); that is attributed to flexible but short, $(CH_2)_2$ groups which tend to leave the chains relatively stiff. When cooled sufficiently fast from the melt to a temperature below Tg, PET solidifies in amorphous form and can be highly transparent. If the amorphous solid is subsequently heated to above Tg, crystallization takes place to about 30% crystallinity; that becomes more predominant if the temperature is raised to above 120ºC. It is a common practice to first preshape PET articles in amorphous state and then reprocess to give them uniaxial (fibre or tapes) or biaxial (film, containers) crystalline orientation. Using injection moulding, PET can yield amorphous transparent objects by cold mould or crystalline opaque objects by hot mould technique.

PET is widely used as a fibre in which form it was introduced initially; it is also used in uniaxially oriented tape form for strapping. Though PET is not widely used as a bulk plastic material but it finds applications including photographic and x-ray films, magnetic tapes, electrical insulation, drafting sheets and food packaging films, including boil-in-bags and retort pouches. PET is also used in the form of biaxially oriented films biaxially-oriented vials, and bottles for containing alcoholic beverages, drinkable liquids as well as for holding carbonated drinks as it is found suitable for pressrized containers.

2.3.11.2 *Polybutylene terephthalate (PBT)*

PBT

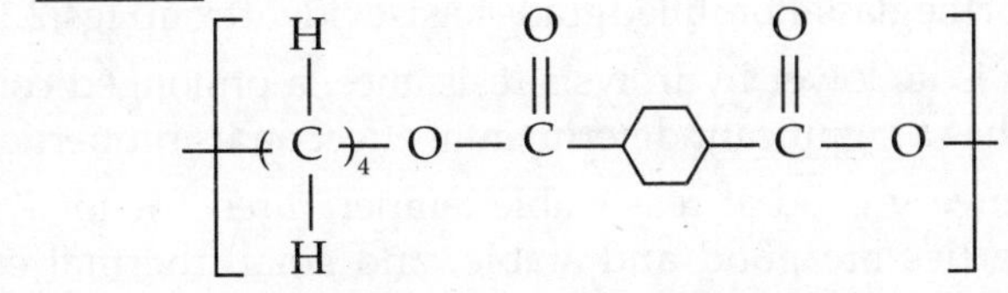

Tg: 50°C (40 — 60)
122°F (104 — 140)

Tm: 245°C (224 — 267)
473°C (435 — 513)

PBT, because of its longer and more flexible $(CH_2)_4$ groups, crystallizes much faster than PET, and it is not normally encountered in amorphous solid form.

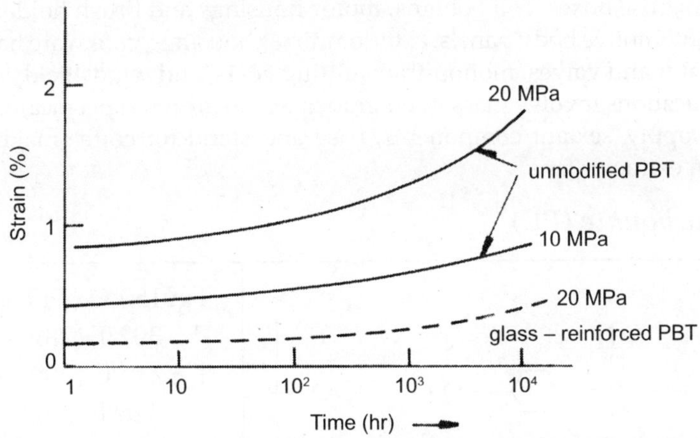

Fig 2.11: Tensile creep at 20°C for unfilled and glass-reinforced grades of PBT.

PBT is used primarily in injection-moulded applications, where fast cycles can be achieved with minimal precautions other than thorough drying of the resin so as to have less than 0.005% of moisture. Suitable grades tend to have rather sharp melting transition, with a relatively low melt viscosity which allows moulding of intricate shapes. The rapid crystallization and the high degree of crystallization, however, are often associated with warpage problems.

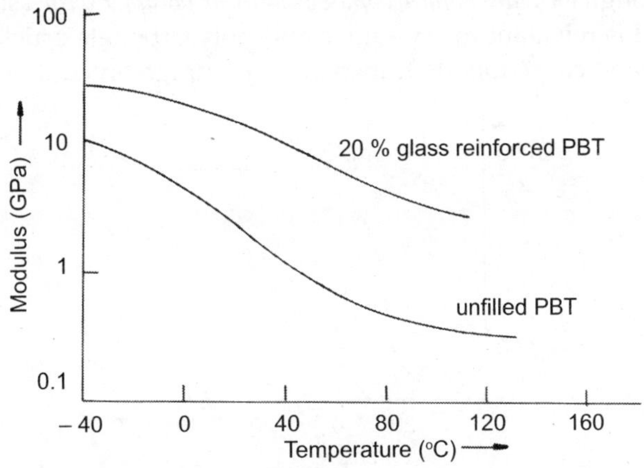

Fig 2.12: Tensile modulus versus temperature for unfilled and glass-filled PBT (polybutylene terephthalate).

PBT is particularly compatible with glass fibres. That is why almost 95% of the total PBT used in mouldings are in glass-filled grades. PBT is characterised by dimensional stability, being largely unaffected by variations in the ambient humidity, and by high rigidity which is retained to high temperatures. This is particularly true for the glass-fibre filled grades, as is evident from Figs. 2.11 and 2.12.

A disadvantage of PBT polyester is its lower hydrolysis resistance; a prolonged contact with water at 95°C or even as low as 50°C has a significant deteremental effect on its properties.

Mechanical properties of PBT are very good at reasonable temperatures, frictional and wear properties are best, electrical properties are good and stable, and small thermal expansion, particularly with filler grades, give very good dimensional stability.

PBT finds use in bodies and stems of push-button switches, valve components for automatic cattle waterers and electric mountings in refrigerators for bear chillers. Other applications include electrical connectors, fuse boxes, coil bobbins, motor housings and brush holders, distributor and ignition coil caps, automotive body panels, exterior mirror housings, cookware handles, pumps and meters housings, rotor and valves, motion-transmitting gears, and windshield wiper blade frame. Many of these applications involve glass filled grades which are also specifically used for terminal boxes for electric supply, sextant components, base and stand for coffee-making machines, and motor-end plates in extractor fans.

2.3.11.3 *Polycarbonate (PC)*

PC

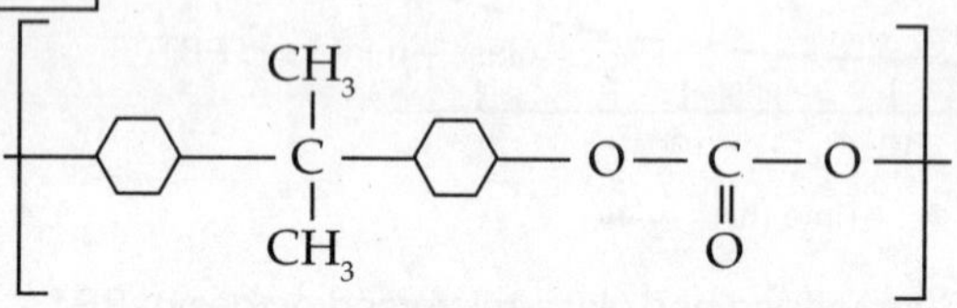

Tg: 150°C (143 — 156)
302°F (289° — 313°)
T_m: 230°C (200 — 267)
446°F (392 — 513)

Basically non-crystalline structure (amorphous)

Polycarbonates involve the simple carbonate linkage O – CO – O or – CO_3–; although only one such material that based on bisphenol A ($HOC_6H_4 . C(CH_3)_2 . C_6H_4OH$) is important commercially. Structurally it is a polyester derived from carbonic acid and the glycol, bisphenol A, but unlike aliphatic polyesters it is reasonably resistant to degradation by hydrolysis, particularly in solid-state, less so in the melt. Although *formally polycarbonate is a linear polymer with a symmetrical structure*, the melt-processed material is reluctant to crystallize and polycarbonate articles and components are *substantially amorphous*, and consequently transparent in unpigmented, unfilled grades.

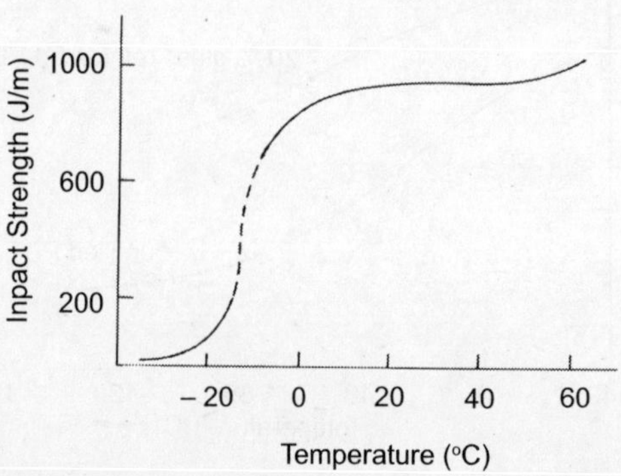

Fig 2.13: Temperature versus impact strength of Polycarbonate.

Polycarbonates are well known for their outstanding toughness (Fig. 2.13) which is superior to that of most rigid plastics, especially transparent ones, good stiffness and excellent creep resistance also contribute to making polycarbonates good engineering plastics for structural applications.

The amorphous nature of polycarbonates results in low moulding shrinkage, allowing close product tolerances. The dimensional stability of polycarbonate parts is very good, with very little moisture absorption.

General-purpose polycarbonate is self-extinguishing and special grades are available that are basically non-flammable. Although chemical resistance of polycarbonate is not bad but aromatic hydrocarbons, esters, ketones, amines, and strong bases like alkaline detergents, ammonia compounds, etc. can severely effect the properties of polycarbonates. At elevated temperatures; hydrolysis may cause chemical degradation. Polycarbonates are resistant to oxidation in air at elevated temperatures.

Polycarbonates have tendency to absorb ultraviolet (UV) radiations; that may lead to discolouration upon exposure to sunlight, however no structural damage is caused.

Grades of PC are available for injection moulding and for extrusion-based processes including blow moulding. As polycarbonate is a polyester, and therefore susceptible to hydrolysis, the material must be dried before processing, and care must be taken to keep the polymer dry in the hopper of the processing unit. Despite relatively low molecular weights of 25,000 to 50,000 the melt viscosities at moderate temperatures tend to be rather high, requiring high injection moulding pressures and appropriate mould design.

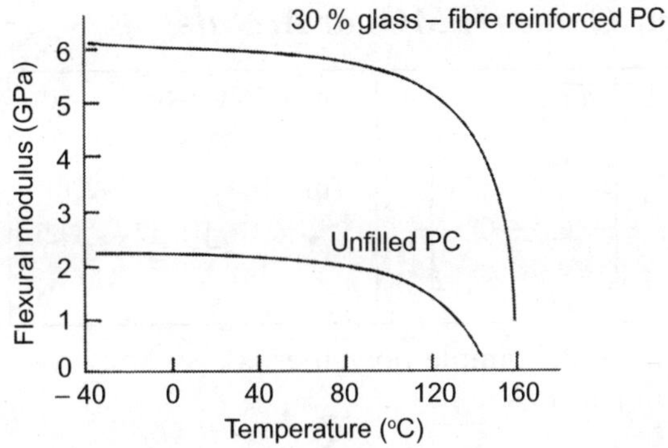

Fig 2.14: Temperature versus Flexural Modulus of Polycarbonate.

Glass fibre-filled polycarbonates have higher rigidity but reduced impact strength; as is evident from fig. 2.14 and table 2.2.

Table 2.2 Effect of glass fibre on properties of polycarbonate

Glass Fibre Content	Flexural Modulus	Izod Impact Strength
(%)	(G Pa)	(J/m)
0	2.4	650–850
10	3.5	250
20	5.6	107
30	9.85	134

Applications involving the use of polycarbonates are numerous. Based on their characteristics some of the well known uses include the following:

Transparency and Toughness: Safety glazing for public places, windshields, bottles — ranging from feeding bottles to 25 litre containers, partition shields in taxis and police vehicles, vandal-resistant street light globes, illuminated signs, vandal-resistant roof-lights, inspection glasses in chemical plants, outdoor air-conditioner housings, etc.

Toughness and High Softening Point: Hair dryer bodies, coffee makers, spot light housings, housings for automotive lenses and light assemblies, housings for power tools and portable appliances, pump impeller parts for small electrical motors and steam irons.

Toughness and Flame Resistance: Aircraft interior fittings, switch boxes for mines/quarries, high voltage plugs and sockets.

Nontoxicity and General Biocompatibility: House wares and food industry items like returnable milk and water cooler bottles, beer pitchers, microwave oven-wares, etc. and for medical applications.

Miscellaneous Applications: Camera bodies, modular phone connectors, capacitor film, ski slalom poles*, drafting films, safety helmets and goggles, gears, etc.

2.3.12 Ether-Oxide Polymers

In this group of plastics a number of industrially important polymers have the words ether or oxide in their names; that is indicative of the presence of an ether oxygen linkage, – O –, in the backbone of the repeat unit. The rest of the repeat unit may feature aliphatic (methylene, ethylene, propylene) or aromatic (phenylene) group. Two important engineering plastics of this class are polymethylene oxide (PMO) and polyphenylene oxide (PPO). The first one of which (PMO) is highly crystallizing while the second one (PPO) is an amorphous thermoplastic. Brief description of both these plastics follows.

2.3.12.1 *Polymethylene Oxides (PMO) or Acetals*

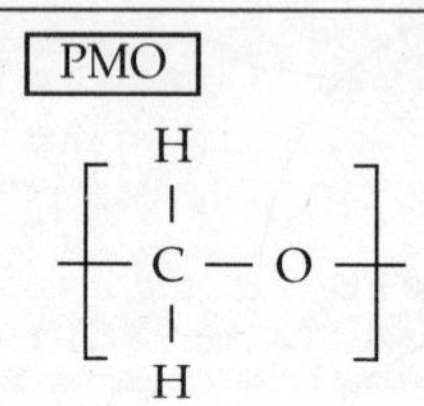

Tg: –75°C (–85 — –50)
–103°F (–121 — –58)
Tm : 180°c (175 — 200)
356°F (347 — 392)

Polymethylene oxide, a structurally simple polyether is also known as *polyoxymethylene* (POM), *polyformaldehyde, polyacetal*, or just *acetal*.

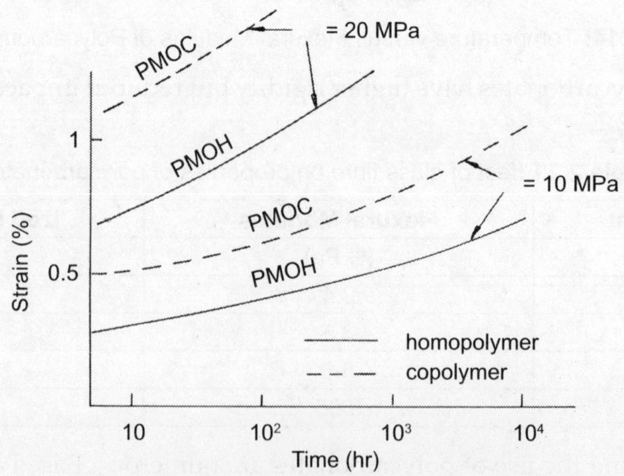

Fig 2.15: Creep behaviour of Acetal under tensile loading.

* Poles to indicate the safe limits for zig-zag skiing path.

Straight polymerization of monomer – CH_2 – O – gives a material with a strong tendency for de-polymerization at high temperature, making it very difficult to process this plastic. Two methods have been adopted to overcome this problem. The first method involves *end-capping*, that is a suitable modification of chain ends by acetylation or methylation and the resulting material is known as PMO homopolymer (PMOH). The second method involves a random copolymerization by introducing a small fraction (about 2%) of ethylene oxide-type repeat units whose C – C bonds are thought to impede de-polymerization and *the term copolymer is used to specify such polymethylene oxides*; materials so produced are called polymethylene oxides copolymer (PMOC). A consequence of these stabilisation methods is that the homopolymer degrades indefinitely if the chain is broken or an end cap removed, whilst the copolymer degrades only to the next comonomer residue; also the copolymer is more stable to hydrolysis. On the other hand, the comonomer restricts crystallization of the copolymer, giving it lower rigidity and strength than the homopolymer; the melting is also lowered from 175°C that of PMOH to 163°C that of PMOC. The consequential effect on creep behaviour of the two types of PMO are shown in Fig. 2.15.

Acetals have a high degree of polymerization (> 1000), and the simple regular chains crystallize to a high degree of about 75–80%. The high density (> 1.4) points to the tight packing of the crystallized chains. Also, the glass transition temperatures of acetals are very low in comparison with those of other crystallizing engineering plastics. Both types of acetals share a number of excellent properties that ranks them among the very best of engineering plastics.

PMO resins are in many ways comparable to nylon 6 and 66. Although acetals have higher density (> 1.4) nylons have better overall mechanical properties than acetals but nylons have one severe limitation in that they are hygroscopic. When moulding nylon the water content has to be extremely low, so that after moulding nylon has very little water and is rigid. The relative behaviour of dry and wet nylon 66 vis-a-vis acetal copolymer (PMOC) are shown in Fig. 2.16.

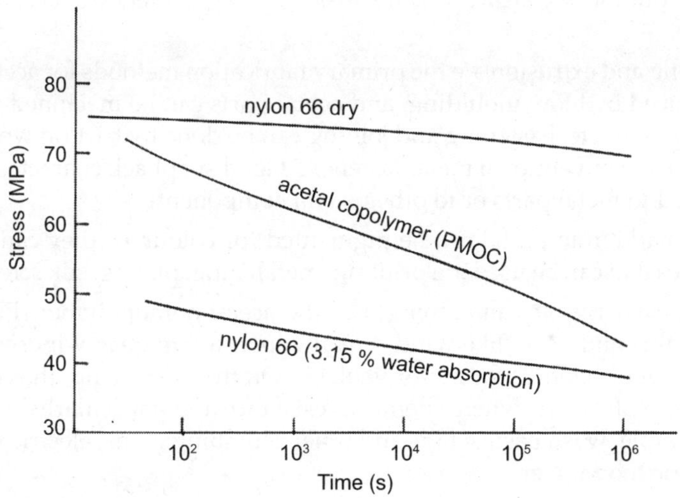

Fig 2.16: Long term strength data: Nylon 66 versus Acetal copolymer.

Acetals have high stiffness and strength, as also good dynamic fatigue resistance as is shown for PMOH in Fig. 2.17. Toughness is good, particularly under repeated impact loading. Friction and wear properties are excellent for such stiff and strong material that makes it suitable for components like bearings, slides, cams and gears. The coefficient of dynamic resistance of PMOC vis-a-vis other materials are as follows:

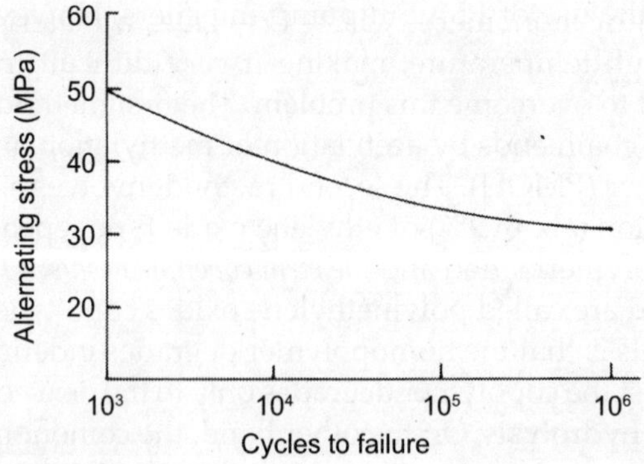

Fig 2.17: Fatique behaviour of Acetal homopolymer (PMOH).

	Steel	0.15
PMOC:	Brass	0.15
	Aluminium	0.15
	PMOC	0.35

Acetal resins can decompose on processing due to overheating, with the formation of toxic formaldehyde gas which at best is unpleasant and at worst can be extremely dangerous.

Acetals have excellent resistance to hot water up to about 82°C; that is why acetal copolymer (PMOC) has been used since 1969 for hot water plumbing applications. An advantage of acetal for such applications over metals is because of less scaling.

Resistance of PMO to radiation is not so good so UV stabilization is required to reduce the effects of sunlight in outdoor exposure. The intrinsic burning tendency of acetals is comparable to that of polyethylene.

Injection moulding and extrusion are the primary fabrication methods for acetals. Small pressure containers are produced by blow moulding, and acetal parts can be machined from stock shapes, i.e. from rods, bars, plates, etc. Fastening and joining can be done by friction welding or ultrasonic welding or by the use of conventional metal fasteners. Good snapback characteristics enable acetal parts to be snap-fitted to metal parts or to other acetal components.

Aesthetic parts made from acetal can be pigmented for colour, or they can be painted. Other decorating techniques that can be used are printing, metallizing, plating, silk screening, and dyeing.

Generally, the greater rigidity and strength of the acetal homopolymer (PMOH) promote its preferential use in cams and gears like wind screen wiper, worm gear, window winding handles for cars, and exterior door handles for motor vehicles. On the other hand, the copolymer (PMOC) is often preferred for applications where chemical resistance and particularly hydrolysis resistance is important, for example, wash basins, taps and other plumbing items, electric kettles, and mincer attachment for electric food mixer.

Applications of acetals generally involve relatively small parts that must perform an important mechanical function, require close dimensional tolerances, and often involve motion and contact. The automotive industry, for example, makes use of acetal gears, bearings, slides, cams and ratchets. The resistance to gasoline justifies the use of PMO for carburettor and fuel pump parts.

Acetal has long been used for making aerosol containers, and the fuel reservoir of disposable lighters.

Resistance of acetal to cold and hot water, in particular, has led to its use in plumbing applications for fixtures or fittings such as shower heads, valves, pumps, sink baskets, etc. Resistance to solvents and most alkalies permit the use of PMO in pump impellers. However, contact with most acids should be avoided.

Other uses of acetals include zip fasteners, roller chain and other conveyor components, cams and counter wheels, etc. on petrol pumps, business machines, etc.

2.3.12.2 Polyphenylene Oxide (PPO)

PPO

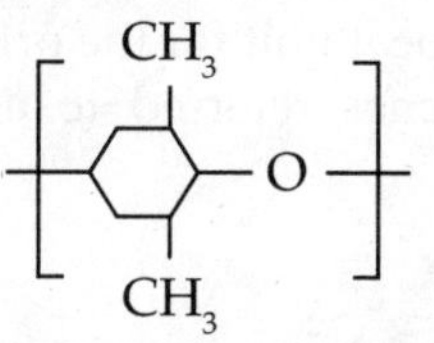

Tg: 210°C (204 —234)
410°F (399 — 453)

Polyphenylene oxide (PPO) is a polyether through – O – group in the repeat unit and is also sometimes termed as *polyphenylene ether* (PPE).

Pure PPO has a high glass transition temperature (T_g) of 210°C (410°F). It is capable of crystallizing between T_g and the melting point, T_m, which is around 260°C (500°F). Such a narrow range appears to contribute to its solidification behaviour that it is normally considered amorphous. Because of its high cost and certain processing difficulties associated with a high melt viscosity, have led to the development of blends (polyalloys) of PPO with polystyrene (PS) or high impact polystyrene (HIPS). The two polymers are very miscible in each other, and the resulting blend features a single glass transition temperature. A balanced blend (50/50) has T_g around 150°C (302°F) and reported T_g values for commercial grades range between 100 and 135°C (230°F – 275°F). These blends are known as *modified polyphenylene oxide* (MPPO).

MPPO has, in general, good mechanical properties with a high stiffness and low creep over a sufficiently long temperature range. Good toughness extends to low temperatures. It has excellent dimensional stability which is associated with its amorphous behaviour, a low coefficient of thermal expansion, and very low moisture absorption. The density of MPPO at 1.06 is among the lowest of engineering thermoplastics. In spite of its amorphous character and the high miscibility of its constituents, MPPO is *not* available in transparent grades.

Electrical properties of MPPO are generally good and are not affected by moisture. Dielectric properties are especially good and stable.

MPPO being self extinguishing and non-dripping finds use in transport vehicles.

MPPO has notable resistance to cold and hot water as also steam. It can be repeatedly sterilized in steam autoclaves. It is, however, attacked by some inorganic chemicals, such as strong acids and alkalies. However, weatherability is considered good.

Injection moulding is the most commonly used process, and MPPO exhibits good flow, low shrinkage, and little warpage associated with its amorphous nature and the presence of polystyrene. Thermoforming of such an amorphous plastic is rather straight forward but involves fairly high temperatuers.

For its high hydrolytic stability MPPO finds its use in hot water systems for valves, pumps, fittings, plumbing fixtures etc.; and for medical supplies requiring steam sterilization.

In the automobile industry MPPO finds application for interior parts like dashboard or exterior parts like grilles, wheel covers, etc. which are often painted or metallized.

Because of its resistance to dilute aqueous solutions, MPPO is used for photographic processing tanks.

Finally it can be said that though MPPO falls short of its more costly counterpart, PPO, in tensile strength and temperature resistance, but because of its higher impact strength, reduced cost and its available properties MPPO is used in such applications as television yokes, tool handles, small appliance housings, business machine housings, sprinkler parts, and decorative grills.

2.3.13 Ketone Polymers

A comparatively new family of polymers in the industry features ketone groups, – CO –, phenyl (aryl) rings, and ether linkages, – O –, in the chain backbone. The combination featuring two ether linkages and three aryl rings, corresponds to the repeat unit for the original PEEK (Polyetherether ketone) commercial plastic which is also sometimes referred to as polyaryletheretherketone (PAEEK). Brief description of which follows.

2.2.13.1 *Polyetheretherketone (PEEK)*

PEEK

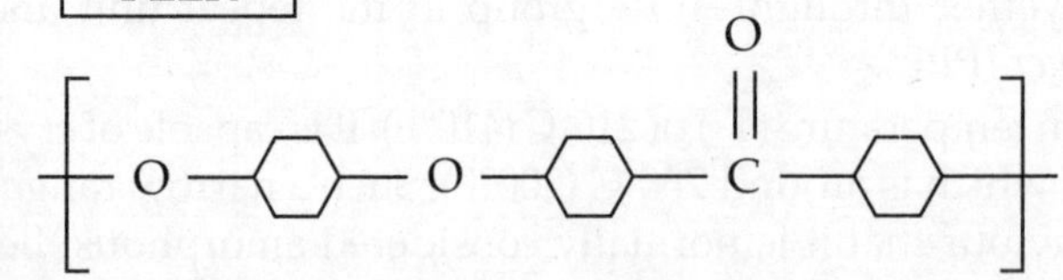

Tg: 143°C (142–144)
290°F (288–291)
Tm: 335°C (332- 342)
635°C (630 - 648)

PEEK with its high glass transition temperature, T_g, and high melting temperature, T_m, combined with its high temperature chemical stability, is rated amongst the most temperature-resistant thermoplastics. It retains good mechanical properties at high temperatures, such as 200°C (392°F), for prolonged period of time.

PEEK is particularly resistant to hydrolysis by steam or high temperature pressurised water and absorbs little water. With its crystallinity rating of 25-50%, like other crystallizing engineering plastics, it is resistant to environmental stress-cracking or crazing.

PEEK is noted for its excellent resistance to nuclear radiation, and its electrical properties are suitable for use as an insulating material. With its low flammability and very low smoke and toxic gas emission, it is very promising for critical applications, such as are encountered in aeronautical and aerospace areas.

PEEK can be injection moulded at high temperatures up to 400°C (752°F), into hot moulds up to 180°C (356°F), to favour crystallization. In spit of low water absorption of the resin, predrying is needed. Extrusion is used to produce fibres and films.

PEEK with its excellent mechanical properties at high temperatures is found suitable for wire and cable covering for use in critical applications in aerospace and nuclear industry. Injection moulded parts include pump impellers, valve parts, electrical connectors, etc. PEEK films can be used for printed circuits or as barrier membranes. Fibres woven into fabrics have been used as filters and conveyor belts in high temperature service.

2.3.14 Fluoro Polymers

Fluoro polymers are a family of carbon chain polymers featuring fluorine atoms that form very strong bonds with carbon atoms (C – F). Those fluoropolymers which have a characteristic plastic-like behaviour are called *fluoroplastics* while those with rubber-like behaviour are called *fluoroelastomers*.

Compared with other thermoplastics, the fluoroplastics offer superior chemical and heat resistance, as also good electrical properties. There is a large number of commercial fluoro thermoplastics but the four more widely used amongst them are:

1. Polytetra fluoroethylene (PTFE)
2. Polychlorotrifluoroethylene (PCTFE)
3. Polyvinyl fluoride (PVF)
4. Polyvinylidene fluoride (PVDF)

Brief description of these four fluoroplastics follows.

2.3.14.1 *Polytetrafluoroethylene (PTFE)*

PTFE 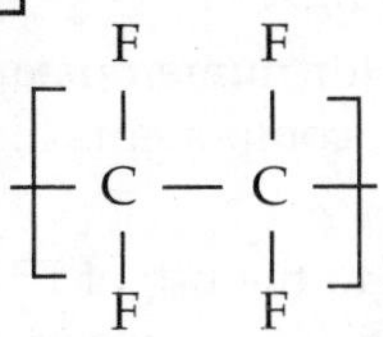	Tg: –130°C (–150 — –112) –202°F (–238 — –170) Tm: 327°C (327 — 330) 621°F (621 — 626)

PTFE is the most distinguished member of the fluorocarbon family, and it contains the highest fluorine content. Since fluorine content regulates the properties, PTFE possesses the maximum number of properties for which the fluroplastics are noted.

PTFE has a very simple and regular molecular structure. Solid PTFE is highly crystalline (50-57%) and, consequently, opaque. Above the crystalline melting point, T_m, the high molecular weight ($5 \times 10^5 - 5 \times 10^6$) polymer becomes very rubbery (gel-like), in a manner similar to that of ultra high molecular weight polyethylene (UHMWPE) with a very high melt viscosity typically 10^6 times higher than that of the melts of most thermoplastics.

PTFE has the greatest chemical resistance of all plastics and is attacked only by molten sodium, and molten sodium hydroxide (NaOH).

PTFE has exceptionally good resistance to ultraviolet (UV) radiations and weathering in general. It is totally unaffected by cold or hot water and will not be oxidised or decomposed at high temperatures in the solid-state; it is stable in air to at least 250°C (482°F) for continuous service. PTFE, however, deteriorates rapidly when exposed to high energy radiations. It has inherent flame retardency; it will not burn and will only decompose at very high temperatures leaving behind little residue.

PTFE is well known for its extremely low coefficient of friction (0.05) which is the lowest of all plastics. It will not stick to, or even be wetted by, practically any substance. For these reasons, it is often referred to as *self-lubricating, antistick* or *nonadhesive plastic*. Depending upon the actual service conditions, it can have high to low resistance to wear and abrasion. Mechanically, PTFE is tough and moderately flexible with high elongation to break point. Toughness is retained over a wide range of temperatures from cryogenic temperatures of – 268°C (– 450°F) to over 250°C (482°F). PTFE, with its very compact crystalline structure, has the highest density (2 – 2.3) of commercial plastics.

PTFE also has outstanding electrical properties as insulator or dielectric material.

Although PTFE is a linear chain polymer and a thermoplastic, processing is not possible by the conventional methods used for other thermoplastics. PTFE is processed more like ceramics or powdered metals; powder metallurgy techniques of processing are employed. Pressure preforms are sintered at temperatures above the crystalline melting point, typically at 380°C (716°F). Complicated shapes are obtained by machining, making them very expensive. Extrusion, particularly of thin sections, is accomplished by lubricating dispersion PTFE powder with petroleum ether, and subsequently drying and sintering the extrudate as in the production of PTFE tubes. Unsintered tape is produced by calendering and is used by plumbers in pipe thread seals.

Rods, tubes, profiles, wire coatings, and fibres can be produced by special ram extrusion process, using long dies, high pressures, and high temperatures of the order of 400 – 450°C (752 – 842°F). Tapes can be sliced (skived) from solid stock; similar to producing of plywood laminations. Films as thin as 25 microns thickness may be produced by skiving.

Because of its excellent properties, PTFE finds application in bridge bearing pads, chemical piping equipment, electronic insulators, arch chutes, seal backup rings, seals, human heart valves, cryogenic valves, and automobile power stearing units. Valves, gaskets, expansion bellows are typical uses of PTFE.

Non-stick coatings including home cookware, bread processing, toffee making, coating on skis, tool blades, snow shovels, conveyor parts and piston rings.

The chemical resistance of PTFE makes it useful for numerous applications in the chemical process industry as liners for reactors, and a variety of components like pump parts, filter media, and for laboratory wares like seals, spigots, etc.

Flame and weather resistance, in particular, leads to the use of PTFE as a binder and coating for glass or other fabrics used for tent-like structures, like for covered stadiums. PTFE is also well known for its use as pipe thread tape sealant.

Electrical and electronic applications include high temperature wire and cable insulation and critical coaxial wire spacers particularly for high frequency applications.

2.3.14.2 *Polychlorotrifluoroethylene (PCTFE)*

PCTFE

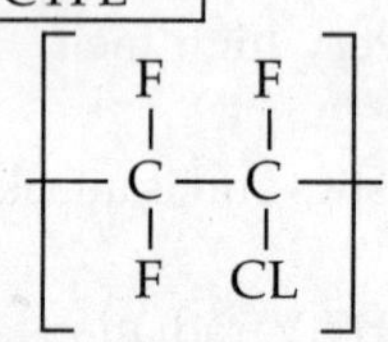

Tg: 40°C (20 — 60)
104°F (68 — 140)
T_m: 210°C (185 — 224)
410°C (365 — 435)

The presence of cl atom in its repeat unit reduces the flexability of the chains, compared to PTFE, and leads to higher glass transition temperature, T_g. Its tendency to crystallize is also reduced to 40-65% and the melting point, T_m, to 210°C (410°F). PCTFE can be made transparent, like amorphous thermoplastics, up to a thickness of 3 mm. The reduced melting point and lower molecular weight of 100,000 to 200,000 makes PCTFE a true thermoplastic.

PCTFE is known for its low permeability to water and gases, making it the least permeable of all plastics. PCTFE has better mechanical properties than PTFE in that it is stiffer and stronger but its frictional properties are much inferior to those of PTFE. It has good chemical resistance, although not comparable to PTFE. It has good resistance to liquid oxygen which makes it useful for application in special areas like aerospace and medical. It has better resistance to high energy radiations than PTFE.

PCTFE has good flame retardancy and is listed as non-burning plastic but it has higher dielectric losses than PTFE, particularly at higher-frequencies.

Being melt-processable, PCTFE can be processed by injection and compression moulding, blow moulding or extruded into tubing and thin films that are transparent. Transparency, combined with high resistance to radiation and zero moisture permeability, makes PCTFE the prime choice for packaging of medical supplies that can be sterilized by radiation after packaging. It is also used for packaging medicines in strip and blister packs for tablets and capsules.

In the chemical process industry, PCTFE is used as gaskets, o-rings, seals, valve seals, and instrument windows.

The non-tracking and high dimensional stability of PCTFE enables its use in super-critical electronic connectors, coil forms, terminals, wire and cable insulation, protective shrink sheaths and electronic flexible printed circuits, etc.

2.3.14.3 *Polyvinyl fluoride (PVF)*

PVF

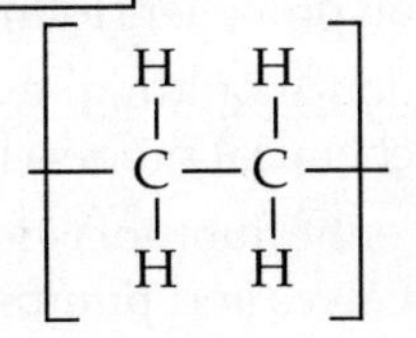

Tg: –20°C
–4°F
T_m: 198°C (195 — 200)
388°F (383 – 392)

PVF involves the molecular chain that is capable of crystallizing to a high degree, but many of its most common products like films and sheets are transparent.

PVF is strong and tough, with good abrasion and staining resistance to fairly high temperatures of 100–150°C (212–302°F). The material is classified as slow burning. It is generally less chemically resistant than fully fluorinated plastics, say PTFE, but show excellent UV resistance and good colour retention and is not affected by water. Excellent weatherability has made PVF a choice material for exterior applications, like coatings for metals such as for sidings, gutters, etc., plywood finishes, architectural sheets, lighting panels, and glazing for solar energy collection. Other applications of PVF include electrical wrapping tape and parting layers for laminates.

2.3.14.4 *Polyvinylidene fluoride (PVDF)*

PVDF

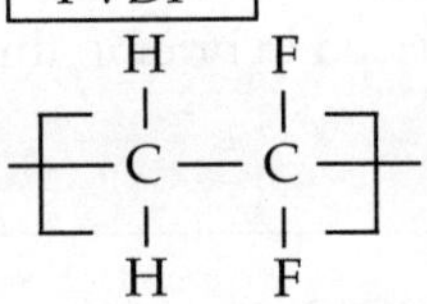

Tg: –35°C (–40 — –30)
–31°F (–40 — –22)
T_m: 170°C (165 — 178)
338°F (329 – 352)

PVDF which is also abbreviated as PVDF or PVF_2 features a repeat unit (CH_2 – CF_2). It is a fluoropolymer which shares some properties including high cost, with PTFE. It is normally highly crystalline (65%), but transparent films can also be produced.

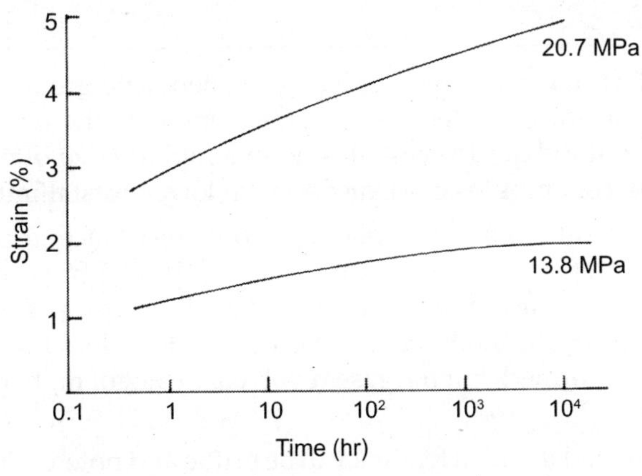

Fig 2.18: Creep behaviour of PVDF at 20°C.

PVDF is strong, tough and abrasion resistant; and in terms of stiffness and creep resistance, it is much better than PTFE. Representative creep data are given in Fig. 2.18.

Resistance of PVDF to ageing/weathering/high energy radiation is exceptionally good; for example, its resistance to gamma radiation is exceptional; the material is little affected even by a does of 300 Mrad from Cobalt-60 source. However, its chemical resistance is not as good as that of PTFE. It burns but is classified as self-extinguishing and non-dripping. The abrasion/erosion resistance of PVDF is exceptional and has led to its use as a coating on propellers for high performance watercraft.

PVDF is melt-processable in conventional injection moulding, extrusion, and fluidized bed coating equipment. It finds a wide variety of uses in chemical process industry.

The high stiffness of PVDF enables it to be the only fluoroplastic to be used in rigid pipes. Common uses are in chemical process equipment in valves and pumps.

Coatings or finishes are applied to metal surfaces such as aluminium siding and roofing. Colours are easily incorporated and are stable.

Moulded parts include seals, gaskets, packings, diaphragms, valve parts, pump impellers, pipe fittings, and coil bobbins. Extruded products include pipes (tubing, ducting), wire insulation, and jacketing, as also semi-finished forms like slabs, rods, etc. that can be used for making thermowelded liners for chemical reactors to machine parts. Monofilaments are woven into fabric for use as filters. Due to its superior mechanical properties, PVDF is used in load-bearing applications that other fluoroplastics cannot sustain.

PVDF is in competition with other materials like UHMWPE for the fabrication of porous pen tips which must be tough, as well as abrasion and ink resistant.

2.3.15 Miscellaneous Heterochain Polymers

There are not many members in this family of thermoplastics; the one which is industrially important is Polyphenylene sulfide (PPS) and is described in brief in this section.

2.3.15.1 Polyphenylene Sulfide (PPS)

PPS

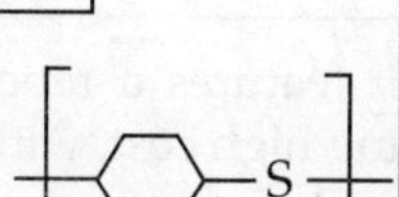

Tg: 88°C (85 — 90)
190°F (185 — 194)
Tm: 288°C (275 — 290)
550°C (527 — 554)

PPS is made up of symmetrical repeat unit consisting of phenylene group and a sulphur atom in the backbone. The phenylene group confers chain rigidity; and both the symmetry and the absense of bulky side groups, allow the chain to crystallize with a high degree of crystallinity at 45–65%. But a rapid solidification in cold moulds can cause a much lower crystallinity at 15%.

PPS is temperature-resistant engineering plastic; also its resistance to chemicals is very good with no known solvents at temperatures below about 180°C (356°F). PPS is not affected by moisture and has good weatherability, as also good resistance to high energy radiations. It has good fire resistance, is self-extinguishing; although if a flame is present PPS does burn, leaving a char but it does not drip. Some smoke is formed, but the gases evolved are not of high toxicity. PPS is basically transparent to microwaves.

PPS is stiff and strong material with a tendency to be brittle and notch sensitive. It is well known for its compatibility with glass fibre. In fact, unfilled PPS is not important commercially. It is used primarily in reinforced grades popularly with 40% glass fibre. PPS has good abrasion resistance and makes it useful as a coating.

PPS is processed mainly by injection moulding; extrusions appear to be impractical due, possibly, to its low viscosity.

Regarding its applications, PPS has been called the poor man's polysulfone (PSU). This is because of some similarity in properties between PSU and PPS, however PSU is almost three times more costly.

PPS is widely used as coatings which can be applied by electrostatic, fluidized-bed, or slurry techniques. Coatings are often baked at 357°C (675°F) to induce cross-linking. Special formulations for *antistick* or *non-stick* coating may be obtained by use of PPS/PTFE polyalloys.

Applications of PPS include its use in chemical process industry for the construction of pump housings and impellers, valves, metering devices; also etch tanks are coated with PPS. Other applications include lamp sockets, coil bobbins, gears, bushings, carburettor parts, as also personal-care and household appliances like propellers or turbines for fan heaters or dryers, coatings for washing machines, microwave-compatible cookware, antistick coatings for cookware, etc. It is also used as an adhesive and a laminating resin.

2.4 Thermosetting Plastics

Thermosetting plastics are polymers that are relatively useless in their raw state. For most of them, upon heating to a certain temperature, a chemical reaction takes place that causes the molecules to bond together or cross-link. After curing, the thermoset material remains stable and cannot return to its original state. Thus, *thermosetting* identifies those materials that get set in their useable state as a result of heat additions. Most of these materials are solids in their raw state and may be available in granule or powder form.

Some newer members of this group of materials, however, are in the liquid state as a result of addition of a catalyst to a liquid or by the mixing of two liquids that react and form a solid-state material. These reactions can take place even at room temperature but are accelerated by raising the temperature. Thus, the present-day thermosets may be categorised into three groups depending upon their mode of formation, viz.,

1. *Temperature activated thermosets*; e.g. formaldehyde like phenolics, aminoplasts, urea and melamine vinyl ester, alkyd, allyl, furan, as well as some unsaturated polyesters, epoxies, and polyimides.
2. *Catalyst-activated thermosets*, e.g. unsaturated polyester liquids.
3. *Mixing activated thermosets*, e.g. polyurethane, some epoxy and other resin systems.

2.4.1 General Characteristics of Thermosetting Plastics

Thermosets are the hardest and stiffest of all plastics and are chemically insoluble after curing. Their properties are less affected by changes in temperature than are the heat-sensitive thermoplastics. The closest non-plastic engineering material to thermosets in properties are ceramics. Common examples of thermoset plastics include phenolics, melamine, urea and epoxies. Moulding compounds made from these polymeric resins always contain additional fillers and reinforcing agents to obtain optimum properties.

Thermosets differ fundamentally from thermoplastics in the following aspects.

1. While thermoplastic resins are most commonly supplied by resin manufacturers in polymerized form, and processors do not normally have to deal with any chemical reaction; only physical changes are involved in manufacturing or *physical processing*.

In contrast major chemical reactions take place in the course of the processing of thermosetting resins, which generally undergo a complex combination of polymerization and cross-linking, converting low or relatively low molecular weight molecules into tight or relatively tight three-dimensional molecular networks. This mode of manufacture may be referred to as *reactive processing*.

2. The very high melt viscosities, inherent in thermoplastics because of very high molecular weights, are avoided by working with comparatively low molecular weight thermosets which generally do not require very high pressure equipment, enabling shaping of thermosets in relatively low pressure equipment of lower cost, for example moulds.
3. The time-dominant cooling stage necessary to stabilize the shape of thermoplastic component is replaced in thermoset resins by a time-dominanting chemical reaction taking place in the hot mould or press.
4. The absence of post-moulding shrinkage in thermosets ensures that very close tolerances can be held, leading to acceptance for precision mouldings.

Over all thermosetting resins can be grouped into eight systems. The detailed classification of these plastics is given in Fig. 2.19, and description of commercially important amongst them follows.

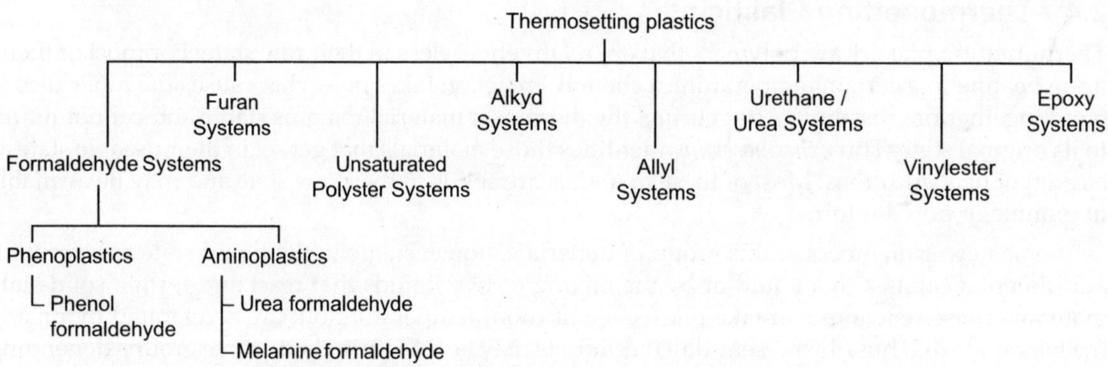

Fig 2.19: Classification of Thermosetting Plastics.

2.5 Thermosetting Resin Systems

Out of the eight thermosetting resin systems listed in Fig. 2.19, the four more widely used industrially are:

1. Formaldehyde Systems
2. Unsaturated Polyester Systems
3. Vinylester Systems
4. Epoxy Systems,

While the remaining systems have limited applications. Therefore, the above listed four systems are described in the following sections in greater details.

The chemical reactions involved in the course of the *reactive processing* of thermosetting resins are generally complex and sometimes incompletely understood or considered confidential by resin manufacturers. While a basic understanding of the major chemicals and reactions involved is helpful, a detailed discussion is beyond the scope of this book.

2.5.1 Formaldehyde Systems

This family of thermosetting polymers involves the reaction of formaldehyde (F) with three hydrogen containing chemicals viz., phenol (P), urea (U), and melamine (M) to form three dimensional networks. These thermosets are frequently compounded with a variety of fillers such as wood flour, α-cellulose fibres, and minerals.

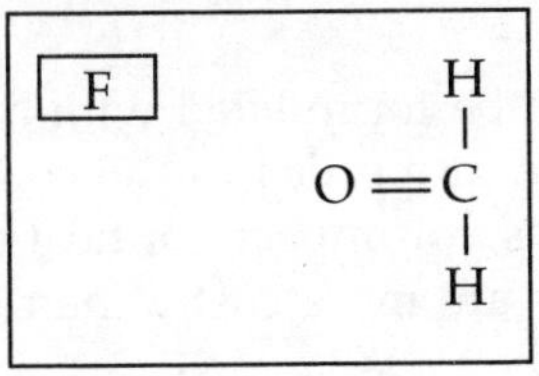

Formaldehyde (F)

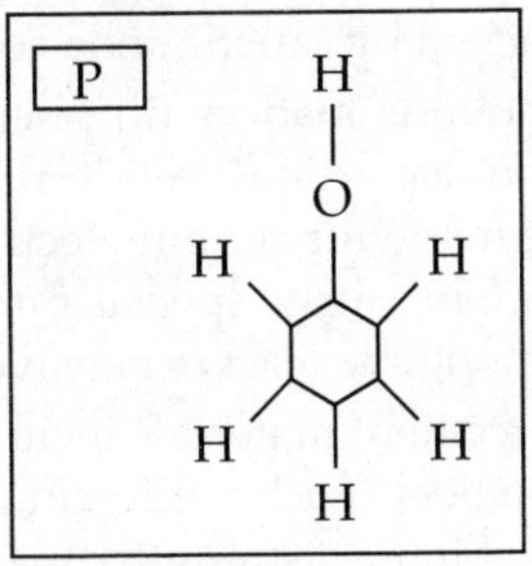

Phenol (P)

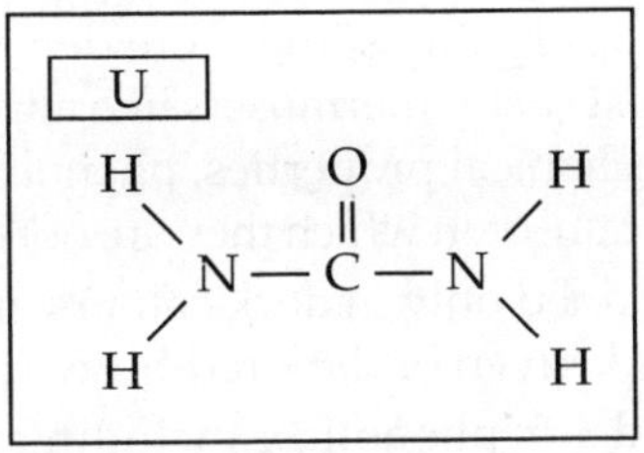

Urea (U)

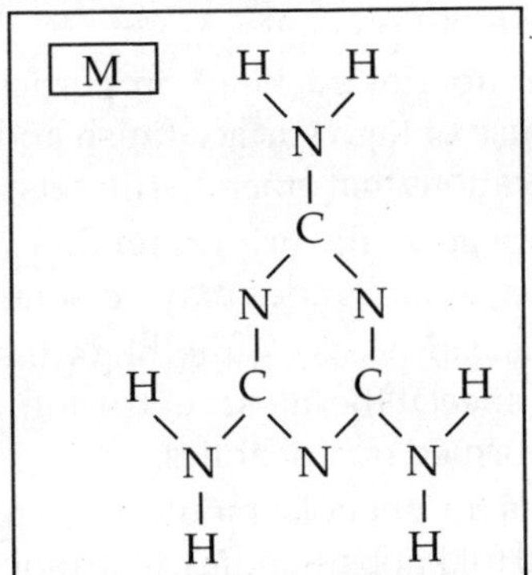

Melamine (M)

This family of thermosetting polymers involves the reaction of formaldehyde (F) with three hydrogen containing chemicals viz., phenol (P), urea (U), and melamine (M) to form three dimensional networks. These thermosets are frequently compounded with a variety of fillers such as wood flour, α-cellulose fibres, and minerals.

For moulding purposes they are generally supplied in powder or granular form or sometimes compacted as preforms. The raw resin can be preheated rapidly using high-frequency (i.e. radio frequency or microwave) heating methods, or hot plasticized slugs can be prepared with an extruder-like equipment.

Moulding temperatures for formaldehyde resins are in the range of 125°–200°C (255–395°F) and moulding pressures are in the range of 14-56 MPa (2000–8000 psi) for compression moulding and can be as high as 140 MPa (20000 psi) for injection moulding.

Apart from their specific applications for individual members, formaldehyde thermosets are widely used as binders, adhesives, coatings, and surface treatment, etc.

These systems of thermosets are usually divided into two groups, viz.,

1. Phenoplasts — Comprising Phenol formaldehyde
2. Aminoplasts — Comprising urea formaldehyde and melamine formaldehyde.

Brief description of these follows.

2.5.1.1 *Phenoplasts/Phenol formaldehyde (PF)*

Phenol formaldehydes (PF) are also known as phenoplasts or phenolics. They are based on phenol (P) and formaldehyde (F) and are among the oldest of the commercial polymers, having been discovered in 1909, and were introduced under the trade name *Bakelite*. Despite age, phenolics are still high-volume engineering materials and continue to find wide ranging applications. Because of their standardization, phenolics are often used as the baseline for comparison amongst the thermosets.

These low cost thermosets have good combination of properties which include rigidity, dimensional stability up to fairly high temperatures, low flammability, and solvent resistance, though they are attacked by caustic soda solution.

Moulding compounds made with phenolic resins can be formulated with a wide variety of fillers that impart an accordingly wide range of possible properties to the compound. Special formulations can be made for certain electrical applications, for impact, for heat resistance, or for chemical resistance. Obviously, special purpose materials are more costly than general-purpose materials that are among the least expensive of all plastics.

Filler and reinforcement materials include wood flour, asbestos, synthetic fibre, chopped cloth, glass fibre, and cotton flock. Each reinforcing medium is selected for a specific purpose. For example, by varying fillers, izod impact strength can be varied between 0.2 and 15.0 ft–lb/in (10.7–802.5 J/m) and tensile strength from 5500 to 7500 psi (38–52 MPa).

Although a lot of phenolics are used in electrical applications, the more demanding electrical devices require superior tracking* properties of the most costly thermosets like alkyds, epoxies, etc. However, because of high surface finish and good mechanical properties, phenolics excel in many structural applications that other thermosets cannot sustain, or in which they are not cost competitive.

Phenolics are generally brittle and they can be obtained only in dark colours, mostly black and brown. However, some grades may be obtained in dark green or dark red-brown.

All thermoset processing methods may be used for phenolics, including screw-injection moulding, and special thermoset extrusion. Machining and polishing are common post-moulding operations performed on phenolics.

Applications for phenolic moulding compounds are numerous, for example PF mouldings are still widely used for knobs, handles, instrument cases, bottle caps and closures. Components for more exacting service conditions include automobile distributor caps, automatic transmission valve bodies, switches, thrust washers, appliance pump impellers and housings, encapsulated electronic components, small motor housings, electronic tool housings, electrical components like connectors, sockets, etc.

Phenolics are also used as binder in grinding and cut-off wheels, brake linings, clutch faces, wood particle boards (chipboards), moulds and cores in foundry work, or adhesive in plywood fabrication and furniture assembly. An important application of PF resins is in decorative laminates like for counter or table tops, etc. where PF-impregnated kraft paper is pressed into heat-resistant, dimensionally stable sheets. Fire-resistant rigid foams are also produced for insulation or structural purposes.

2.5.1.2 *Aminoplasts*

Aminoplasts comprise resins that are based on urea (U) and formaldehyde (F) or melamine (M) and formaldehyde. Both urea and melamine feature amino (NH_2) group, hence they are given the name amino resins. Aminoplasts, in contrast to phenoplasts, can be made translucent or in light colours and are thus suitable where aesthetic considerations are important.

Both materials offer good electrical properties, such as arc resistance, have high surface hardness, excellent gloss and appearance, and resist solvents. They are also non-burning. The chief drawbacks of aminoplasts are their low impact strength and great dimensional change upon exposure to humidity.

Urea Formaldehyde (UF): Urea formaldehyde plastics have lower maximum service temperatures and lower water resistance than phenol formaldehyde (PF), but find wide-spread use as they are available in light pastel colours and are of lower cost. The electrical behaviour of UF is superior to that of PF in that it does not suffer failure by tracking, i.e. degradation of polymer by electrical breakdown leaving a conducting, carbon track on the surface.

* *Tracking* is degradation of a polymer by electrical breakdown leaving a conducting, carbon track on the surface.

Urea formaldehyde is of increasing interest as it is based on intermediates which are not derived from oil, thus price escalation is less of a problem than for most plastics. Urea is obtained by reaction of carbondioxide and ammonia, the ammonia carbamate formed being dehydrated to urea.

$$CO_2 + 2NH_3 \longrightarrow NH_2.CO.ONH_4 \quad \text{(ammonia carbamate)}$$

$$\downarrow$$

$$NH_2.CO.NH_2 \quad \text{(Urea)}$$

Urea and formaldehyde react to form mono and dimethylol ureas which are the usual pre-polymers for UF-plastics.

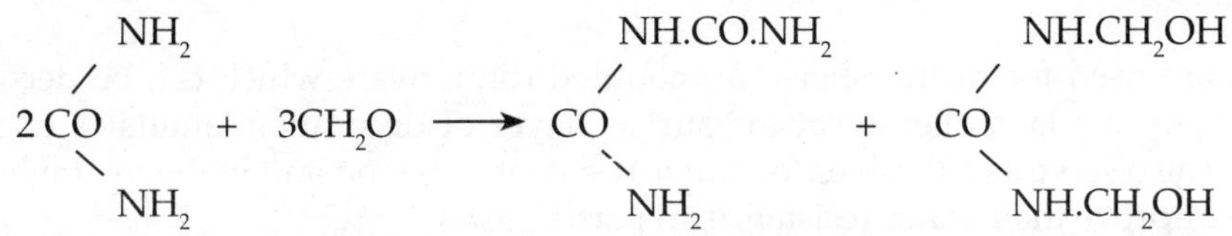

urea formaldehyde monomethylol Urea dimethylol Urea

The attractive features of urea formaldehyde plastics include low cost and a wide colour range — including pastel shades — in contradistinction to PF. They are less heat-resistant than PF, being serviceable only to 70°C (158°F), although this and the water resistance, which is not very good, can be improved by modification with melamine. Notable properties of UF plastics are good electrical properties, including resistance to tracking, which is important in electrical plugs and sockets, and lack of tainting by food-stuffs with which they come in contact — a property that is important for caps and closures. The resistance to glycol and alcohol solvents is also relevant in this last application.

For use in caps and closures a number of considerations favour UF mouldings; important in these are the *feel* of the component and its visual appeal, on both these counts UF scores very highly. On technical considerations, UF caps and closures are rigid and strong and reasonably immune to attack by a wide variety of fluids, including food stuffs, which themselves are not contaminated by contact with this plastic.

Wood flour is normally used as a filler except for very light shades where α-cellulose is employed. The bulk of UF mouldings are found in caps and closures for bottles and other containers and in electrical fittings. However, substantial quantities of UF are still used for applications as varied as buttons and toilet seats. Other important use for UF resins is as wood adhesives in plywood and chipboard manufacture.

Another very important use of UF is as foam (U-foam) for cavity wall insulation. The foam has a high closed-cell content but is very friable; a density of less than 1 lb/ft^3 (16 kg/m^3) can be achieved. UF foams are also used in floral decoration.

Melamine Formaldehyde (MF): Melamine is a colourless crystalline solid with a melting point of 354°C (669°F); and is obtained from urea. Urea is heated in the presence of ammonia at 250–350°C and a pressure of 4–20 MPa; and the following reactions follow:

$$2NH_2.CO.NH_2 \rightarrow NH_2CN + NH_2COONH_4$$

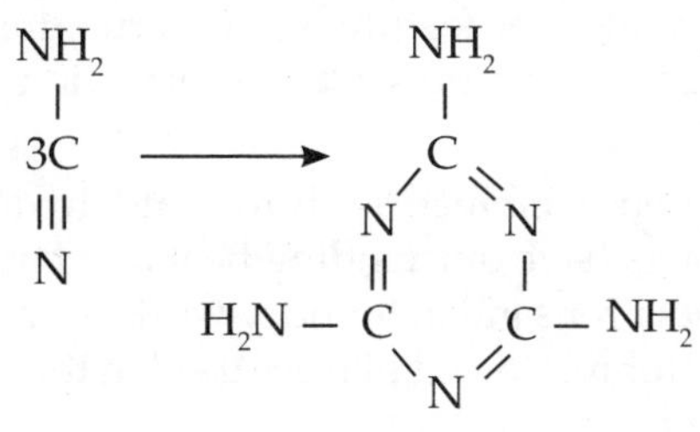

Earlier melamine was mainly obtained from dicyandiamide (NH_2. C. NH. CN) but this route is no longer of commercial importance.
$\quad$ ||
$\quad$ NH

For the preparation of melamine-formadehyde resins, formaldehyde (CH_2O) is generally used as the aqueous solution, formalin.

MF resins are noted for their excellent water resistance, better heat resistance, better stain resistance, and are harder than Urea formaldehyde. In fact MF mouldings compare favourably with their UF counter-parts in all properties except the price.

Melamine formaldehyde mouldings are much more resistant to staining by tea and coffee than UF mouldings; largely because of this factor the major use of MF mouldings is for the manufacture of tableware.

MF resins are used for dishwasher-safe moulded dinnerware which can be decorated with moulded-in paper overlays; they form the surface layer of decorative laminates (e.g. Formica, Arborite). They also serve as adhesives for water-resistant plywood and in the treatment of papers and textiles to improve their water resistance, in particular.

Glass-reinforced MF laminates are used in good quality electric boards, having heat resistance to 200°C (392°F) and good electrical properties, including tracking resistance.

Electrical insulating properties, which are similar to those of UF mouldings under normal conditions, are maintained better at elevated temperatures and in damp conditions.

Other applications of MF resins include circuit breakers and switch-gears, utensil handles, small applicance housings, and applicance knobs. Melamine-impregrated overlays are also used to impart decorative details to items like utensil handles.

2.5.2 Furan Systems

Furfural and *Furfural alcohol* are used to make a variety of polymers which are collectively known as *furan polymers*.

Furan and furfural alcohols are produced from furfural by the following reactions:

(i)

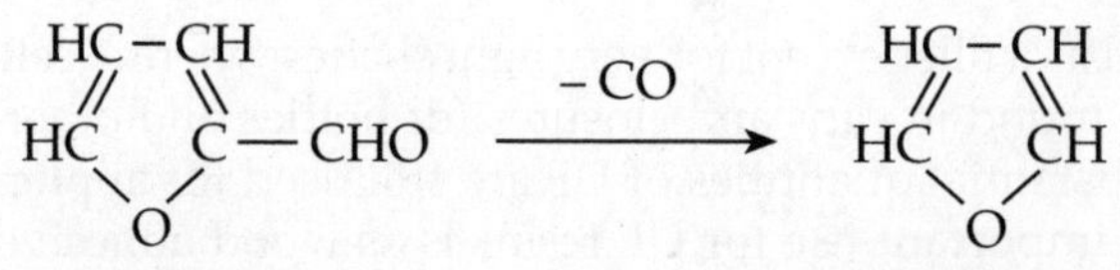

Furfural $\qquad$ Furan

(ii)

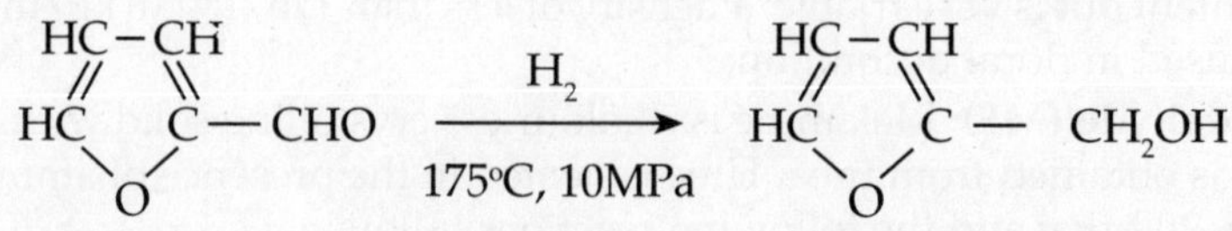

Furfural $\qquad$ Furfural alcohol

Furan, furfural and furfural alcohol feature a ring structure which can be opened to yield polymeric molecules. With unsaturation present, the formation of three-dimensional molecular networks is achieved.

Furan resin systems are often combined with formaldehyde (CH_2O) related thermosetting resins. Such cured furan polymers have outstanding heat and chemical resistance and find use in cement for chemical plants, binder for sand in foundry work, abrasive particles in grinding wheels; as well as adhesives and matrix for reinforced plastics used in the chemical process industry, where corrosion resistance is important.

2.5.3 Unsaturated Polyester System

Maleic acid (MLA)

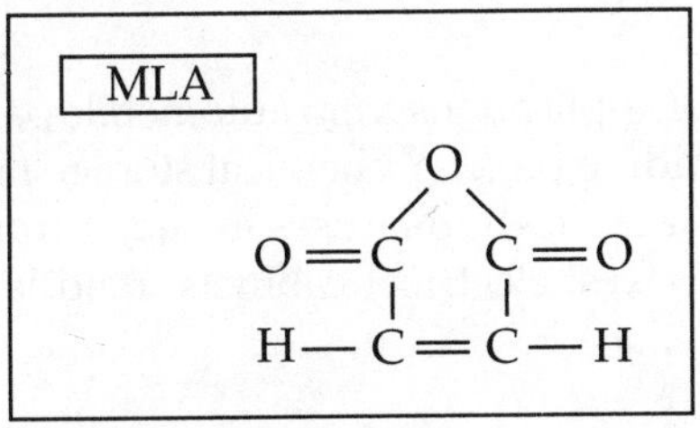

Phthalic acid (PTA)

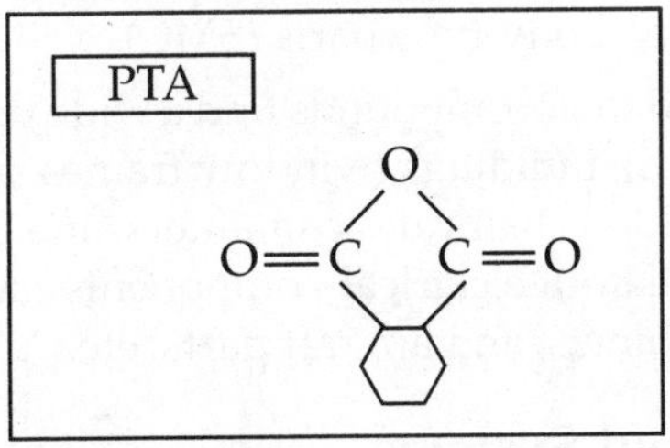

Ethylene glycol (EG)

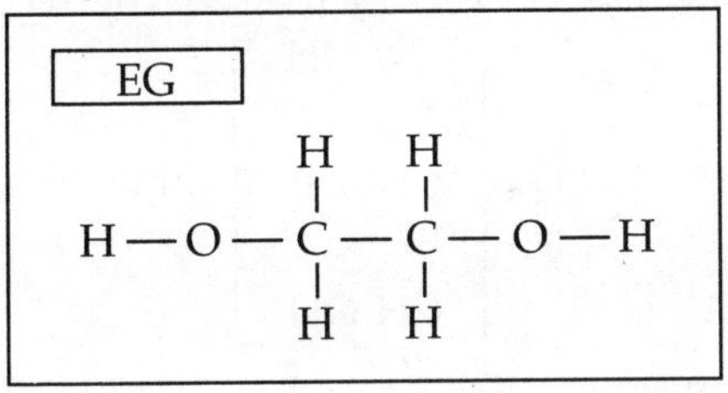

Styrene (S)

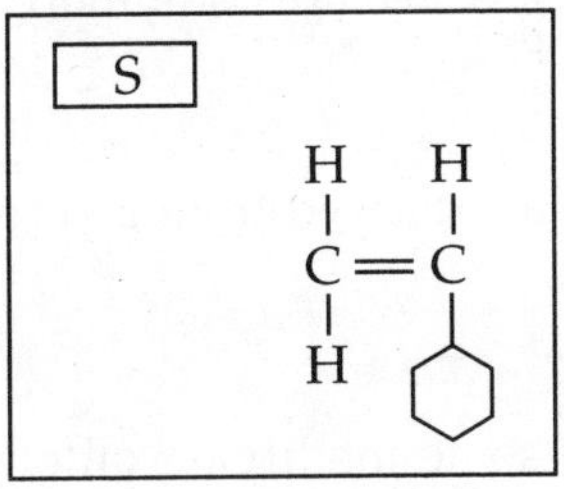

The first major constituent of unsaturated polyester resin (UPR) systems also called simply polyester resins consists of relatively short polymeric chains resulting from the controlled reaction between a difunctional acid or anhydride, e.g. maleic acid or anhydride (MLA), or phthalic anhydride (PTA), and difunctional alcohol or glycol, e.g. ethylene glycol (EG). Characteristic ester linkages are formed, hence the name polyester. At least some of the acid features double bonds between adjacent carbon atoms (resulting in unsaturations), which are required for the subsequent cross-linking of the systems into three-dimensional molecular networks. This first constituent is called *unsaturated polyester*. The typical molecular weight of the corresponding molecules is 1000-3000.

The other major constituent is a monomer capable of polymerizing as well as reacting with the unsaturations in the polyester molecules, to form interconnections and thus molecular networks. The proportion of monomer used in commercial systems ranges between 30 and 50%. Low viscosity monomers also serve as diluent for the higher viscosity polyester molecules. The most commonly used monomer is styrene which is also the least expensive. Other monomers used include para methyl styrene (PMS), methyl methacrylate (MMA) and diallyl phthalate (DAP).

Resin systems are normally supplied with an additive (inhibitor) that prevents or retards premature cross-linking, allowing a suitable shelf life.

Heat or radiation (UV, etc.) alone can often trigger the cross-linking reaction, but generally additives known as catalysts and accelerators are used.

UPR systems may be divided into two groups, viz., those that can cure at room temperature and those that require higher curing temperatures.

Room temperature curing systems or *Catalyst activated systems* are particularly convenient for use as a matrix for reinforced plastics using primarily glass rovings, cloth or mat as the reinforcements; and the fabrication and curing of products at ambient temperatures is particularly convenient for large parts. These so called *fibre reinforced* plastics (FRP) are now widely used to make pipes, tanks, containers, structural elements like beams, posts, springs, hulls, architectural panels, etc.

The *temperature-activated systems* are relatively stable at room temperature. They are compounded with all ingredients (catalyst and accelerator, as well as extenders, fillers, etc.) and are then subjected to the high temperature of a heated mould typically at 150°C (302°F). *Bulk moulding compounds* (BMC) and *Sheet moulding compounds* (SMC), which are essentially mixtures of UPR, glass fibre and filler, that belong to this category are *pre-polymers* designed with economy in mind, and provide the present major growth area of GRP (Glass-Reinforced Polymer) technology. They are processed by injection moulding (BMC), or compression moulding (BMC and SMC) into relatively small parts (BMC) or large body-type parts (SMC).

UPRs and their compounds find a wide variety of applications from automobile parts (car bonnet and frontends), buildings (window frames and cladding panels), chemical storage tanks, chemical piping, moulded chairs and containers, and even car bodies. Other uses include furniture (outdoor tables and chairs), electrical components (switch boxes, electrical cabinets, road lamp housings, television cabinets, and aircraft parts, etc.

2.5.4 Alkyd Systems

Oil-modified polyesters, which are commonly known as *alkyds* (which term is derived from alcohol (alk) and acid (yd)), are derived from glycerol ($C_3H_8O_3$) or 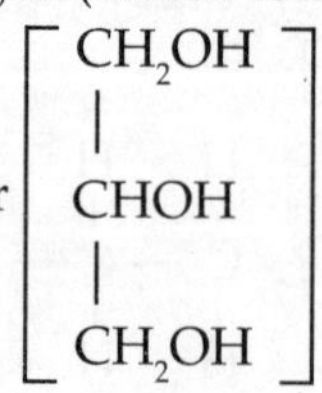, phthalic acid anhydride 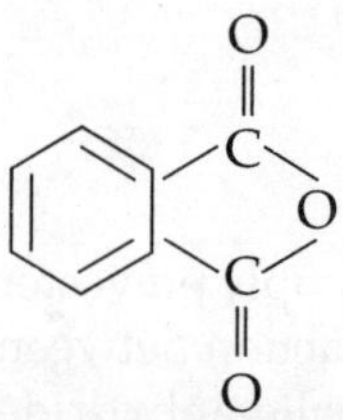 and some modifying oil of plant origin like linseed oil, soybeen, and tung oils.

Other materials blended with alkyds are cellulose nitrate (for fast drying an toughness), chlorinated rubber (for fire-resistance and toughness), silicone resins (for heat–, and water resistance). Also added are phenolic resins (fast drying and corrosion resistant) and amino resins (for hardness, durability and alkali resistance).

Alkyds are network polymers which find extensive use as surface coatings (paints, varnishes, lacquers) and account for almost half of all the resins consumed in the protective coating field.

Alkyd thermosetting moulding compounds made from unsaturated polyester resin (UPR) systems use DAP (diallyl phthalate) monomers for achieving cross-linking.

DAP

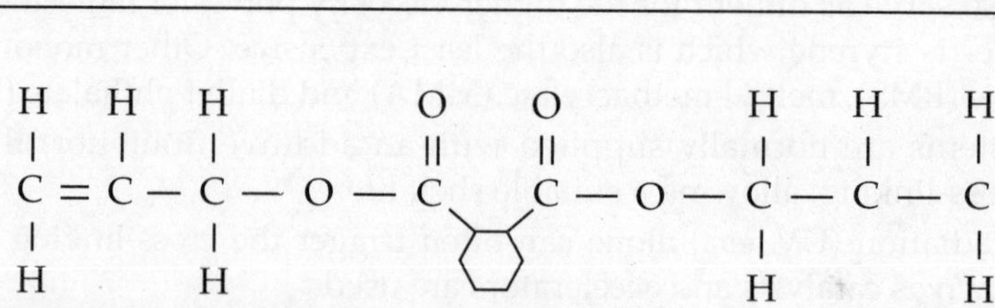

Like most thermosets, alkyds arc hard and stiff and retain their mechanical and electrical properties at elevated temperatures.

Most attractive with alkyds is their arc resistance and track resistance in the presence of an arc. Low moisture absorption and retention of electrical properties when wet permit alkyds to offer excellent performance in automotive ignition equipment. The non-tracking characteristic of alkyds permits their use in arc chutes and in heavy duty switch gear. Low change in dielectric loss factor enables alkyds to be used in TV tuner segments, because tuner capacitance remains unchanged throughout the operating temperature range of the TV set.

The raw material for alkyde moulding compounds comes in four basic form viz., granular, rope, putty, and bulk. Granular materials offer the lowest mechanical properties of all alkyds. Rope alkyd is a putty-like material containing reinforcement with glass or synthetic fibres. Primarily used in transfer moulding, ropes form the workhorse of the alkyd family and offer a good balance of impact strength and other mechanical properties, while retaining excellent electrical properties. Also, ropes cure fast and can be moulded into intricate shapes.

Alkyds are available in many colours, but are seldom used in decorative applications, because of high cost and less smooth surface finish when compared to aminoplasts or phenolics. Many of the alkyd moulding materials are self-extinguishing.

Major applications of alkyds include electrical and electronic insulators, arc chutes, automotive coil tops, distributor caps and switch insulators that rely on their high dimensional stability in the presence of temperature and humidity changes.

2.5.5 Allyl Systems

The manufacture of allyl resin monomers involves the reaction of unsaturated alcohol like allyl alcohol (AA), with a difunctional acid like phthalic anhydride, and isophthalic acid resulting in the formation of most important *allyl esters* for the preparation respectively of diallyl phthalate (DAP) and diallyl isophthalate polymers (DAIP).

2.5.5.1 *Diallyl Phthalate (DAP)*

Allyl alcohol (AA)

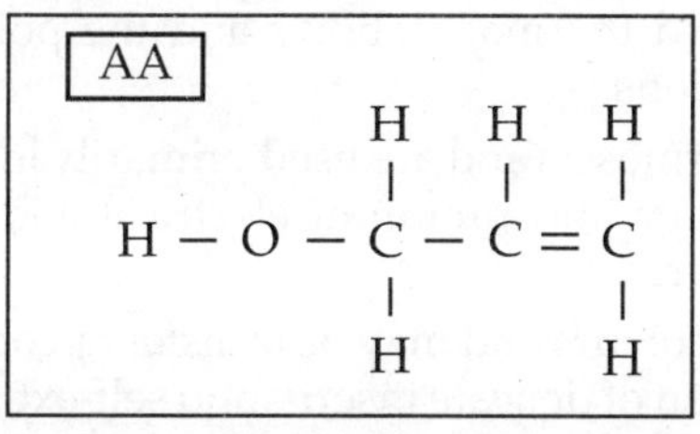

Phthalic Acid (PTA)

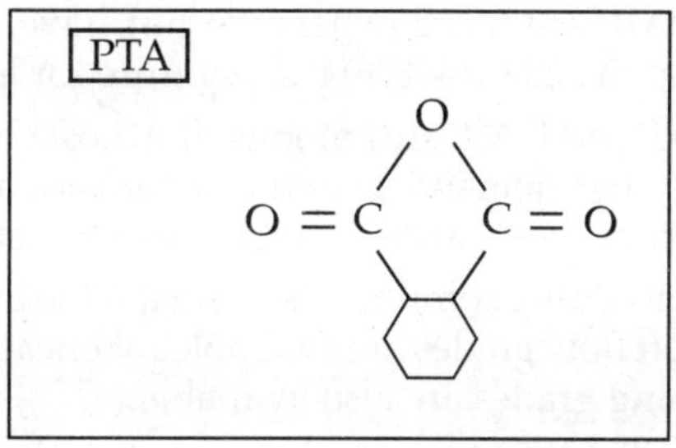

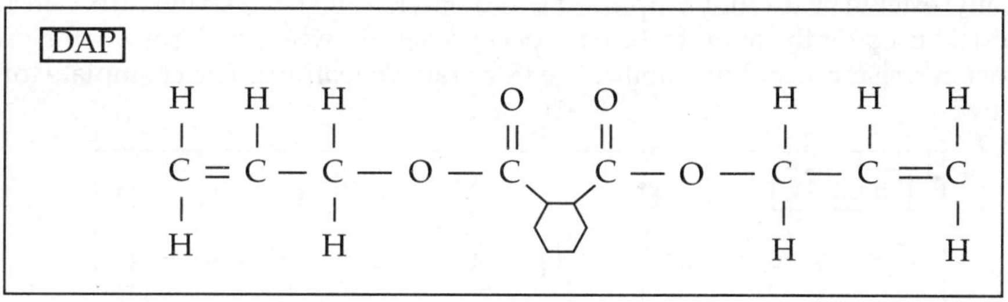

DAP is prepared by reaction of allyl alcohol and phthalic anhydride.

Like other thermosets, DAP requires fillers and reinforcing agents to impart good mechanical properties to the moulding compounds. Typical fillers and reinforcing agents used in DAP compounds include cellulose asbestos, glass fibres and synthetic fibres such as nylon. Filler selection is determined by the particular property combination desired.

Mouldings based on DAP exhibit good electrical insulation characteristics and dimensional stability under conditions of dry and wet heat. The polymer has good thermal stability and can withstand a temperature of up to 180°C (356°F) for long periods. However, the polymer is prone to attack by strong alkalies though it is resistant to acids other than strong oxidising acids. Mouldings based on DAP are expensive and are used mainly for electrical components for use under adverse operating conditions.

Resistance to fungi enables DAP to function well in special military communication gear.

2.5.5.2 *Diallyl Isophthalate (DAIP) Polymers*

Allyl alcohol (AA) Isophthalic Acid (IA)

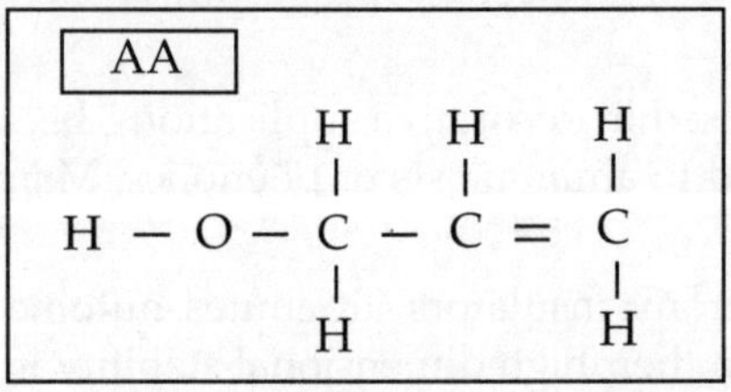

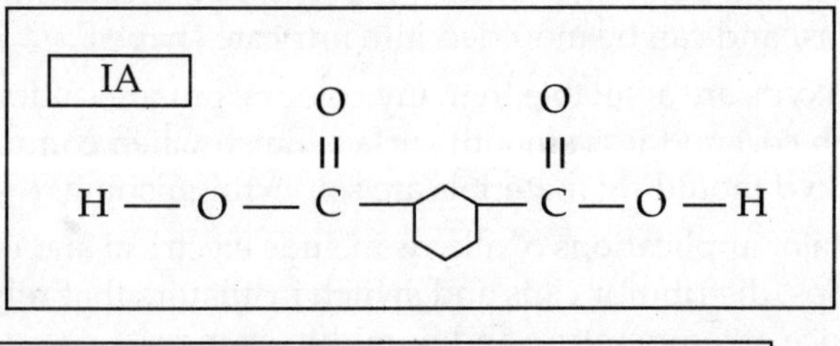

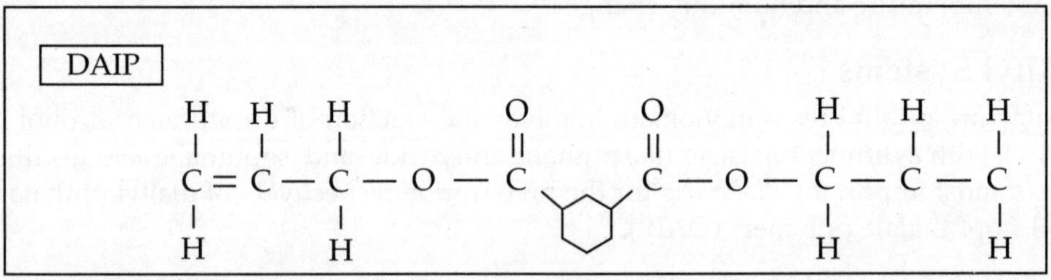

DAIP monomer is prepared by estrification of allyl alcohol and isophthalic acid.

DAIP polymers are very similar to DAP in preparation, properties and applications. Mouldings based on DAIP are more expensive but have enhanced thermal stability and the polymer can withstand up to 220ºC (428ºF) temperature for long periods.

Both DAP and DAIP are amongst the most costly thermosets and are used primarily in speciality applications that demand extreme dimensional accuracy and premium electrical and chemical resistance properties, particularly at elevated temperatures.

Allylic moulding compounds are available in many colours and may be transfer or compression moulded. Soft flow grades are available for encapasulation of delicate inserts, and self-extinguishing or non-burning grades are also available.

The first allyl ester to be usefully employed was diethylene glycol bis-allyl carbonate (DEGBAC) and is still widely used for the manufacture of spectacle lenses, where its light weight, dimensional stability, abrasion resistance and dye-ability are its attractive features. The chemplate for DEGBAC is shown below:

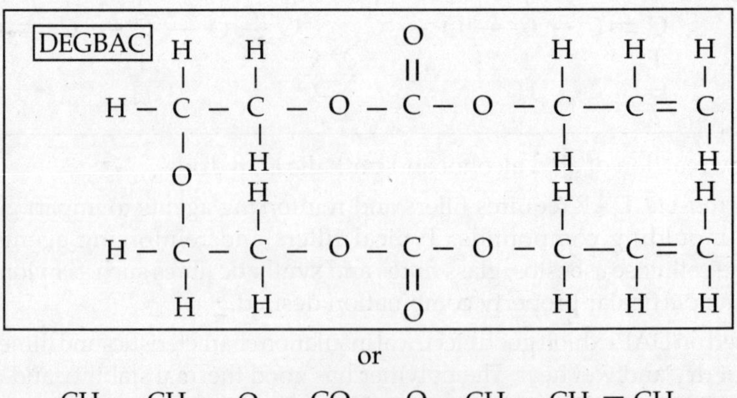

or

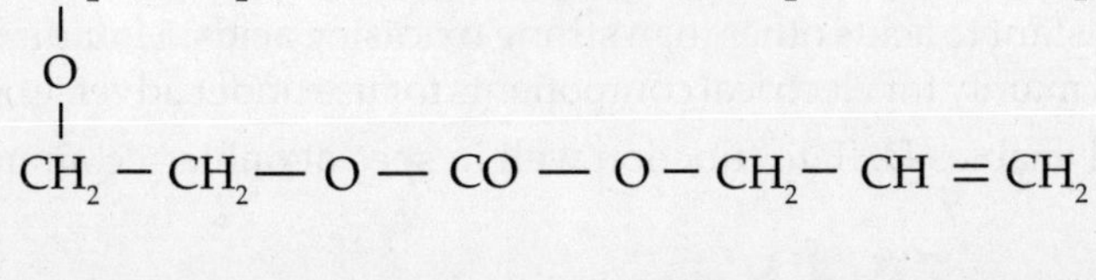

2.5.5.3 *Epoxy Systems*

Epoxide Group

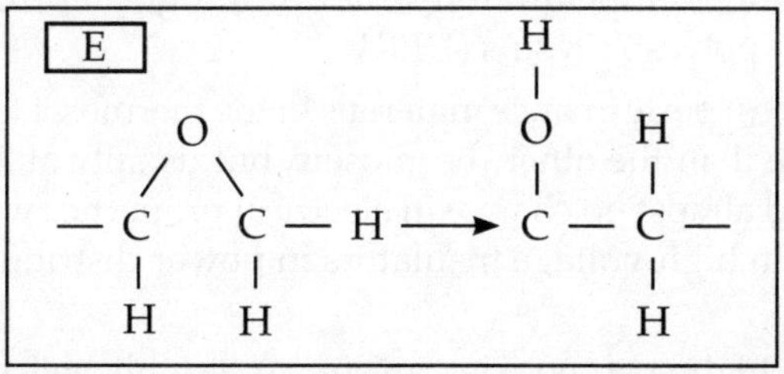

Epoxies are defined as cross-linked or thermosetting polymers in which the cross-linking is derived from reactions of the epoxy group with active hydrogen atoms of acid anhydrides or amines.

ECH — Epichlorohydrin

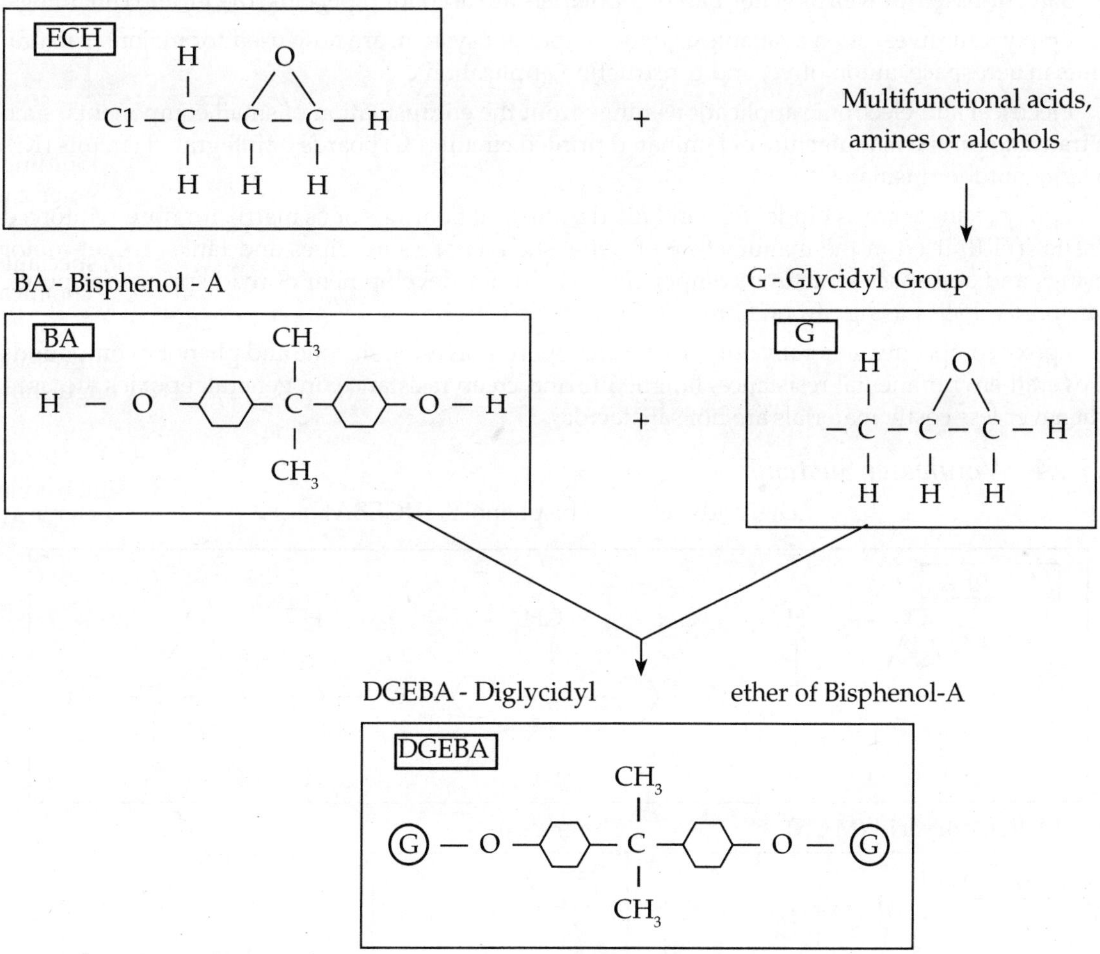

The most common epoxy is based on bisphenol-A and, in its simplest form, is often referred to as diglycidyl ether of bisphenol-A (DGEBA). It is sometimes used in a slightly polymerized form that features hydroxyl (OH) groups.

As with other thermosets, epoxies must be combined with fillers and reinforcing agents, such as glass, to provide the mechanical properties needed for moulded parts. Glass reinforced epoxy parts offer higher strength, high temperature resistance, and greater durability than similar parts made from, say, unsaturated polyester resins (UPRs).

Epoxy resins are the highest performance materials in the thermoset family. They possess greater chemical and heat resistance than the other thermosets, but usually at 2-3 times the cost of UPRs. Out door weather ability and almost no change in electrical properties with outdoor exposure have enabled epoxies to be used in high-voltage insulators in power distribution lines, where ceramics were used earlier.

Despite their relatively high cost, epoxies are firmly established in a number of important industrial applications. The largest single use is for surface coatings and as adhesives, which account for about 50% of epoxy resin output. Protective coatings are widely used for industrial, architectural, marine, and other applications like flooring coatings and road surfacing for heavy wear.

Solvent-based, as well as water-based or powders are used for pipe, tank, drum, and can linings.

Epoxy adhesives, often available as two-component system, are now used for major structural joints in aerospace, automotive, and construction applications.

Electrical and electronic applications range from the encapsulation of small components, such as transistor or the manufacture of laminated printed circuit (PC) boards or integrated circuits (IC), to large outdoor insulators.

Epoxy resins serve as binder for sand filled industrial floorings or as matrix for fibre-reinforced plastics (FRP) used in the manufacture of items such as pressure pipes and tanks, rocket motor casings and aerospace structural components, as also for development of road springs for public transport vehicles using carbon fibres.

Epoxy composites generally outperform analogous polyester, silicone and phenolic compounds in overall environmental resistance, fatigue life and creep resistance. In general, epoxies are used whenever less costly materials are not satisfactory.

2.5.5.4 Vinylester Systems

Diglycidyl ether of bisphenol A (DGEBA)

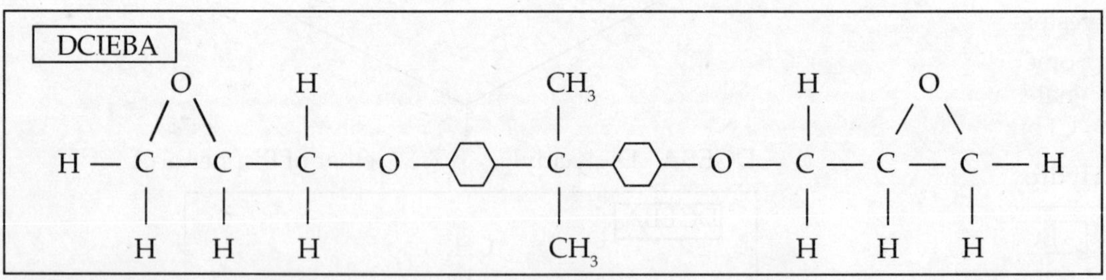

Methacrylic acid (MAA)

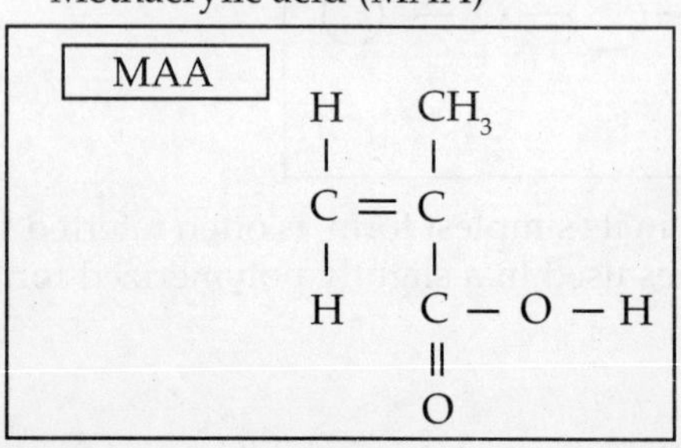

Styrene (S)

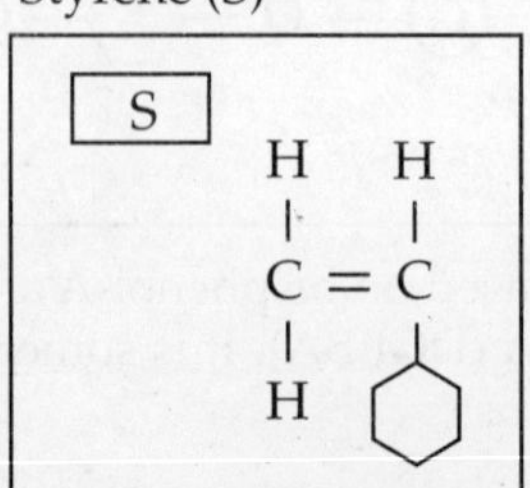

Vinylester resins (VER) consist of short linear molecules resulting from the reaction of epoxide group terminated molecules such as diglycidyl ether of bisphenol-A (DGEBA) with unsaturated acids, such as methacrylic acid (MAA). They feature ester linkages and vinyl-type unsaturations, hence they are termed *Vinylester resins.* They are cross-linked into three-dimensional networks by using styrene monomers, as is done in the case of regular unsaturated polyester resins (UPR).

Major use of vinylesters is as matrix for fibre-reinforced plastics (FRP). They are well known for their very good adhesion to glass and for the high corrosion resistance they impart to the composites. Chemical process equipment like piping, tanks, coatings, etc., constitutes a large market for these resins.

2.5.5.5 *Polyurethanes/Urethane Resin Systems (PU)*

Toluene di isocyanate (TDI)

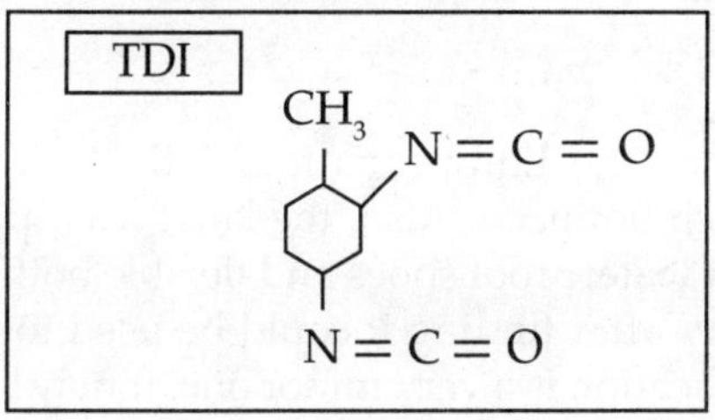

Polybutylene ether glycol (PBEG)

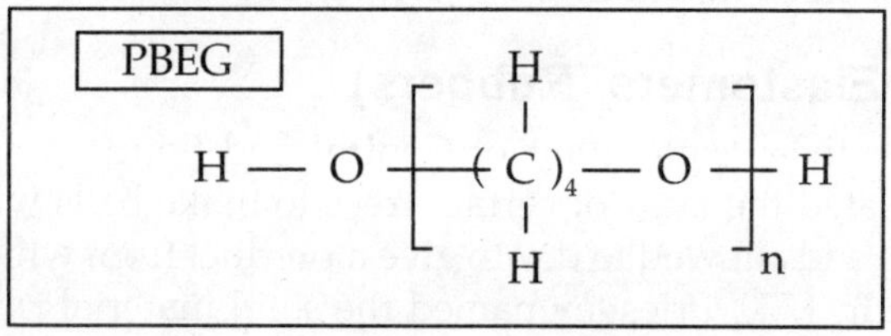

Polyurethanes are defined as polymers which contain urethane groups (– NH – CO – O) in the main polymer chain. However, in the industrially important polymers the urethane group is not usually the principal group present, other groups such as ester, ether, amide, and urea groups are generally contained in the polymer chain in appreciable number.

The urethane group results from the interaction of an isocynate and a hydroxyl compound. When a di-isocynate and a diol (glycol) react together a linear polyurethane is obtained whilst a di-isocynate and a polyol (polyhydric alcohol) lead to a cross-linked polymer. Thus, di-isocynates and diols and polyols are the principal raw materials used in the manufacture of polyurethanes.

Polyols are generally divided into *polyether types*, which have a greater resistance to hydrolysis (i.e., water-associated degeneration), and *polyester types* which have a better resistance to oil and hydrocarbons in general. Polyethers are now used in about 90% of commercial polyurethanes whilst polyesters are used in most of the remainder.

Some hydroxyl-terminated hydrocarbon polymers (OH end-groups) can react with suitable isocynates to form three-dimensional molecular networks (thermosets). They include unsaturated (C = C) butadiene (BOH) and saturated isobutylene (IBOH) types, as shown in chemplates.

Hydroxyl terminated Butadiene (BOH)

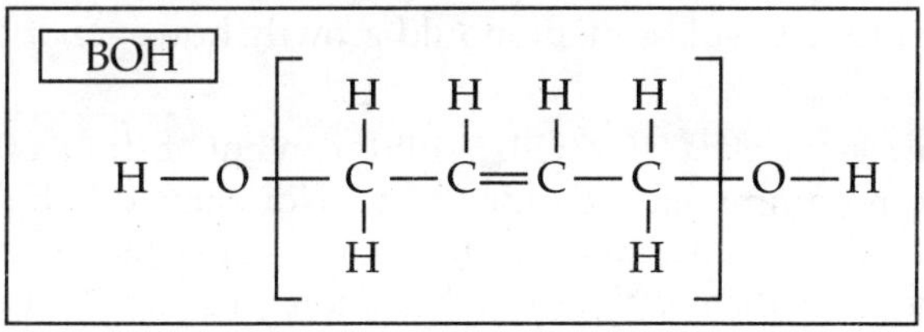

Hydroxyl terminated isobutylene (IBOH)

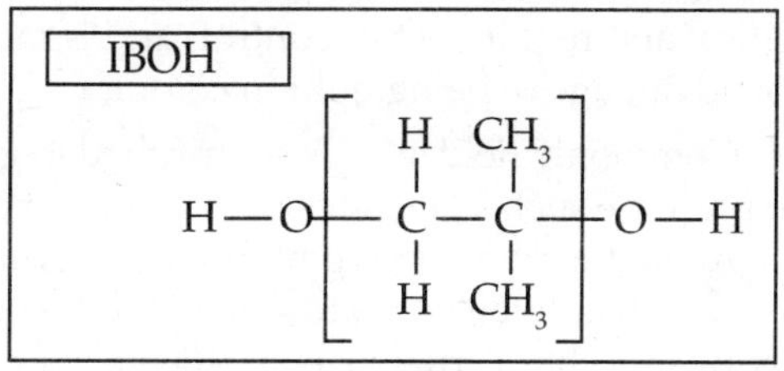

The properties of polyurethanes can vary widely, depending on the exact system considered. Among the mechanical properties, abrasion resistance is generally outstanding.

Many of the parts that need good mechanical properties are made of solid polyurethane by a moulding process that may involve a thermoplastic resin with high moulding pressures (e.g., conventional injection moulding) or a reactive system with much lower moulding pressures but usually a longer cycle time. The stiffness can range from that of a soft rubber to that of an engineering plastic. Examples of such applications include pump liners and impellers, gears, sprocket wheels, bushings, shock mounts, O-rings, seals and gaskets, solid tyres or wheel covers, roller covers for printing paper industry, sports boots for skiing and hockey, conveyer belts, and chute liners.

A later development known as *Reaction* Injection Moulding (RIM) has been developed for processing polyurethanes where in the reactants are injected into a mould where they are made to react by maintaining the mould at a high temperature. It permits, particularly, the economical moulding of large thin parts, such as automobile body components. The products are very slightly cellular and relatively flexible unless a stiff reinforcement is used and the process is then called RRIM.

Other applications of polyurethane as commercially significant plastic include foams (flexible, rigid, and integral), elastomers, surface coatings, adhesives and binders. For these applications, they are discussed in detail in the relevant sections of the text.

2.6 Elastomers (Rubbers)

During their exploration of Central and South America the Europeans found that the natives coagulated the latex of certain trees to make balls which bounced. Also, the latex was spread on fabrics and allowed to dry to give a product from which waterproof shoes and flexible bottles were made. In 1770, Priestley named the solid material *rubber* after finding it could be used to rub out pencil marks; the name has remained although this application is a very minor one. It may be noted that the term *rubber* is now commonly used also as a generic name for any material with rubber-like properties; to avoid confusion the latter materials are now called *Elastomers. Elastomers can thus be defined as rubbery materials capable of large or fairly large elastic deformation.*

2.6.1 Natural Rubber

For the production of natural rubber, the Hevea trees on plantations are tapped, i.e. the bark is cut and the latex that flows is collected in small vessels. The latex is a stable dispersion of rubber in water, proteinous material being the dispersing agent; a typical analysis might be as follows:

(i) Rubber hydrocarbons: 35%

(ii) Non-rubber solid (proteins, lipids, quebrachitol, inorganic salts, etc): 5%

(iii) Water: 60%

The bulked latex is then diluted to a solid content of about 15% and coagulated by the addition of acetic or formic acid. The coagulum is converted into sheet on rollers, then washed and dried. Two types of drying processes are used. In the first method, the sheets of rubber are dried in wood smoke at about 60°C for four days; the resultant brown product is known as *smoked sheet*. In the second method, the sheets are dried in air at about 40°C, for 7 days; the product is *pale crepe*. Compared to smoked sheet, the pale crepe is more expensive to produce (since the drying conditions are more critical and require better control) and is more prone to subsequent mould growth; however, it has the advantage of being light in colour.

Commonly field latex is centrifuged to give *concentrated latex* with a solids content of about 60%. To preserve such a latex over a long period of time, ammonia and other bactericides are added. The coagulated material being highly elastic that could not be shaped by moulding or extrusion; thus recourse had to be made to somewhat labourious *solution methods* of fabrication. In 1820, Hancock discovered that if rubber is masticated (ground) it becomes plastic or capable of flow and hence easier to fabricate. However, there was another problem in that rubber products become hard and stiff in cold weather and soft and sticky in hot weather. The situation was further aggravated by the poor ageing properties shown by the products.

In 1839, Goodyear found that rubber that was heated with sulphur retained its elasticity over a wider range of temperature than the raw material and had better resistance to solvents. Subsequently, Hancock also found that when plastic masticated rubber was heated with sulphur a strong, elastic material was regenerated. The interaction of rubber and sulphur was termed *vulcanization*. The term now also includes other reagents which have similar effects, although sulphur has remained by far the most important vulcanizing agent. Further, the term now encompasses other elastomers besides natural rubbers.

From a chemical point of view, vulcanization is principally a process of cross-linking where by the discreet linear rubber chains are converted into a three-dimensional network. In this way, the chains are prevented from slipping past one another when the sample is stressed and they return to their original positions when the stress is removed.

The discoveries of mastication and vulcanization represent the foundations of the rubber industry.

2.6.2 Types of Elastomers

Elastomers are of the three types, viz.,

1. Vulcanizable or Conventional Elastomers
2. Reactive System Elastomers
3. Thermoplastic Elastomers

Brief descriptions of these types of elastomers follows.

2.6.2.1 *Vulcanizable or Conventional Elastomers*

This method for producing a material with good elastomeric properties (rubber) involves the formation of chemical cross-links between high-molecular-weight linear molecules. The starting polymer (raw rubber) must be of the amorphous type, and its glass transition temperature, T_g, must be well below the room temperature to ensure a rubbery behaviour, and the chemical cross-links must not be too numerous.

Conventional vulcanization of the most common rubbers generally involves sulphur-based chemical systems and relies on the presence of carbon-carbon (C = C) double bonds or unsaturations in the chains. Other chemical systems can be used (e.g., organic peroxides, etc.) but they are generally more costly. The incorporation of the vulcanization ingredients into the raw rubber is carried out at moderate temperatures in an operation generally referred to as *compounding*. The compounds are then shaped into products (by extrusion, moulding, etc.) and heated to a relatively high temperature typically 150°C (302°F) to induce the cross-linking. Vulcanization temperature and duration are functions of the chemical systems used.

A number of fluids are often employed to help reduce the stiffness of elastomers or for other purposes, including ease in processing and reduction in cost. They include components referred to as *pepticizers (digesters)*, *plasticizers, lubricants, factices* (rubber-like substance obtained by vulcanizing unsaturated vegetable oil), and oils such as paraffin waxes, mineral, and other oils.

A number of other additives are often compounded into rubbers. They include antioxidants, anti ozonants, oil extenders to reduce cost or soften the rubber, fillers to reduce cost or stiffen the rubber, and reinforcing fillers.

A typical tyre tread compound has the following composition:

Natural rubber smoked sheet	:	100
Sulphur	:	2.5
Sulfenamide	:	0.5
MBTS (Mercapto benzothiazole disulphide)	:	3
Zinc oxide	:	3
Stearic acid	:	3

PBNA (Phenyl beta naphthylamine)	:	2
Carbon black (HAF)	:	45
Mineral oil	:	3

The major ingredients of a typical shoe sole compound are :

SBR (Styrene butadiene rubber)	:	100
Clay	:	90

While a typical electric cable cover features,

Polychloroprene	:	100
Kaolin	:	120
Carbon black (FEF)	:	15
Mineral oil	:	12

in addition to the vulcanization and protective agents.

Suppliers of raw rubber suggest suitable compounds for specific applications and processing methods, as also the corresponding vulcanization conditions. Experienced rubber processors develop their own recipes that sometimes involve blends or raw rubbers.

2.6.2.2 *Reactive System Elastomers*

Elastomers (rubbers) can also be produced in a single step from low molecular weight reactive chemicals. Mechanism is similar to that employed for thermosetting plastics. However, because relatively lower degree of cross-linking is required for elastomers, chain polymerization is favoured over cross-linking in the overall chemical reaction.

Certain types of polyurethane and silicon elastomers fall into this category. The starting chemicals are normally in liquid form and thus easy to cast, and the reactions can be triggered by heat, catalysts, or mixing as in the case of thermosetting plastics.

2.6.2.3 *Thermoplastic Elastomers*

Thermoplastic elastomers (TPE) are materials which possess, at normal temperatures, the characteristic resilience and recovery from extension of cross-linked elastomers but which exhibit plastic flow at elevated temperatures and can be fabricated by the usual techniques employed for thermoplastics. These effects are associated with linear polymers containing segments which give rise to interchain attraction. At normal temperatures these interactions have the effect of conventional co-valent cross-links and confer elasticity but at elevated temperatures the secondary forces are inoperative and the material exhibits thermoplastic behaviour.

Thus, a thermoplastic elastomer has two intermingled polymeric systems each with its own phase and softening temperature. In the useful temperature range one phase (the soft phase) is above its softening temperature, T_g, whilst the other (the hard phase) is below its softening temperature. The hard phase thus acts to anchor the chains of the soft phase and the material is elastomeric. At temperatures above the softening temperature of the hard phase, the reactions imposed by the hard phase disappear and the material behaves as a viscous fluid. On cooling, the hard phase re-solidifies and the material regains its elastomeric characteristics.

Thermoplastic elastomers (TPE) differ markedly from the previous two classes of elastomers in that processing of TPEs does not involve any chemical reaction. The various types of thermoplastic elastomers can generally be made in a wide range of stiffness and thus bridge the gap between soft plastics and elastomers. A few of them are, however, available in very soft, highly elastic grades.

TPEs have many advantages over conventional vulcanizable elastomers, but they are still generally inferior in a number of important properties, such as creep, set, and resilence.

Due to their thermoplastic nature, TPEs can be processed by many processes, including injection moulding, extrusion, blow moulding, film blowing, and rotational moulding.

2.6.3 Commercial Elastomers and Related Polymers

Commercial elastomers may be grouped into seven types, viz.,

1. Diene and Related Polymers
2. Elastomeric Copolymers
3. Ethylene-Related Elastomers
4. Fluoro Elastomers
5. Silicone Polymers
6. Miscellaneous Elastomers
7. Thermoplastic Elastomers

Brief description of these different types of elastomers follows.

2.6.4 Diene and Related Elastomers

Many of the elastomers in this group are widely used as commodity rubbers which are often associated with the term *diene* that is itself related to the presence of carbon-carbon (C = C) double bonds which are suitable for conventional vulcanization. The important members of this group include the follwoing:

1. Polyisoprene (PI)/Natural Rubber
2. Polybutadiene (PB)
3. Polychloroprene Rubber (PCR)
4. Polyisobutylene (PIB)/Butyl Rubbers.

2.6.4.1 *Polyisoprene (PI)/Natural Rubber*

Cispolyisoprene (PI – C)

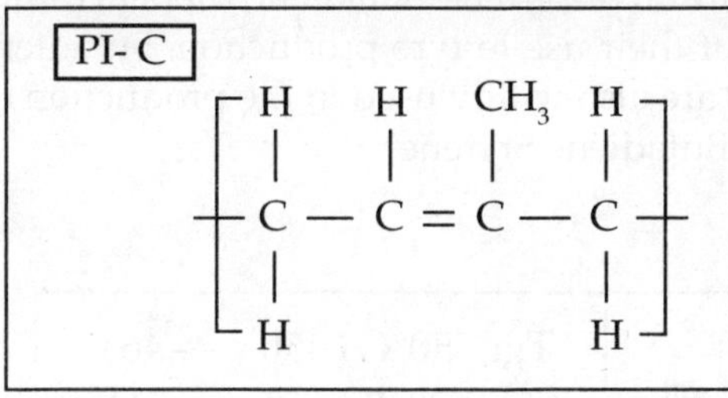

Tg: –70ºC (–70 — –68)
–94ºF (–94 — –90)

Transpolyisoprene (PI – T)

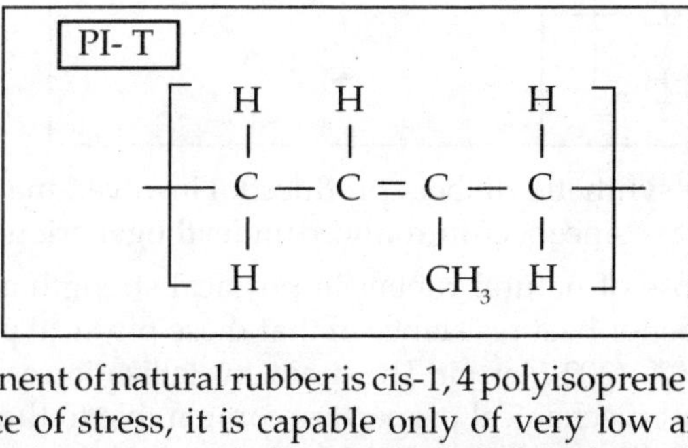

Tg: –70ºC (–75 — –70)
–94ºF (–103 — –94)
Tm: 68ºC (60 — 74)
155ºF (140 — 165)

Major component of natural rubber is cis-1, 4 polyisoprene (PI-C) and is now produced synthetically. In the absence of stress, it is capable only of very low and slight crystallization at temperatures between its melting point of about 28ºC (82ºF) and its glass transition temperature. When highly stretched, however, it crystallizes readily, and this strain-crystallization is associated with its excellent mechanical resistance, even in the absence of reinforcing filler (gum rubber).

Trans-1, 4 polyisoprene (PI-T) with a molecular weight of 30,000 a major component of balata or gutta-percha has about the same glass transition temperature as cis-1, 4 polyisoprene but it crystallizes quite readily below its melting point of about 68°C (155°F). It is a gocd dielectric and was one of the first plastics used.

Natural rubber is well known for its high tensile and tear strengths, high resilience; resistance to cold flow; and resistance to wear, abrasion and fatigue.

Natural rubber does not have a good intrinsic resistance to sunlight, oxygen and ozone, heat ageing, as also oils and fuels.

Natural rubber in unfilled form is widely used for products requiring very large elastic deformations or very high resilience, such as pharmaceutical or medical devices, tyres, gaskets, and shock absorption devices.

2.6.4.2 *Polybutadiene (PB)*

PB

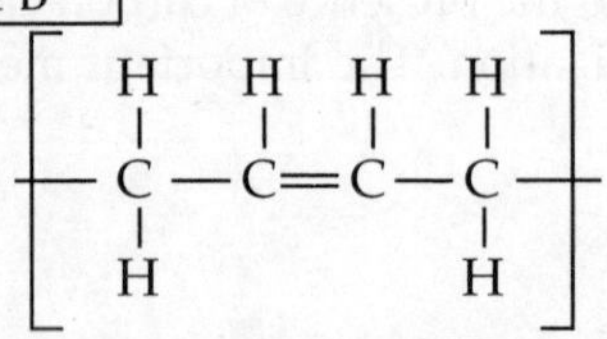

Tg: – 90°C (–108 —–85)
–130°F (–162 — –121)

Polybutadiene rubbers are superior to natural rubber with respect to resilience, heat build-up and abrasion or wear resistance. These properties are particularly significant in automotive tyres. However, PB rubbers have lower tensile strength and tear resistance; and PB tyres have relatively poor road adhesion in wet conditions. Due to these reasons and to help ease in processing, butadiene rubbers are generally used in blends with natural or styrene-butadiene rubbers (SBR); such blends usually contain less than 50% PB. Because of their use in tyre production, butadiene rubbers are produced in significant tonnage. PB rubbers are also widdy used in the production of high impact polystyrene (HIPS) and ABS (Acrylonitrile) Butadiene Styrene).

2.6.4.3 *Polychloroprene (CR)*

CR

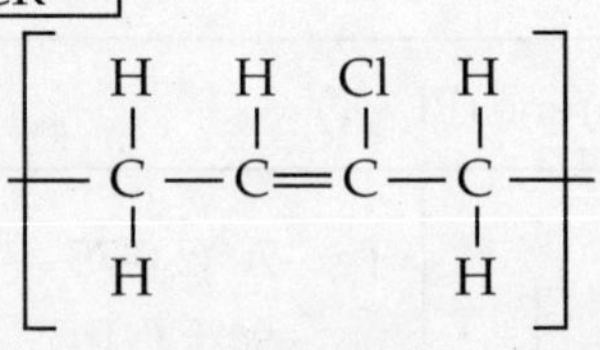

Tg: –50°C (–50 — –46)
–58°F (–58 — –51)

Polychloroprene was one of the very first synthetic rubbers produced. However, many types and grades of polymers and copolymers of CR have since been introduced under the generic name *neoprene*.

CR rubbers are broadly similar to those of natural rubber in physical strength and elasticity. However, polychloroprenes show much better heat resistance in that these physical properties are reasonably well maintained up to about 150°C (302°F) in air. The strain crystallization associated with its molecules is responsible for its excellent mechanical properties, comparable to those of natural rubber but the normal grades readily crystallize and become stiff when cooled below-10°C (14°F).

Polychloroprene rubbers have high order of oil and solvent resistance but aromatic and chlorinated solvents cause some swelling. The presence of chlorine-substituted double bond makes the polymer have good resistance to most chemicals, oils, oxygen and ozone. The high chlorine content makes it generally self extinguishing and have good weather resistance.

Polychloroprene rubbers are used for automotive products like fuel hoses, cable sheathing, protective boots, etc. They are also used for bridge pads, roofing membranes, fabric coatings, gloves, adhesives, and corrosion-resistant coatings.

2.6.4.4 *Polyisobutylene (PIB)/Butyl Rubbers*

Polyisobutylene (PIB)

PIB

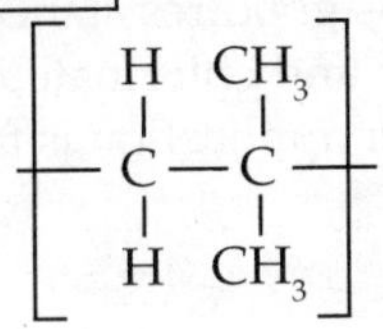

Tg: –65°C (–70 — –50)
–85°F (–94 — –58)

The physical properties of PIB are very dependent on molecular weight. Polymers with average molecular weight of about 15000 are sticky viscous fluids whilst those with molecular weight of 100,000–200,000 are rubber-like, resembling un-milled *crepe rubber*.

PIB is remarkably different from most common rubbers in two respects viz., it has a very low gas permeability and very high damping properties. Since the molecules do not contain unsaturations (C = C) they cannot be cross-linked by most of the conventional methods. On the other hand, networks have good weathering, ozone, and heat ageing resistance. PIB is capable of strain-crystallization and can have good mechanical resistance. Resistance to oils and fuels is, however, not good.

A further limitation of PIB is its tendency to cold flow and, as a result, the polymer finds little use in self-supporting form. Applications are, therefore, restricted to adhesives, fabric and paper coatings, and blends with other polymers. Low molecular weight PIB is also used in caulking compounds.

If isobutylene is copolymerized with small fractions of isoprene (1-2%), the isoprene unsaturations can be used for conventional vulcanization. Such isobutylene-isoprene copolymer (IIR) is referred to as *Butyl Rubber*.

A major limition of butyl rubber is its poor compatability with other common rubbers and generally low adhesion properties. Partial chlorination or bromination brings a marked improvement and also allows faster curing.

Vulcanized butyl rubber broadly resembles vulcanized natural rubber in physical characteristics, but shows better resistance to abrasion, repeated flexing and tear; it, however, has lower resilience. Vulcanized butyl rubber has relatively good resistance to degradation by heat, oxygen and ozone. The outstanding property of butyl rubber, however, is its very low permeability to gases. Because of this last characteristic, butyl rubber is very widely used for tyre inner tubes. Other applications include automobile damping fixtures, radiator hose and window seals, and cable insulation. It is also used for curing bags or bladders, steam hoses, and shock and vibration absorption devices, like engine mounting pads.

2.6.4.5 *Ebonite*

Ebonite

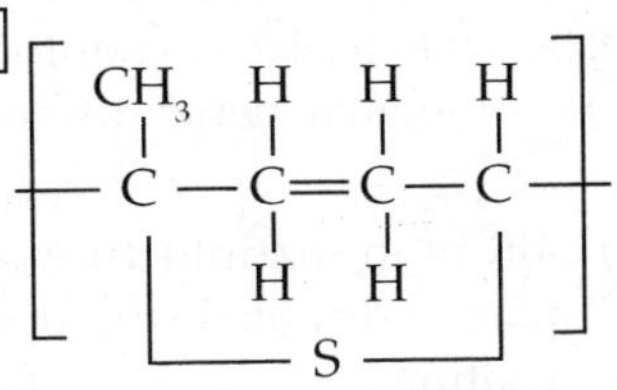

Ebonite (Vulcanite or *hard rubber*) is the hard product obtained when natural rubber (or other diene is heated at about 150ºC (302ºF) with approximately half its weight of sulphur until reaction is substantially complete.

Typically, the rubber (often cheap reclaimed rubber) is mixed on a mill with sulphur, diphenyl guanidine (accelerator), magnesium oxide (activator) and ebonite dust (filler) to give a soft stock which is then compression moulded at 170–180ºC (338–356ºF) unit vulcanization is complete.

The structure of ebonite is basically similar to that of a *soft* vulcanizate prepared by the use of a low accelerator to sulphur ratio although, of course, there are far more sulphur-containing structures per unit volume and the material is essentially saturated.

Ebonite shows very low water absorption and has good resistance to most aqueous reagents, except nitric acid. It is, however, swollen by aromatic and chlorinated hydrocarbons. Ebonite has good electrical insulation characteristics and the major application is far car battery cases.

2.6.5 Elastomeric Copolymers

The important members of this group are:

1. Styrene - Butadiene Rubber (SBR)
2. Acrylonitrile - Butadiene Rubber (NBR)
3. Ethylene - Propylene Rubber (EPR)

Brief descriptions of these elastomers follows.

2.6.5.1 *Styrene Butadiene Rubber (SBR)*

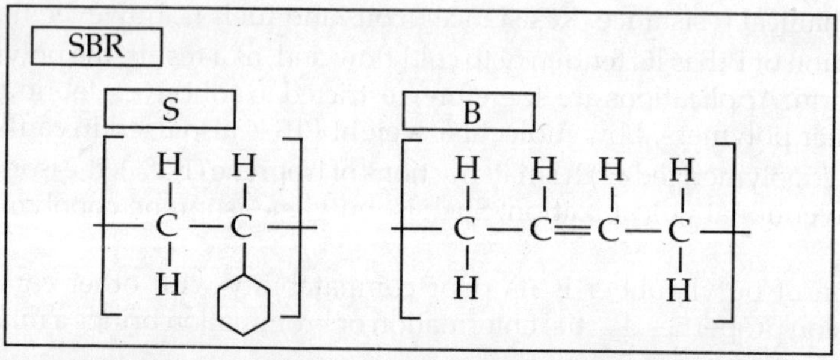

SBR is a random copolymer of butadiene (67–85%) and styrene (15–33%). The glass transition temperature of a typical 75/25 composition is about – 60ºC (– 76ºF).

SBRs have characteristics very similar to those of natural rubber. They are compounded and processed in much the same way and may be vulcanized with either sulphur or peroxides. The styrene-butadiene chain is irregular and there is no tendency to crystallize on stretching (unlike natural rubber); thus in gum form (i.e. without reinforcement) SBR has low tensile strength. However, reinforcement with carbon black leads to a product with properties which resemble those of natural rubber and the two products are interchangeable in most applications. SBR shows a high heat build up on flexing which makes it unsatisfactory for the manufacture of large tyres.

SBR is one of the least expensive rubbers and generally processes easily. When compared to natural rubber, which it initially sought to replace in reinforced form, it is inferior in most mechanical properties except wear. But it is somewhat better in heat ageing, ozone resistance, weathering resistance and resistance to oils.

The most important use of SBR is in manufacture of car tyres but it finds numerous other applications including footwear, wire and cable insulation, industrial rubber products, adhesives, paints, carpet backing and paper coating.

2.6.5.2 *Acrylonitrile-Butadiene Rubber (NBR)*

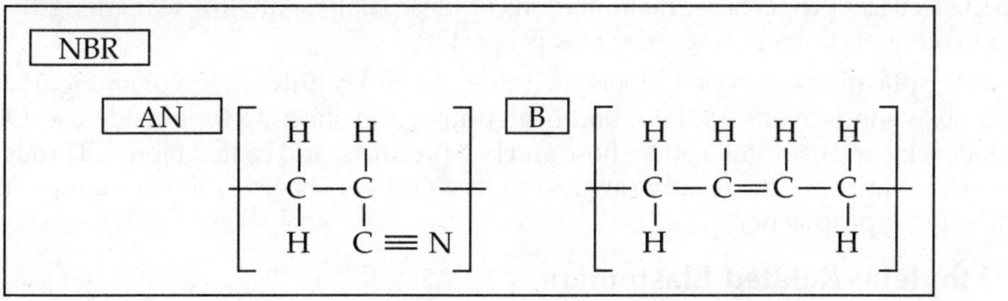

Acrylonitrile-butadiene rubbers or Nitrile rubbers (NBR) are compounded and processed in much the same manner as natural rubber. They can be vulcanized with either sulphur or peroxides. As with SBR, gum form has low strength but reinforcement with carbon black leads to rubbers with tensile strength approaching that of natural rubber.

NBR, a copolymer with 20-25% of acrylonitrile, was developed as a general purpose, oil-resistant rubber. The presence of polar nitrile groups (C ≡ N) is responsible for the drastically increased resistance to oils, fuels, and solvents, which is superior to that of polychloroprene rubber (CR). Resistance to heat ageing is quite good, and abrasion resistance is high. Cost is moderate. However, NBR does not have good resistance to ozone and weathering. Also, low temperature flexibility is limited in some grades. Because of its polar nature, NBR shows high dielectric losses.

Applications for NBR include fuel and oil tubing and hose, gaskets and seals, conveyor belting, and printing rolls and pads.

2.6.5.3 *Ethylene - Propylene Rubber (EPR)*

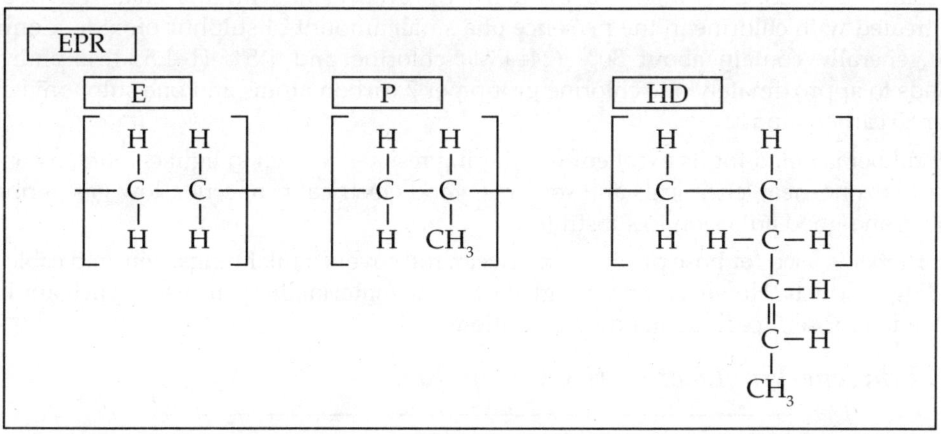

Since neither ethylene (E) nor propylene (P) repeat units have unsaturations (C = C), E–P copolymers, referred to EPR are saturated and completely amorphous. This makes them resistant to ozone, weathering, and heat ageing, but does not allow conventional vulcanization with sulphur-based systems and are vulcanized by heating with peroxides.

The incorporation of a small fraction of a third monomer containing unsaturations like 1, 4— hexadiene (HD) allows conventional vulcanization with sulphur systems. Such polymers are known as *terpolymers* or *EPT rubbers*. These rubbers retain the excellent ozone and weathering resistance and the good heat ageing of EPR rubbers.

EPR rubbers are made from low cost monomers and have low specific gravity of 0.87, they therefore have considerable commercial potential as general-purpose rubbers. Being saturated polymers, they have outstanding resistance to degradation by heat, oxygen and ozone. They also have good electrical properties. The limitations of these rubbers are low tear strength, difficulty in bonding to other elastomers, and poor tack properties.

Typical applications of these rubbers include solid and cellular (foam or sponge) weather strip for automobile windscreens, radiator and heater hose, and sheeting for outside use. Other uses of EPR include wire and cable insulation, hose and belt products, and coated fabrics. Through blending, they are used to improve the weathering resistance of other rubbers or the low temperature impact resistance of polypropylene.

2.6.6 Ethylene-Related Elastomers

Two of the important members of this group are:

1. Chlorosulfonated Polyethylene (CSPE)
2. Ethylene-Vinylacetate Copolymers (EVA)

Brief descriptions of these elastomers follows.

2.6.6.1 *Chlorosulfonated Polyethylene (CSPE)*

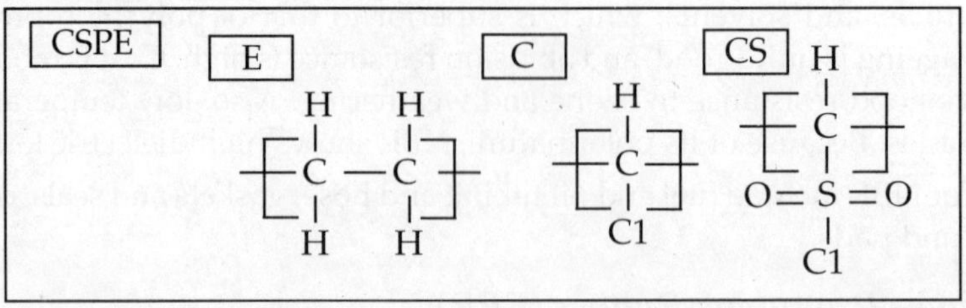

In the preparation of CSPE commonly a low density polyethylene with molecular weight of about 20,000 is treated with chlorine in the presence of a small amount of sulphur dioxide. Commercial products generally contain about 30% (24–43%) chlorine and 1.5% (1–1.5%) sulphur, which corresponds to approximately one chlorine group per 7 carbon atoms and one sulphonyl chloride group per 85 carbon atoms.

CSPE rubber is noted for its excellent weathering resistance even in light colours. It has a good resistance to ozone, heat, chemicals and solvents, good electrical properties, low gas permeability for a rubber, and good adhesion to substrates.

CPSE rubber is used for hose products, footwear, roll covers, tank linings, wire and cable covers, and building products. However, rather high cost of the material limits its use to such applications as are intended for service in demanding conditions.

2.6.6.2 *Ethylene-Vinylacetate (EVA) Copolymer*

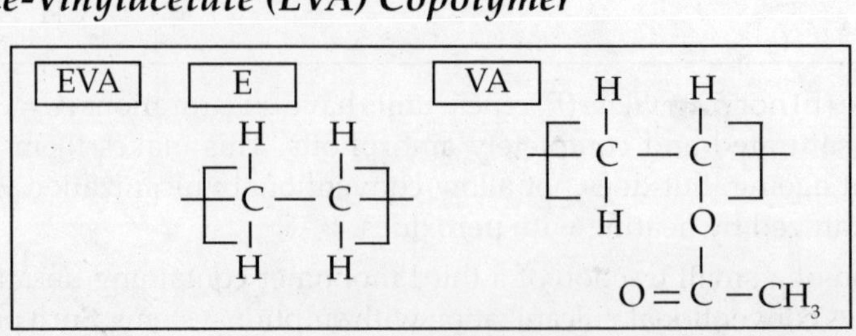

EVA copolymers with vinyl-acetate content in the range of 5-20% by weight are widely used. Compared to low density polyethylene, these copolymers show improved clarity, low temperature flexibility, impact strength, flexibility and stress crack resistance. They have greater oil and grease resistance but are more permeable to gases and water vapours; they are also less heat stable.

Copolymers with about 5% vinyl acetate are used for film extrusion; such a film finds use in food packaging. Copolymers with about 10% vinyl acetate are used for extrusion and injection moulding of such items as medical tubing, squeeze bulbs and gaskets. Copolymers with about 20% vinyl acetate find use as coatings and hot melt adhesives.

2.6.7 Fluoro Elastomers

There are a number of fluoro elastomers which are used industrially but the most well known are:

1. Polyvinylidiene fluoride (PVDF)
2. Polytetra fluoro ethylene (PTFE).

Both these elastomers are discussed collectively in this section.

PVDF

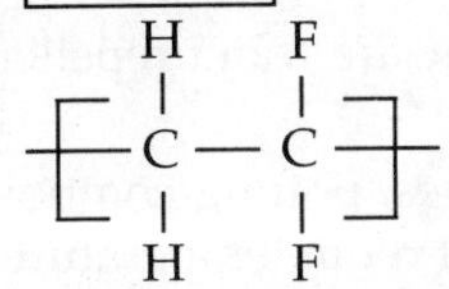

Tg: –35°C (–40 — –30)
–31°F (–40 — –22)

PTFF

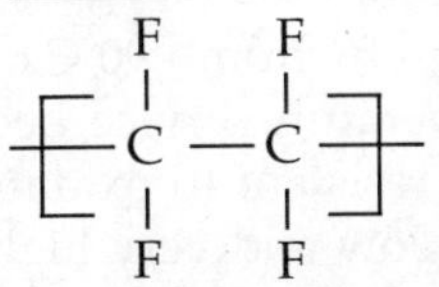

Tg: –130°C (–150 — –112)
–202°F (–238 — –170)

Fluoro elastomer chains, in principle, are made by suppressing the crystallinity associated with the chains of fluoroplastics, as discussed earlier in the section on thermoplastics, and to provide a mechanism for cross-linking sometimes by terpolymerization where in three repeat units are involved. However, details of the processes involved are usually not disclosed.

There are a number of monomers used for the development of fluoroelastomers but the ones used for PVDF and PTFE are vinylidiene fluoride and tetra-fluoroethylene respectively.

Fluoro elastomers are generally expensive, but possess a number of outstanding properties. They offer exceptionally good resistance to chemicals, in general, and to oils and solvents, including chlorinated solvents, in particular. They can withstand high tempratures. They have excellent weathering (i.e. resistance to UV radiation) and ozone resistance. They have good barrier properties, i.e., low permeability to gases and vapours. Their mechanical properties can be quite good, but they are not normally suitable for low temperature elastomeric applications.

Fluoroelastomers being specialty polymers are used for only demanding applications in the chemical, petroleum, automotive, and aerospace industries.

Specific applications include mechanical seals, packings, O-rings, gaskets, diaphragms, expansion joints and connectors, hose liners, roll covers, as also wire and cable insulation.

2.6.8 Silicone Polymers

Polydimethyl Siloxane (PDMS)

PDMS

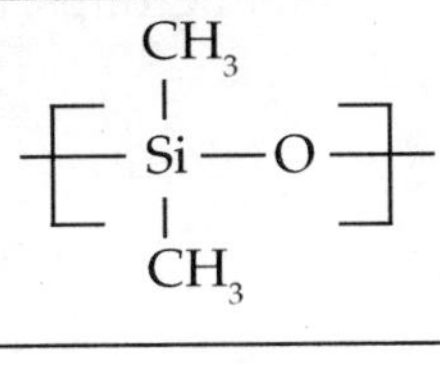

Tg: –125°C (–125 — –123)
–193°F (–193 — –189)
Tm: –55°C (–58 — –53)
–67°C (–72 — –63)

Silicone polymers or Polysiloxanes are based on polymeric chains featuring the very stable, alternating combination of silicone (Si) and oxygen (O) atoms in the backbone and a variety of organic side groups attached to the silicone atoms. The most common side groups are two methyl (CH_3) groups, and the polymer so produced is called Polydimethylsiloxane (PDMS).

Silicone elastomers are based on linear polymers which need to be cross-linked in order to obtain characteristic elastic properties.

High molecular weight (300,000–700,000) high consistency resins containing unsaturations are suitable for conventional rubber processing. These heat-curable elastomers (HCE) are cross-linked through high temperature vulcanization (HTV), normally through the use of peroxides.

The outstanding property of silicones is their thermal stability. In general, these polymers can be heated in air to about 200°C without much change in properties.

Another property of silicone polymers is that they are water repellent and have, therefore, the non-stick behaviour.

The silicones include resins, elastomers, adhesives, potting compounds, fluids and oils. One of the most familiar applications of silicones is that of room temperature vulcanized (RTV) rubber. RTV rubber is used widely in making flexible moulds and patterns for polyurethane, polyester, or epoxy parts. The moulds can be used for casting of models or prototypes. Adhesives and caulking materials that cure when exposed to air are used for sealing around edges of bath tubs and windows.

Most of the silicone-rubber moulded parts remain useful from – 90°C to 250°C (– 130°F — 482°F). The silicone rubbers are especially useful for high-temperature service, because they do not employ plasticizers to maintain rubberiness. They are highly resistant to oxidation and deterioration by heat ageing; in addition to their serviceability at very low and very high temperatures. Silicones are physically inert, odourless, tasteless, and non-toxic. Because, they are tolerated by body tissues so they are used as body implants.

Typical applications of silicone elastomers include power cable insulation, high voltage leads and insulator boots, ignition cables, spark plug boots, etc. In the electronic industry, semi-conductors are often encapsulated in silicone.

Mechanical applications include O-rings, gaskets, seals for aircraft doors and windows, freezers, ovens and applicances; diaphragms, flapper valves, protective boots and bellows, and printing rolls.

Silicone membranes are also used in artificial organs, such as kidneys.

The inclusion of a small number of B – O –Si bonds in PDMS structure leads to the development of fusible elastomers and *bouncing (silly) putty*. Bouncing putty is capable of highly elastic or perfectly viscous behaviour depending on the speed or temperature of loading. It can as well be drawn into thread. When dropped on a hard surface the material shows high elastic rebound; it shatters, however, when given a sharp blow.

Silicone coatings applied to clear plastic-like acrylics or polycarbonates and cured at moderate temperatures, provide scratch, wear, and abrasion resistance, as well as improved resistance to weathering and environmental stress-cracking.

Silicones are now well established as polymers. However, they are comparatively expensive; they are speciality material and, in terms of tonnage, are manufactured on a relatively modest scale.

2.6.9 Miscellaneous Elastomers

There are a number of other elastomers which are from different groups; three of the better known among them are:

1. Acrylic Rubber (AR)
2. Epichlorohydrin Rubber (ECHR)
3. Polysulfide Rubber (SR).

2.6.9.1 *Acrylic Rubber (AR)*

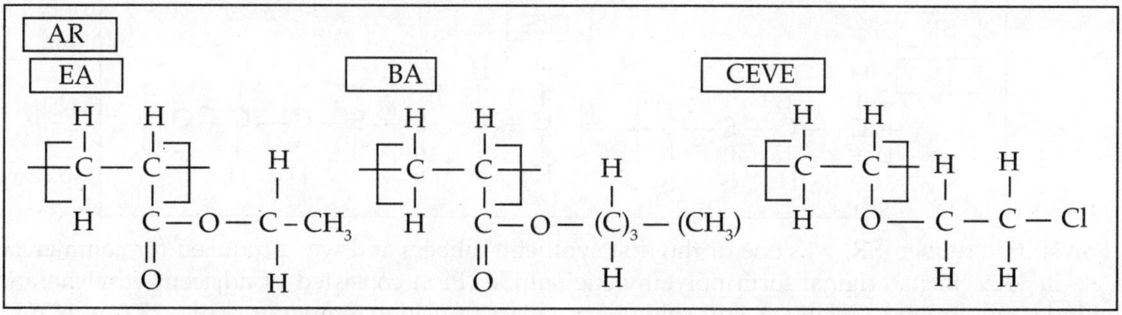

Acrylic rubber (AR) often consists primarily of polyethylacrylate (PEA) copolymerized with a small amount of about 5% of 2-chloro ethylvinyl ether (CEVE), for example, which serves as a cure site. The glass transition temperature, T_g, of PEA is about –22°C (–70°F), and AR is not suitable for low temperature application. The carbon black reinforced materials age very well in air up to about 200°C (392°F). A similar polyacrylate, polybutylacrylate (PBA), can be used, which has a low glass transition temperature of about – 45°C (– 49°F).

The important characteristics of acrylic rubbers are good resistance to high temperatures, lubricating oils, including sulphur-bearing oils, which often cause the hardening of other oil-resistant rubbers, such as nitrile rubbers (NBR). Major applications of AR include oil seals for automobiles, gaskets, and hoses.

2.6.9.2 *Epichlorohydrin Rubber (ECHR)*

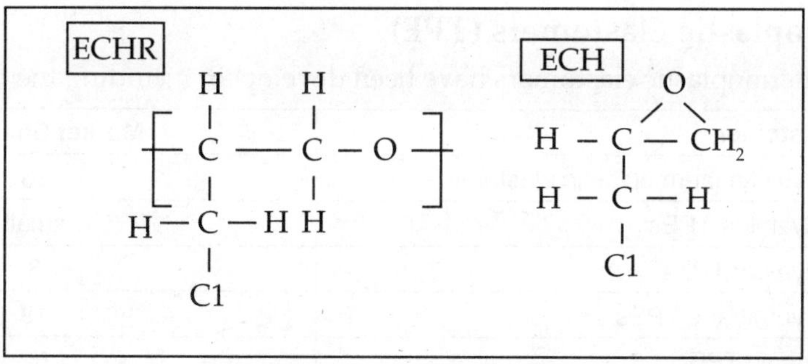

ECHR results from the polymerization of epichlorohydrin (ECH) with a repeat unit Polypichlorohydrin CH (CH_2Cl) CH_2 On (M_w = 700,000) (PECH). The relatively high glass transition temperature of PECH limits its low temperature use, but copolymers with more flexible ethylene oxide (EO) repeat units with glass transition temperature of – 67°C (– 88.6°F) are available which are suitable down to about – 40°C (– 40°F).

The presence of chlorine contributes to its excellent resistance to oil and fuels and to a good flame resistance. It is resistant to heat ageing and ozone. It has low temperature flexability and low permeability to air.

The largest use of ECHR is in the automotive field as fuel, lubricating fluid, air and vacuum hose. Other uses of this elastomer include seals, gaskets, diaphragms, wire and cable covers, and protective boots.

2.6.9.3 *Polysulfide Rubbers (SR)*

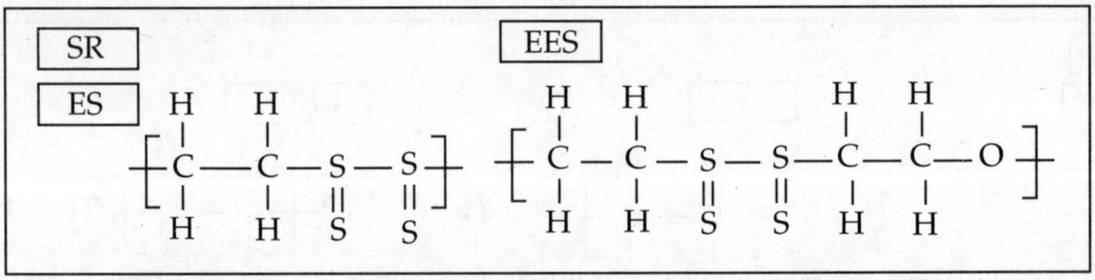

Polysulfide rubber (SR) was one of the first synthetic rubbers and was produced for commercial use in 1929. In its original form polyethylene sulfide (PES) consisted of adjacent ethylene and sulfide units, giving a relatively stiff chain with a glass transition temperature of – 27°C (– 16.6°F). The low temperature flexibility is increased by the closer positioning of ethylene oxide units with glass transition temperature of – 67°C (– 88.6°F) in the polyethylene ether sulfide (PEES) elastomer.

The mechanical properties of polysulfide elastomers are not generally good, but they are distinguished by outstanding resistance to oils and solvents. The resistance improves as the sulphur content of the polymer increases. Products of high sulphur content are also less permeable to vapours and gases. Polysulfide elastomers are also noted for oxygen and ozone resistance and have good low temperature flexibility, as also good weatherability. However, these rubbers are attacked by strong oxidizing acids and strong alkalies.

Polysulfide rubbers find applications that involve contact with fuels, paints and printing inks, and outdoor exposure, i.e. as caulking material, mastics* and putty. However, the current consumption of polysulfide elastomers, compared to that of other elastomers, is very small, the use is restricted mainly to special applications involving resistance to oils and solvents.

2.6.10 Thermoplastic Elastomers (TPE)

Several types of thermoplastic elastomers have been developed, including the following:

S.No.	Elastomer	Market Share (%)
1.	Polyolefin thermoplastic elastomers	18
2.	Polyamide TPEs	small
3.	Polyester TPEs	6
4.	Polyurethane TPEs	19
5.	Styrenic TPEs	52

Brief description of these elastomer follows.

2.6.10.1 *Polyolefin Thermoplastic Elastomers*

Polyolefin TPEs are generally blends of poly-propylene with up to 65% ethylene-propylene rubber and it is supposed that short propylene blocks in the latter co-crystallize with segments of the polypropylene chains to give micro-crystalline regions which act as cross-links.

These elastomers are characterized by low cost, good mechanical properties and excellent resistance to oxygen, ozone and polar solvents. They are, however, attacked by non-polar and chlorinated solvents, especially at elevated temperatures. The main application of these elastomers is in the automotive field for such parts as radiator grills, headlight surroundings and rub strips.

* **Mastics:** Waterproof filler and sealants used in buildings.

2.6.10.2 *Polyamide Thermoplastic Elastomers*

Polyamide TPEs are block polymers made up of alternating ternating hard and soft segments. The hard segments are polyamide blocks and the soft segments are either aliphatic polyether or polyester blocks. The properties of these elastomers depend on the type and length of the two constituent blocks. By appropriate selection of the segments, mechanical, thermal, and chemical properties can be regulated within a broad range. Generally polyamide TPEs are high cost materials characterized by high tensile strength, good recovery and flex fatigue properties, good abrasion resistance and ability to withstand service temperature up to 150°C (302°F).

Applications of polyamide TPEs include hoses, gaskets, and seals and high performance athletic goods such as ski-boots and footballs.

2.6.10.3 *Polyester Thermoplastic Elastomers*

Polyester TPEs are block copolymers made up of alternating hard and soft segments. The hard segments are alkylene terephthalate blocks and the soft segments are poly (alkylene ether) terephthalate blocks.

Polyester TPEs are tough materials which exhibit useful mechanical properties over the range of – 40 - 130°C. As polar polymers, they have good oil and fuel resistance. However, they are attacked by chlorinated solvents and by concentrated acids and bases. Despite their rather high cost relative to most other rubbers, polyester TPEs have found various uses, including automotive parts, gear wheels and footwear soles.

2.6.10.4 *Polyurethane Thermoplastic Elastomers*

Polyurethane TPE is a block polymer that is made up of alternating polyester or polyether blocks (soft segments) and polyurethane blocks (hard segments).

The properties of polyurethane TPEs are quite similar to those of cast elastomers but the absence of primary cross-links results in higher compression set and more pronounced loss of strength with increasing temperature. Also, although the thermoplastic elastomers are insoluble in most common solvents, they are soluble in highly polar solvents such as dimethyl acetamide. A major use of polyurethane TPEs is in the manufacture of skiboots. Polyurethane TPEs may also be produced in fibrous form to give elastic fibres (spandex fibres). Such fibres are made either by solution spinning or by reaction spinning.

Spandex fibres have replaced natural rubber elastomeric fibres in various applications such as foundation garments on account of their greater strength, improved ageing characteristics and easier dyeability. They are used in stretch-fabrics.

2.6.10.5 *Styrenic Thermoplastic Elastomers*

Styrenic TPEs are styrene-butadiene block copolymers. The characteristic properties of these elastomers stem from the inherent incompatibility of the polystyrene and polybutadiene blocks. In the bulk material the two different blocks aggregate into separate domains leading to a dispersion of polystyrene domains (as the lesser component) in a countinuous matrix of polybutadiene. At ordinary temperatures the polystyrene (PS) domains are rigid and immobilize the ends of the polybutadiene (PB) segments, in effect serving both as cross-links and filler particles. Thus, the system has the strength of a filler-reinforced vulcanized amorphous elastomer. At temperatures above the glass transition temperature of polystyrene (PS) 100°C (212°F), the PS domains are readily disrupted and the material may be melt processed. The maximum service temperature of components fabricated from the copolymers is about 60°C (140°F).

The chemical properties of the block copolymers are similar to those of vulcanized PS–PB rubber. Thus the block copolymers are soluble in a range of solvents including chlorinated and aromatic hydrocarbons, esters and ketones.

Styrene-butadiene TPEs find use in such applications as footwear soles and heels, elastic bands and adhesives.

2.7 Special Polymeric Products

There are various types of specialty products produced from polymeric materials. Industrially more important amongst them are:

1. Polymeric fibres
2. Polymeric films
3. Cellular Polymers, i.e., Foams and sponges.

Brief description of these polymeric products follows.

2.7.1 Polymeric Fibres

There are many naturally occurring polymeric fibres, which can be subdivided into fibres of animal origin (e.g., wool, fur, silk, etc.) and fibres of vegetable origin (e.g., cotton, linen, sisal, etc.). These fibres are either protein or cellulose-based.

Polymeric fibres which are produced by chemical reactions are referred to as *man-made, synthetic*, or *artificial fibres*. They are not just polymeric materials, but are to a large extent, products in finished or semi-finished form.

Polymeric fibres are used in a variety of applications, including filters, cords, cables, and also as fabrics.

Fabrics frequently require that the synthetic fibres undergo a process which gives them a texture similar to that of most natural fibres. Fibre-forming polymers are normally of the crystallizing type. Besides crystallization, strong orientation of molecules and crystalline units is an essential feature.

There are six major types of conventional fibres, viz., rayon, acetates, olefinics, vinylics, acrylics, nylons and polyesters.

Rayon or *Viscose Rayon* fibres are marketed as regular or cross-linked types.

Acetate or *Triacetate* fibres are based on cellulose acetate.

Olefinic fibres include polyethylene (PE) and poly-propylene (PP).

Vinylics are based on polyvinyl chloride (PVC), but often involve copolymerization with vinyl acetate (VA) or Vinyl-idene chloride (VDC), for example.

Acrylics or *Modacrylics* are based on polyacrylonitrile (PAN), but generally involve copolymerization with polyvinyl chloride (PVC), for example.

Polyamides or *Nylons* involve aliphatic polyamides.

Polyesters involve polyethylene terephthalate (PET).

In addition a number of special-purpose or high performance fibres have been commercialized, including polyurethanes, aramids, extended chain polyethylenes, PIB (polyisobutylene) and PEEK (poly ether ether ketone).

2.7.2 Polymeric Films

Polymeric materials are widely used in the form of film and sheet. The term film is generally associated with relatively soft wide product of uniform thickness between 40 μm and 400 μm. Thicker products are usually referred to as sheet particularly for relatively stiff materials. Dry-cleaning garment covers which are amongst the thinnest soft material films, consist of low-density polyethylene (LDPE), about 13 μm thick. The thickness of ultra-thin high density polyethylene (HDPE) film can be as low as about 7 μm, and the term *tissue* is sometimes used in this case.

Polymer films can be manufactured by several processes, including flat extrusion on chill rolls which is basically a *Calendering process,* and *tubular or bubble extrusion blowing*. *Mechanical skiving* from polymer blocks is occasionally used for materials which are difficult to process. Uniaxial or biaxial molecular orientation, which can be promoted by a flat stretching device or through the bubble process, tends to improve certain important properties, as in the case of oriented polypropylene (OPP).

Many polymers are used in film form. The commercially important among them include *regenerated cellulose* or *cellophane* (CEL) that has long been used, often coated with a thermoplastic polymer for heat sealing. Cellulose nitrate (CN) and cellulose acetate (CA) were also among the early films. Low density and high density polyethylenes (LDPE, HDPE) are by far the most common film materials, Polypropylene (PP) is generally used as oriented polypropylene (OPP).

Ionomers (IO) are olefin-related film materials, polyvinyl chloride (PVC) is commonly used in plasticized form. Polyvinylidene chloride (PVDC) often known by its trade name *Saran*, is generally used in copolymer form, for example with 10-15% of acrylonitrile (AN). Ethylene-Vinyl acetate (EVA) and ethylene-vinyl alcohol (EVOH) copolymers are also used. Polyethylene terephthalate (PET), a thermoplastic polyester, is widely used in film form. Polystyrene (PS) is sometimes used in biaxially oriented form. Polycarbonate (PC), polysulfones (PSU), polyethersulfones (PESU), polyimides (PI), polyether imide (PEI) and several fluoro-polymers such as PTFE (polytetra fluoro ethylene), and FEP (fluorinated ethylene-propylene), are also used for engineering applications.

Composite polymeric films are also produced by a combination of parallel layers of different materials designed to offer a set of properties not possessed by anyone of the individual materials. The coating or lamination or materials, such as paper or aluminium foil with polymers has long been used. Multi-layer all-polymer films are now generally made by coextrusion. Typically for a food packaging application, different layers will have to serve the major roles of mechanical strength (PET, etc.), sealing (PE, etc.), and barrier (PVDC, EVOH, etc.). Since different polymers generally do not form a good interlayer bond, additional tie-layers of adhesives must be included. Scrap of composite films is sometimes reground and incorporated as an additional *fill layer*. Complex extrusion dies, capable of handling over ten layers, are used for either flat or tubular extrusion.

Applications

Polymer films are extensively used for food packaging wherein it may be *barrier packaging* or *mechanical packaging*. In barrier packaging the film must provide a suitable control of the in and out movement of gases and vapours, while in mechanical packaging it is mainly the mechanical strength of the film that matters.

Barrier packaging may involve cases where the entire package is flexible (bags, pouches) while in other cases it may be relatively rigid (cups for yogurt, cans for soups, tubs for margarine, etc.) with a mechanical closure (cover, lid), or a sealed film closure.

Barrier packaging is a vast field involving packaging of liquid food stuffs (fruit juices, wines, hard liquors, soft drinks and beer). Semi-fluid stuff may include tomato ketchup, sauces, jelly, Jam, marmlade, etc.

Animal products include fresh (wet) fish, fresh red meat, fresh or frozen poultry. PVC film is found most suitable for such purposes.

Dairy products include milk which is now sold in rigid plastic containers (PC, PET) as well as flexible film pouches (LDPE).

Barrier packaging also involves packaging of bakery products, dry foods (cereals, biscuits, snack), confectionery, as also tobacco products (OPP).

Mechanical film packaging application include thick-gauge industrial sacks for powdered or granular products (LDPE), garbage or trash bags (LDPE); general merchandise, paper-like thin gauge bags (HDPE); supermarket grocery bags (LDPE); very thin gauge dry-cleaning garment covers (LDPE); gift or florist wrap (HDPE).

Heat-shrinkable polymer films are also used for fitting closely to the contours of an object like packaging of sweet-meat boxes. Shrinking of the films (PVC, LDPE, PP, etc.) is achieved by hot air convective or radiation heating.

Polymeric films are also used as construction coverings for weather protection of food grain bags, wet concrete; as well as industrial liners and roof liners.

In agriculture and horticulture, polymer films are used as crop silo covers, water reservoir liners, seed-bed covers, green-house covers (LDPE).

In electrical and electronic industry, polymer films are used as capacitor dielectrics (PS) in membrane switches, as wire insulation tape (PVC, PI), and as audio/video tape (PET).

Polymeric films are also used as selective diffusion membranes for chemical separation, water desalination, blood dialysis, etc.

Miscellaneous uses of polymer film include photographic and cinematographic substrates (CN, CA, PET), moisture-proof envelopes (HDPE), high altitude balloons or sail material.

2.7.3 Cellular Polymers (Foams, Sponges, etc.)

The polymer material in which gas is incorporated to form voids and cells, the resulting lighter material is referred to as *cellular, blown,* or *expanded polymer, sponge*, or more popularly as *foam*.

Depending upon the elastic properties of the polymeric material, referred to as the matrix, the foams can be classified into three types, viz., elastomeric foams, flexible foams, and rigid foams.

Elastomeric Foams are those in which the matrix is an elastomer or rubber capable of large elastic deformations.

Flexible Foams involve a soft plastic matrix, such as plasticized polyvinylchloride (PPVC), low density polyethylene (LDPE), and certain polyurethane (PU) plastics.

Rigid Foams involve rigid matrices, such as polystyrene (PS), unsaturated polyesters, phenolics (PF), and certain polyurethane (PU) plastics.

The manufacture of foamed parts has been restricted mainly to the cheaper polymers although the range is increasing. Foams are often classified by density; *low density foams* have a density lower than 0.1 Mg/m^3 and are used in packaging, seating, and heat insulation. *Medium density foams*, with density ranging between 0.1 Mg/m^3 and 0.6 Mg/m^3, are used for footwear, furniture, and impact resistance pads. *High density foams*, sometimes referred to as *structural foams*, have densities between 0.6 and 1 Mg/m^3 and are used in many *rigid applications*, for example washing machine tubs, television panels, and business machine covers.

Structural foams consist of a low density cellular core sandwiched between solid integral skins. Foamed parts are manufactured mainly using either injection moulding or an extrusion process with a polymer melt in which an inert gas is present. This inert gas can be generated either by chemical blowing agents which decompose on the application of heat, or by the addition of a compressed gas, usually nitrogen or fluorocarbon, into the melt before injection. This solution is injected into a mould cavity under pressure and is then allowed to expand in the cavity. A solid skin is formed by the solidification of the melt before expansion can take place. This skin forms an insulating layer for the core to expand by foaming.

Structural foams of thermosetting polyurethane can also be produced by reaction-injection moulding (RIM) process under low pressures (about 3.5 atm. or 50 psi) with good surfaces; it is therfore particularly suitable for large parts.

Experience with foams has shown that the stiffness and strength vary with the square of the foam density, for densities in the range of 0.4 to 1 Mg/m^3.

In general, the advantages of a structural foamed part over an unfoamed plastics component are:

1. increased rigidity for same polymer weight, or
2. reduced weight,
3. possibility of moulding thick and thin section, and
4. reduced cost.

Applications: Cellular polymers have a wide field of applications as their mechanical properties, on weight basis, can be excellent. Rigid foams are widely used in structural applications, often as core material in conjunction with rigid shells, and in sandwich construction.

Because of their outstanding thermal insulation properties, closed-cell foams are widely used as insulation board and for the packaging of frozen or perishable foods, such as ice cream, fish, etc.

Closed-cell foams with their very low density are ideal buoyancy material, the upward thrust per unit mass is exceptionally high. They are also much safer than empty ballasts. Cellular polymers are particularly useful for the protection of sensitive objects against shocks and vibrations and for cushioning purposes. They are widely used in rigid, semi-rigid, flexible, elastomeric form for numerous applications in packaging, automotive occupants protection, cushioning, bedding, and fruit and egg packaging.

Suitable cellular polymers can be very efficient for acoustic insulation or as dampening material.

Open cellular polymeric structures are also used in filtering and humidifying applications.

2.8 Reinforced Plastics and Composites

Reinforced plastics are polymers to which some other material, called filler or reinforcement, is added to achieve desired mechanical, chemical, or electrical properties not possessed by the parent plastic in its individual capacity. Reinforced plastic mouldings include such products as storage bins, loud speaker horns, machinery housings, truck cabs, and aircraft panels.

When this addition of reinforcing material is done on a more systematic or planned way, the product obtained is called a *Composite*.

Reinforced plastics are usually classified into three types:

1. Reinforced Thermoplastics
2. Reinforced Thermosetting Plastics
3. Laminated Plastics

Brief description of these three types of reinforced plastics follows.

2.8.1 Reinforced Thermoplastics

Reinforced thermoplastics (RTP) or fibre-reinforced plastics are polymer compositions in which fibrous reinforcements are embedded in the thermoplastic resin. Almost all thermoplastic polymers that are generally reinforced are nylon, polystyrene, SAN (Styrene copolymerized with acrylonitrile), ABS (Acrylonitrile-Butadiene Styrene), polycarbonate, polypropylene, acetal, polyethylene, polyurethane, and polysulfone. The most common reinforcing fibre is glass, although many other specialty fibres such as boron fibres, carbon fibres, and ceramic fibres are also used. Fibre glass is still the industry's workhorse reinforcement, particularly in large volume end products.

The addition of fibres to any polymeric material serves the same purpose as steel rods added to concrete. The fibres or rods increase and improve the tensile and flexural strength. *Fillers* on the other hand, do not increase these physical properties, but act merely as *extenders**. The resin actually bonds the reinforcements together, transmitting loads to the fibres so that optimum fibre stresses can be developed before failure. A reinforcing fibre must have a strong adhesive bond to the plastic resin that is used. Coupling agents are generally added to the reinforcing fibre to increase this adhesive bond. Two basic finishes have provided optimum bonds for fibre glass, that is, chrome finishes and *silane*** finishes.

The tensile strength of a thermoplastic polymer can be at least doubled by the addition of glass reinforcement, and deformation under load is greatly decreased. Unlike thermosetting reinforced plastics, fibre glass reinforced thermoplastic compounds can be pelleted and used in conventional moulding equipment. Reinforced thermoplastics produce increases in strength and rigidity and marked decrease in the coefficient of thermal expansion. The most significant effect in thermoplastics is the retention of impact strength at very low temperatures. Deflection temperature is improved most markedly in nylon. Other beneficial effects may include increase in hardness and abrasion resistance and decrease in mould shrinkage, creep, and dimensional changes with humidity.

Some of the undesirable effects of fibreglass as reinforcement in thermoplastics include opacity, occasional fibreglass appearance on the surface, reduced gloss, some difficulty in electroplating, and loss of mechanical flexibility. Fibreglass reinforced thermoplastic material is generally not extruded because that results in non-random orientation of the glass fibres.

2.8.1.1 *Processing of Fibreglass Reinforced Thermoplastics*

There are four basic methods of processing thermoplastic polymers with fibre glass reinforcement. These are:

1. Reinforced pellets method
2. Super reinforced pellet concentrates method
3. Fibreglass rovings method, and
4. Prechopped fibreglass method.

Brief description of these processing methods follows.

1. Reinforced Pellets Method: In making fibreglass reinforced thermoplastic pellet materials, continuous glass roving is first sized to increase interfacial adhesion between the glass and resin. After the sizing operation, the fibreglass is impregnated and coated with thermoplastic resin and finally chopped into pellets. These pellets contain core bundles of parallel glass fibres impregnated and surrounded by a sheath of thermoplastic resin. The pellets contain from 20 to 30% fibreglass by weight. The fibreglass in pellets can be obtained in 3 mm lengths or in 10 to 13 mm lengths. The longer glass fibres give higher impact strength. In this type of reinforced pellets, there is virtually no limit to the types of plastics available in reinforced grades. Fig 2.20 shows the essential elements of a set up for production of components from reinforced pellets by injection moulding process.

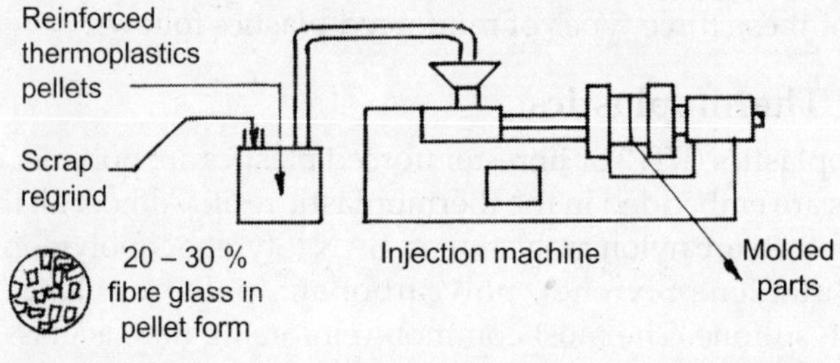

Fig 2.20: Set up for Injection Moulding from reinforced pellets.

* *Extenders* are low cost plasticizing additives added mainly to reduce the cost of compounds, when the level of stiffness is not important, e.g., wood flour.

** *Silanes* are compounds with the general formulae $Si_n H_{2n+2}$ and are given the generic name silanes, for example SiH_4 is silane.

2. Super-Reinforced Pellet Concentrate Method: The super-concentrated pellets have 80% fibreglass by weight and are blended with virgin resin powder or pellets to achieve a 20% glass content. This type of concentrate can be processed very easily with a screw injection moulding machine; a set up for which is shown in Fig. 2.21.

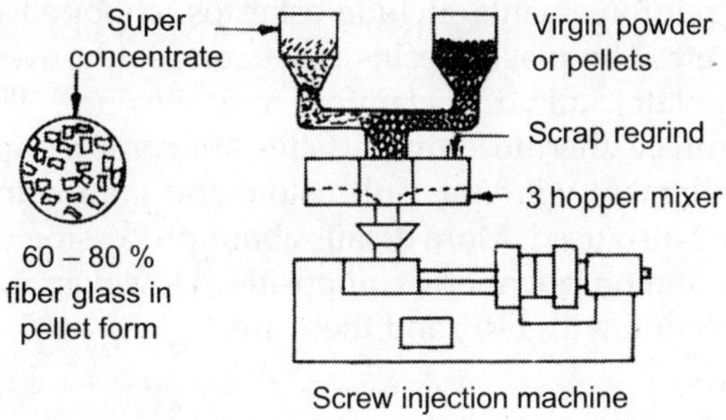

Fig 2.21: Set-up for screw Injection Moulding from super reinforced pallet concentrate.

3. Fibreglass Rovings Method: The third method is to use fibre glass rovings and virgin polymer powder. The fibreglass rovings are chopped and blended with powdered resin in the screw moulding press as shown in Fig. 2.22. It may be noted that in this process a screw-type injection moulding machine is necessary in order to mix the virgin resin powder thoroughly with the fibreglass.

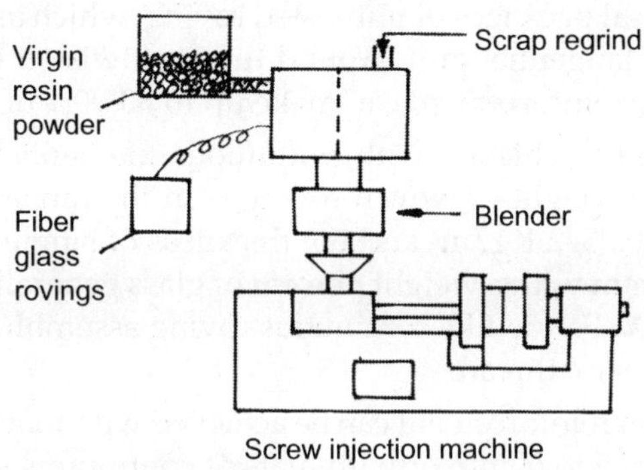

Fig 2.22: Set-up for Screw Injection Moulding using fibreglass rovings.

4. Pre-Chopped Fibreglass Method: In this method, the setup for which is shown in Fig. 2.23, the pre-chopped fibreglass is mechanically metered into the resin flow. The method is better adaptable to bigger moulding machines that have large diameter screw. The polystyrene beads or powder is the best type of material to use in this process. It is not economically feasible to compound reinforced thermoplastic polymers at the press, unless the operation can be set up to run the year around.

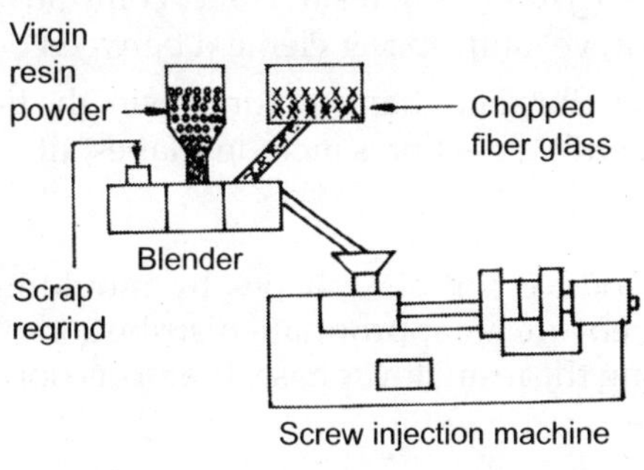

Fig 2.23: Setup for Screw Injection Moulding for pre-chopped fibreglass reinforced polymers.

2.8.2 Reinforced Thermosetting Plastics

Reinforced thermosetting plastics, also known as Fibre Reinforced Plastics (FRP), are plastic compositions in which fibrous reinforcements are embedded in a thermosetting resin. The reinforcements are usually fibreglass in the form of rovings, mats, fabrics, chopped fibres, and short milled fibres. Other reinforcements include asbestos, chopped paper, macerated* fabrics, jute, sisal,** cotton, nylon, etc. The plastic resins most commonly used are polyester, phenolics, epoxies, DAP (diallyl phthalate), alkyd, melamine, and silicones. The fabricating methods of forming or moulding reinforced thermosetting articles are compression moulding, hand lay-up moulding, bag moulding, filament winding, pultrusion, and laminating. Transfer moulding and screw injection moulding are also used. More details about processing of reinforced thermosetting plastics are given in the section on polymeric composites. However, forms of fibre reinforcement need consideration in connection with FRF; and these are:

1. Fibre reinforcements
2. Fibre preforms
3. Fibre prepregs.

Brief description of these forms follows.

2.8.2.1 *Fibre Reinforcements*

Fibre Reinforced thermosetting plastics having continuous filament reinforcement are manufactured in the form of wound cylindrical packages of untwisted roving, which are held stationary for centre pull, or mounted on reels for tangential pull. Wound tubes of twisted yarn are normally used for fabric weaving. Uniaxial glass reinforcement can make up to 80-90% of the FRP weight.

Biaxial reinforcement can be achieved with continuous filaments by using woven roving or yarn-based woven fabric. The weight of woven roving is in the range of 0.3–0.75 Kg/m^2, that of woven fabric in the range of 0.05–2 Kg/m^2 and the thickness of single ply of woven fabric-based FRP is in the range of 0.1–2.5 mm with a weight percent of glass generally between 45 and 55% that may occasionally go up to 75%. Fabric-like continuous roving assemblies are also available, which involve stitching with a polymeric thread.

Multidirectional or random reinforcement can be achieved with mats of discontinuous chopped strands ranging in length from 5 to 50mm or with mats of continuous strands also known as *swirl mats*. Such mats have weights in the range of 0.3 to 0.9 Kg/m^2. The glass content in corresponding FRP products is generally between 25 and 35% and occasionally up to 50% by weight. Overlay mats with lighter surfacing, also known as *Veils*, are used to form resin-rich surface layers, containing no more than 20% by weight of glass, which help improve appearance and chemical resistance.

A special form of FRP, known as *honeycomb structure*, is now widely used in aerospace and other advanced applications. Such structures, initially made of aluminium, are now increasingly made of reinforcing fibres embeded in a thermosetting resin. Honeycomb structures in thickness around 10 mm are used primarily as a light, yet stiff, spacer element between bonded FRP skins.

To improve the bond between fibre and thermosetting resin, the fibres are treated with coupling agents. The couple agents used with glass fibres include silanes, titanates and zirconates.

2.8.2.2 *Fibre Preforms*

Fibre preforms are made to produce complex shapes by moulding in processes such as *Resin Transfer Moulding* (RTM). They ensure an appropriate distribution of the reinforcement, and help minimise its displacement during the resin flow phase. Preforms normally consist of a mat of fibres

* *Macerated* means softened

** *Sisal* is a tree with large fleshy leaves used for making fibres, ropes, etc.

held together by a binder and having the shape of the mould cavity. There are two methods of producing fibre preforms, viz.,

1. Dry Preforming Technique
2. Wet Preforming Technique

1. *Dry Preforming* has two variants viz.,
   (i) Direct Fibre Method
   (ii) Plenum Chamber Method.

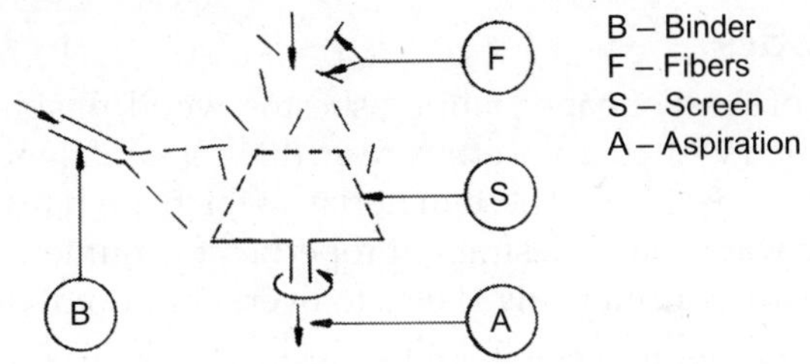

Fig 2.24: Schematic of Dry Preforming Technique.

In *Direct Fibre Method,* the roving is cut into 25–50mm lengths of chopped strands that are blown through a flexible hose onto a rotating preform screen as shown in Fig. 2.24. Suction holds the strands in place while a binder is sprayed on the preform. The wet preform is heated in an oven to cure the binder.

In the *Plenum Chamber Process* of making preforms, the fibre glass roving is fed into a cutter on top of the plenum chamber. The chopped strands are directed on a spinning fibre distributor to separate the chopped strands and distribute them uniformly in the plenum chamber. From the plenum chamber, the glass strands are sucked onto the rotating preform screen. As the strands are falling, a resinous binder is sprayed on them. The wet preform is heated in an oven to cure the binder.

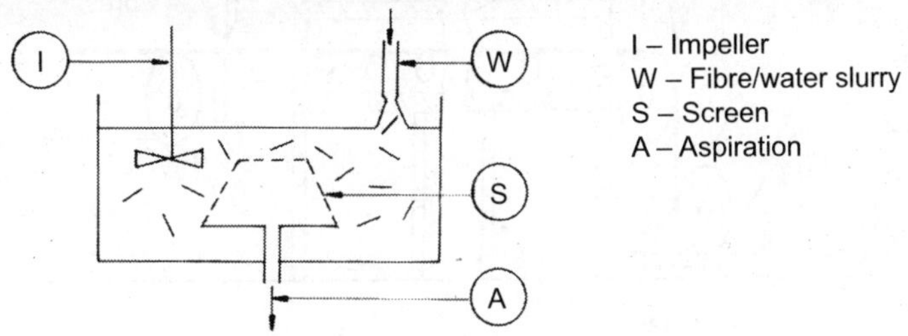

Fig 2.25: Schematic of Wet Preforming Technique.

2. *Wet Preforming Technique* which is also known as *Water Slurry Method,* shown schematically in Fig. 2.25, involves the suction of stirred suspension of chopped strands in water, onto the perforated preform screen. A small amount of cellulosic fibres, flexible in wet state but quite rigid in the subsequent dry state, is used as the binder. The use of preforms allows glass loadings of about 35% by weight.

The completed preform, which looks like a thick white blanket, is placed in the cavity or over the male portion of the metal die. A resin mix comprising resin, catalyst, fillers, and pigments is measured out and poured over the chopped strand fibreglass preform. The press is then closed, and under these conditions, the liquid resin is forced through the preform, thus forcing out the air before it. The curing of the liquid plastic resin takes place immediately. This process is used for items like fibreglass bucket-type chair seat wherein the fibreglass preform is placed in the mould cavity, and liquid polyester resin is poured over it. The press is closed and the chair seat is moulded into shape.

2.8.2.3 *Fibre Prepregs*

Prepregs are any form of reinforcement that is obtained in the form cf pre-impregnated ribbons, tapes, or sheets, in which a thermosetting resin is pre-cured to a semi-solid state, which facilitates the fabrication of parts. That has to be followed by a time-consuming operation of curing at an elevated temperature and under an appropriate pressure.

Thermoplastic-based pre-pregs have also been introduced that involve shorter heating, fabrication and cooling stages.

2.8.3 Laminated plastics

Laminated plastics consist of sheet of paper, fabric, asbestos, wood, or similar other materials which are first impregnated or coated with thermosetting resin (binder) and then combined under heat and pressure to form commercial materials. These materials, which are hard, strong, impact-resisting, and unaffected by heat or water have desirable properties for numerous electrical applications. The final product may consist of from a few sheets to over a hundred, depending on the required thickness and properties. Although most laminated stock is made in sheet form, rods and tubes as well as special shapes, are also available. The material has good machining characteristics which permit its fabrication into gears, handless, bushings, furniture, and many other articles.

The principal resins used in laminated products include phenolics, melamine, polyesters, epoxy and silicones. *Phenolics* are low cost and have good electrical properties. *Melamines* are more costly but offer flame resistance and have excellent electrical properties. The *polyesters* are low in cost with average properties. The *epoxies* are high in chemical resistance and are extremely moisture resistant. *Silicones* are used primarily for their retention of mechanical and electrical properties even at very high temperatures, when their high costs can be justified.

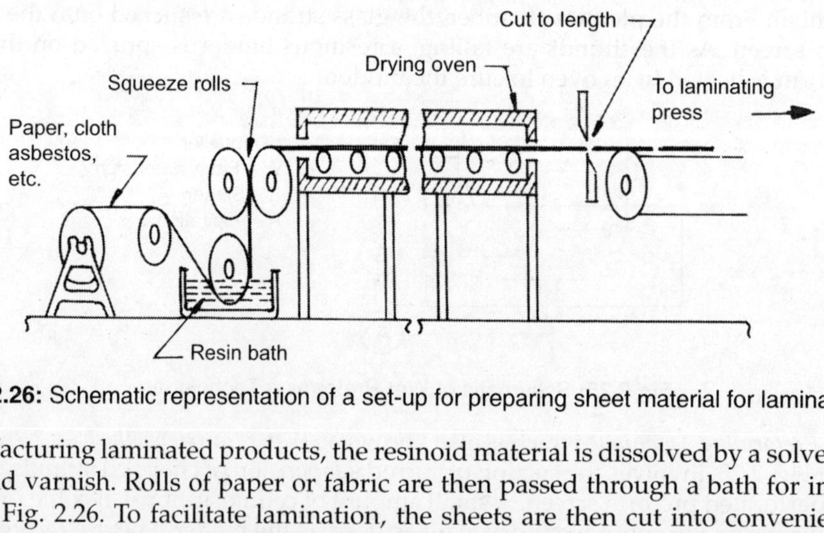

Fig 2.26: Schematic representation of a set-up for preparing sheet material for lamination.

In manufacturing laminated products, the resinoid material is dissolved by a solvent to convert it into a liquid varnish. Rolls of paper or fabric are then passed through a bath for impregnation, as shown in Fig. 2.26. To facilitate lamination, the sheets are then cut into convenient sizes and stacked together in numbers sufficient to make up the thickness of the final sheet. The stock is then compressed in a laminating press, wherein heat and pressure are applied to the layers causing the resin to flow and harden the laminate into one solid mass.

The laminates are of two types, viz., *high pressure* and *low pressure laminates.* The high pressure laminates are cured at or above 1000 psi (7MPa) and more commonly in the range of 1200-2000 psi (8.4-14 MPa) and heated to around 300°F (150°C). If the pressure is under 1000 psi (7 MPa), the product is called low pressure laminate. The laminates made with little or no pressure, such as in hand lay-ups, are sometimes called *contact pressure laminates.*

Solid laminated rods are made by winding the impregnated filler web on a very thin mandrel, which is withdrawn before moulding. The central hole is filled up when pressure is applied in the metal mould; but flash develops along the mating lines of the two halves of the mould. That is subsequently ground off by centreless grinding.

In making a tube, the mandrel is left in the tube, and the tube is moulded in the same manner as the rod.

Parts with irregular shape, such as gears and bearing are often formed by cutting an uncured impregnated sheet to a pattern and then stacking and moulding. Where small parts are required, as for spacers, cams, contact arms and levers, it is economical to saw them or punch them from stock sheet, rods or tubes.

Uses: Paper-base laminates are used in electrical products. Fabric-base materials, which are stronger and tougher, are well suited for stressed parts. Gears made of a canvas base are quiet in operation. Asbestos and fibreglass cloth are recommended for heat resisting and low water-absorption uses. Thin sheets of wood are laminated to produce a light material equal in strength to some metals and resistant to moisture. Safety glass is, in effect, a laminated plastic product, because it uses thermoplastic layers between the glass sheets to make it non-shattering.

2.8.4 Polymeric Composites

Fibre-reinforced plastics, referred to FRP, *structural composites*, or simply *composites* are associated with products in which a polymeric matrix is combined with reinforcing fibres. The polymeric matrices used in FRP are primarily of the thermosetting type and include unsaturated polyesters, vinylesters, and epoxies. The reinforcing fibres include glass, carbon/graphite and aramid fibres; however glass is, by far, the most common type.

There are five general classes of polymer composite materials, viz.,

1. Laminar Composites
2. Fibre Composites
3. Flake Composites
4. Separate Particles Composites, and
5. Filled Composites.

Basic physical nature of all these types of composites is illustrated in Fig. 2.27.

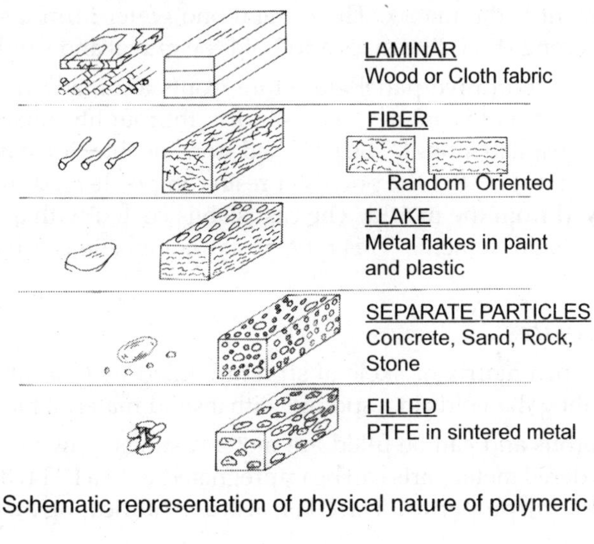

Fig 2.27: Schematic representation of physical nature of polymeric composites.

Brief descriptions of all these composites follows.

2.8.4.1 *Laminar Composites*

This was the first man-made polymeric composite. It is made up of layers of different materials like in plywood, bonded or fastened together with different plastic resin. Plywood is made with the grains of alternating plies at right angles that are bonded together with an adhesive to make a solid, rigid article. However, polymeric laminar composites consist of layers of resin-impregnated fabrics, paper, cloth, glass, etc. which possess high strength-to-weight ratios. If highly directional materials like wood, woven fabrics, or bundles/layers of continuous filaments are employed, the resulting laminate is highly directional in its mechanical properties.

2.8.4.2 *Fibre Composites*

Practically every type of plastic, rubber, elastomer, and ceramic has been reinforced with fibres. Fibres are arranged in a random, uni-directional, and bi-directional patterns. In selecting a reinforcing fibre for a composite, the different aspects to be considered include,

(a) fibre orientation,

(b) length, shape, and composition of the fibres,

(c) mechanical properties of the matrix, and

(d) integrity of the bond between fibres and matrix.

2.8.4.3 *Flake Composites*

A very pleasing decorative effect can be produced by using small metallic flakes in paints and plastics. The flakes produce a brilliant metallic highlight. The highly reactive flake pigments are particles of either aluminium or copper and copper alloys. The aluminium pigments reflect and produce brilliant blue-white highlights. The copper-based pigments produce a bronze or gold colour.

Numerous problems have been encountered in using metallic pigments in plastic materials. In injection moulding, for example, the orientation of the flakes in the plastic material is upset during the moulding process at flow and knit lines, producing an unsightly appearance. This condition can, however, be improved by using a large fan type gate in the mould.

2.8.4.4 *Separate Particle Composites*

The oldest and most widely used particle composite is cement concrete wherein gravel and sand are the particles and cement is the matrix. The cement and water form a slurry that hardens by chemical reaction into a strong stone-like mass and hold the gravel and sand firmly embeded in it.

In the plastic industry, decorative panels for store fronts are made from sand and polished pebbles with polyester resin as the matrix. The sand is thoroughly mixed with the catalyzed polyester resin and cast into a slab mould. The polished stone pebbles are partially embeded on the surface of the sand-resin mixture. After the polyester resin composite mixture has polymerized and become hard, it is removed from the mould. The cast slab is coated with a silicon resin to protect the composite from the outside elements. This type of slab composite is used only for a decorative item for building fronts.

2.8.4.5 *Filled Composites*

A filled composite is an open matrix or skeletal structure filled with another material. It may be described as similar to filling the voids in a sponge with a solid material like plaster.

Materials that are porous and can be filled are metal castings, powder metal parts, ceramics, graphite and foams. Powdered metal parts can be impregnated with a PTFE fluoroplastic to provide a good combination of bearing properties. The PTFE acts as a lubricating medium. Porous castings of aluminium can be sealed with liquid plastic resin such as polyester or epoxy. The liquid resin is forced into the porous metal by air pressure or vaccum. The liquid resin fills all of the voids and is polymerized to a 100% solid.

2.8.5 Fibre Moulding Compounds

Combination of discrete fibres and temperature activated thermosetting resins, such as unsaturated polyesters, directly suitable for compression or injection moulding in hot moulds are oftern referred to as moulding compounds. Some of the more important comercially important amongst them include the following:

1. Bulk moulding compounds (BMC)
2. Pelletized moulding compounds (PMC)
3. Sheet moulding compounds (SMC)
4. Thick moulding compounds (TMC)
5. Cross moulding compounds (XMC).

These are briefly described in this section.

2.8.5.1 Bulk Moulding Compounds (BMC)

Bulk moulding compound (BMC) which is also known as *Dough Moulding Compound* (DMC) is obtained by blending about 25% (by weight) of chopped glass fibres and a thick thermosetting resin in a blade-type mixer. The compound achieves the consistency of moulding clay or putty and is often extruded into rods or ropes for easy cutting into suitable moulding charges.

The mixing process tends to reduce the length of the chopped fibre strands. But what is lost in properties can be gained in processing as the material can flow well into fairly complex shapes like ribs and bosses, etc. BMCs are well adapted to moulding by compression moulding, transfer moulding and even injection moulding.

2.8.5.2 Pelletized Moulding Compounds (PMC)

These moulding compounds are in the form of dry free-flowing pellets, in contrast to the generally *tacky* BMC or DMC. Such compounds are intended primarily for injection moulding.

2.8.5.3 Sheet Moulding Compounds (SMC)

These moulding compounds involve the combination of about 25%, by weight, of relatively long (1-5 cm length) chopped glass fibres with a high viscosity thermosetting polyester resin matrix. The paste-like consistency of the resin is partly associated with the use of magnesium oxide (MgO) as an additive or filler. The manufacture of sheet moulding compounds is normally a continuous operation and the rolled material can be stored or shipped for subsequent moulding.

The Procuction of SMC, illustrated schematically in Fig. 2.28, generally involves a polyethylene carrier film (F) on which a layer of resin paste (P) is applied by means of a dotor blade (D). The carrier film is supported by a wire mesh belt, B. Glass fibre roving, G, is cut with a rotory chopper, C, into strands that fall in a random fashion on the paste layer.

A polythylene cover film, F, also coated with resin paste is overlaid, and the glass-resin system is compacted between the supporting belt, B, and a rolling device, B′, to ensure a good penetration of the fibre strands into the resin. The SMC sheet is then wound up into rolls that, after a resin-maturing stage, can be used in compression moulding. A typical production rate is about 3000 kg/h/m of width.

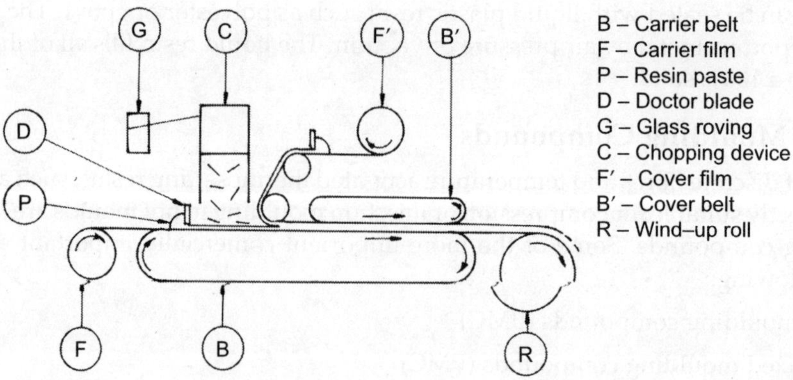

Fig 2.28: Schematic illustration of the set-up for manufacture of Sheet Moulding Compounds.

Sheet moulding compounds are particularly suited for the manufacture of large, relatively thin parts such as automotive body components. The sheet material can be cut to the desired shape and placed in the mould, after stripping off the polythelene carrier and cover films. To accomodate larger variable thicknesses, the sheet material can be layered. Sheet moulding compounds are compression moulded into heated moulds at about 150°C (300°F). The curing time ranges from about 1 to 5 minutes, depending on the part thickness.Moulding pressure is relatively high at 1.52 to 15.2 MPa (220-2200 psi) and thus sturdy steel moulds and large presses are required for large parts like automotive body components. The cost of a moulding press for SMC, however, may be only about 1/5th that of a corresponding metal stamping press.

Contraction of the resin after curing and cooling may give a part with somewhat uneven surface; this problem can be tackled by the use of the so-called low-shrink resin additives. A technique referred to as *in-mould coating* (IMC) has also been developed to inject a lacquer into the mould, at an intermediate stage, to obtain a high-class finish that may, for example, be desirable for automotive body components.

2.8.5.4 *Thick Moulding Compounds (TMC)*

Thick moulding compounds are prepared by a special technique that permits the production of SMC-like compounds up to a thickness of 5 mm. In this case, the chopped glass fibres are added to the resin at the nip of two counter rotating rolls, from which the resulting sheet-like material is stripped, and subsequently formed and compacted into the thick final compound. Thick moulding compounds can be compression moulded into thick products and they can also be used as feed material for injection moulding.

2.8.5.5 *Cross Moulding Compounds (XMC)*

Cross moulding compound is a special moulding compound that has been developed for applications such as large panels, which involve little deformation of an initially flat moulding compound during moulding, but call for high mechanical properties in the product. It features a high loading of continuous glass strands laid in a cross-pattern (hence the name XMC), supplemented by either chopped strands or continuously swirled strands to provide resistance in directions other than the major fibre directions. For placement of continuous fibre, glass strand helically wound on a large cylindrical mandrel is used.

3 TESTING OF PLASTICS

For selecting a plastic for a specific application, it is essential to know whether its properties are adequate to serve satisfactorily for the intended service. Though it may not be possible to discuss every aspect of all useful plastics but most of them are discussed where their following characteristics are known.

1. Tensile strength
2. Compression strength
3. Flexural strength
4. Shear strength
5. Impact strength/Toughness
6. Hardness
7. Heat Deflection Temperature
8. Heat Resistance
9. Softening point
10. Flow Rate (Melt Index)
11. Specific Gravity
12. Coefficient of Thermal Expansion
13. Thermal Conductivity and Specific heat
14. Optical Properties
15. Flammability.

Brief description of these properties of plastics and their relative ratings, where available, are included in this chapter.

3.1 Tensile Strength

The tensile properties determine the behaviour of a material when subjected to forces that tend to pull it apart. This gives an important measure of the strength of the material.

Tensile properties are determined on a standard shaped specimen similar to the one shown in Fig. 3.1. A standard tensile test specimen as per ASTM* D-638 has an overall length of 8½″ (216 mm) and the central section is 2¼″ (57 mm) long with constant cross-sectional dimensions of 1/2″×1/8″(12.7 mm × 3.2 mm). In the centre section a 2″ (51 mm) length is marked off and is called the *gauge length*. It is this portion of the sample (within gauge marks) that is tested.

---

* ASTM — American Society for Testing and Materials.

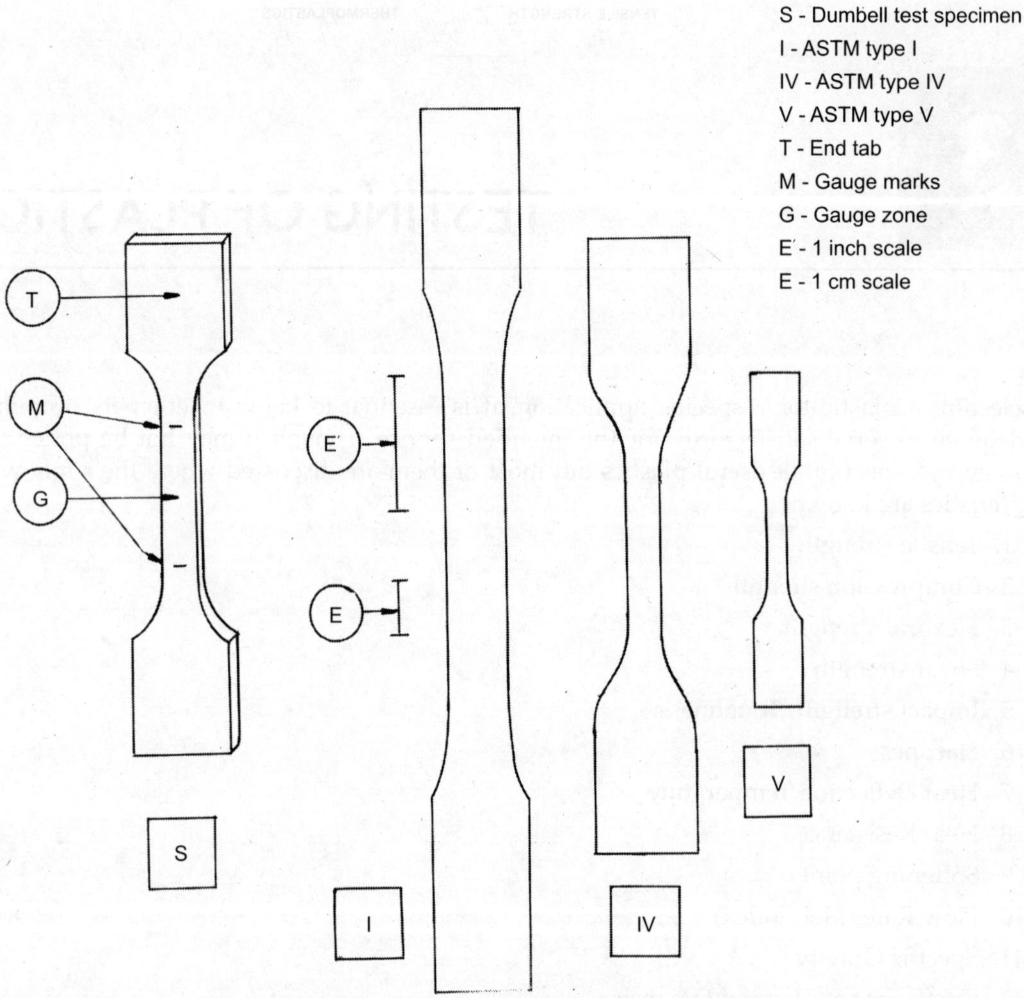

Fig 3.1: Standard tensile test specimens.

Because of the known effect of speed on the mechanical behaviour of polymeric materials, the strain rate must be specified. For the test specimen of Fig. 3.1, it corresponds to specified movable jaw speeds. Typical testing speed of 5 mm/min for a test specimen with 100 mm gauge length, is referred to as a *moderate speed* and the corresponding test as *short term test*.

The tensile behaviour of polymeric materials can vary widely as a function of polymeric system, as well as the temperature or speed of testing and other factors. A full stress-strain curve is the ideal way of reporting the tensile behaviour of a material, but specific quantities are generally reported, which include Young's (tensile) modulus E, the tensile yield stress σy, when applicable, the ultimate tensile strength σ_{μ}, and also the elongation at $(\in_y)$ and at break $(\in_u)$.

The modulus of elasticity or Young's modulus E is a measure of material stiffness and is commonly reported in mega or gigapascals (MPa or GPa). The stiffest unreinforced plastic corresponds to $E \approx 3.5$ GPa but for most engineering plastic E lies in the range of 2-3.5 GPa. Tables 3.1 and 3.2 show the relative tensile strengths of some industrially important thermoplastics and thermosetting plastics respectively.

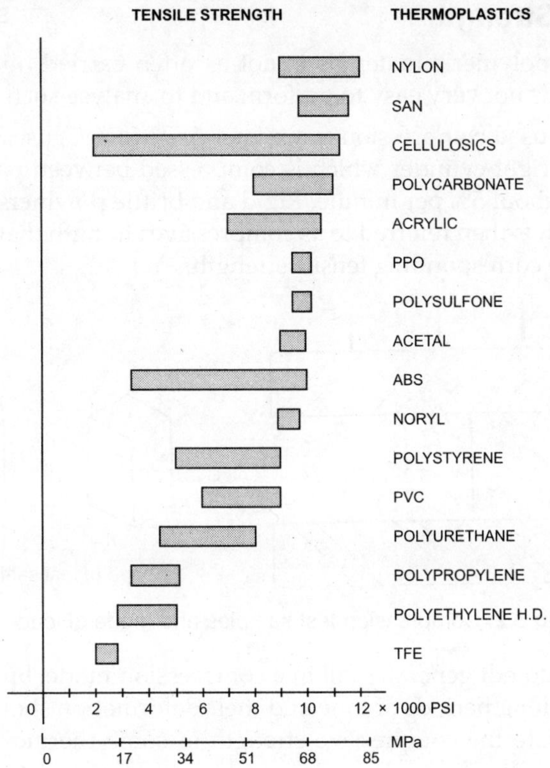

Table 3.1: Average tensile strength of some thermoplastic materials at room temperature.

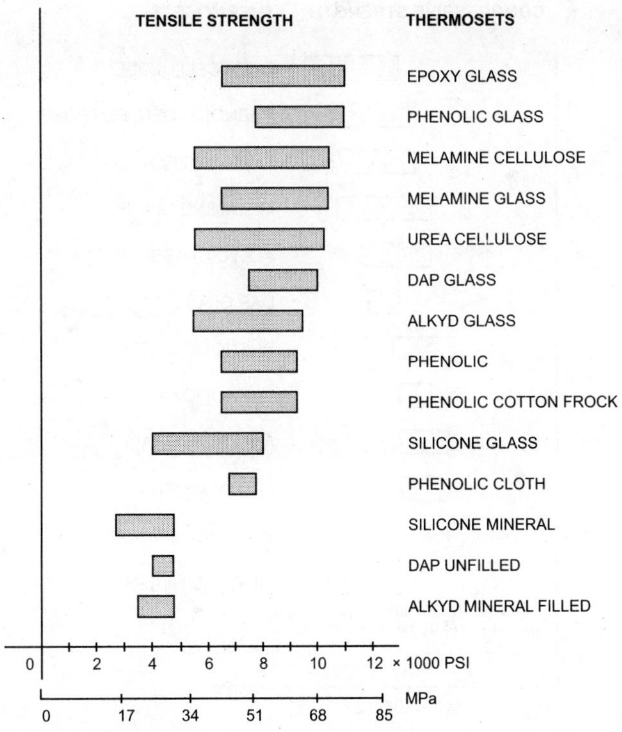

Table 3.2: Average tensile strength of some thermosetting plastic materials at room temperature.

3.2 Compressive Strength

Compression testing of polymeric materials is not as often carried out as is tensile or flexural testing, partly because it is not very easy to perform and to analyse such test data.

The most commonly used compression test, as per ASTM D695, uses 1″ (25 mm) square prism or 1″ (25 mm) diameter right cylinder which is compressed between parallel plates as shown in Fig. 3.2 at a slow rate of about 5% per minute. Rigid and brittle polymers may fracture at a certain compressive stress, which is then referred to as compressive strength that is generally significantly more (2-3 times) than the corresponding tensile strength.

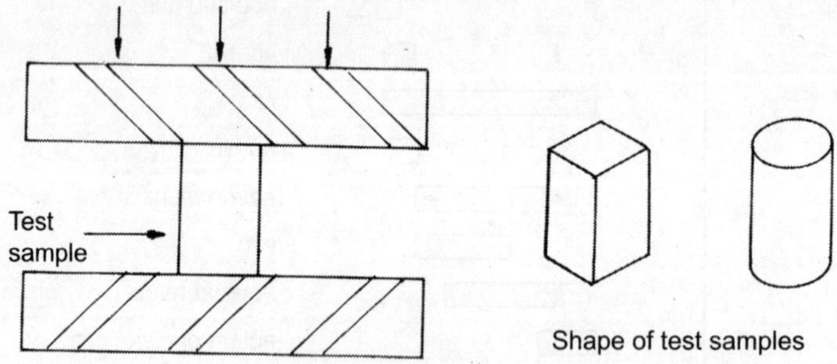

Fig 3.2: Compression test samples and mode of loading.

Polymeric materials do not generally fail in a compression mode, but they are often subjected to compressive loads for long periods of time and their deformation should not become excessive. Tables 3.3 and 3.4 illustrate the compressive strength of some thermoplastic and thermosetting plastics respectively.

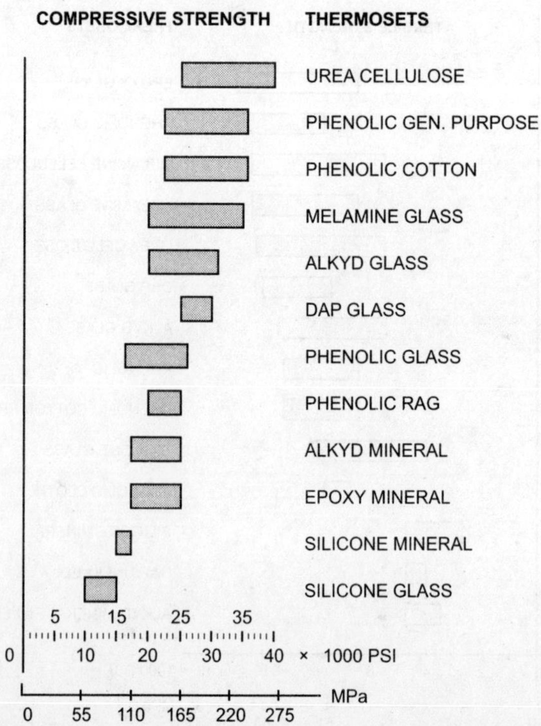

Table 3.3: Average compressive strength of some thermosetting plastic materials at room temperature.

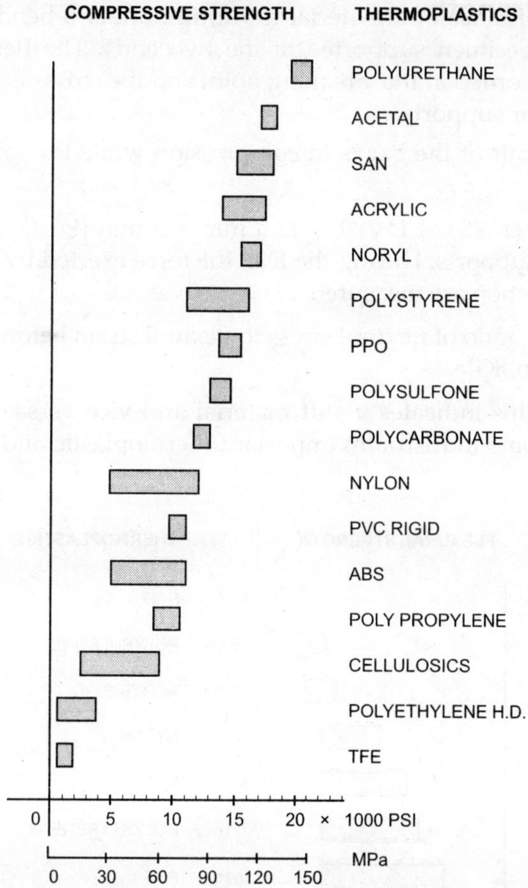

Table 3.4: Average compressive strength of some thermoplastic materials at room temperature.

3.3 Flexural Strength

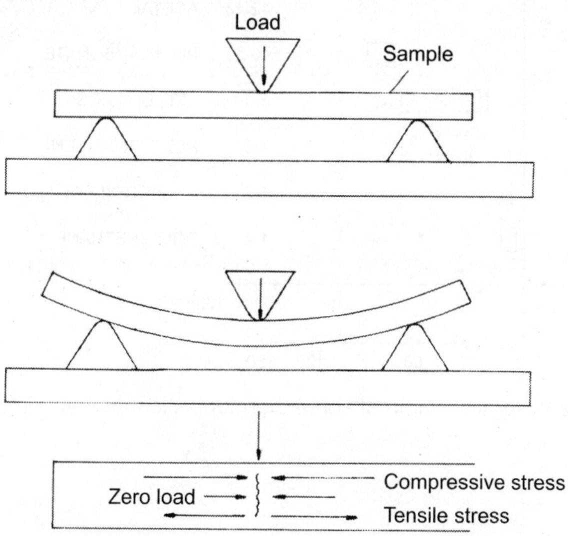

Fig 3.3: Set-up for flexural test and the nature of stress in testpiece during testing.

Flexural strength is the resistance of a material to rupture under a bending stress applied at the centre of a rectangular specimen supported at the two ends. The flexural strength in KPa is calculated from the load recorded at the breaking point and the cross-sectional area of the bar and the distance between the bar supports.

In this test the upper half of the bar is in compression while the lower half is in tension, as illustrated in Fig. 3.3

The test specimen as per ASTM D-790 is 12.5 mm × 3 mm (½"×1/8") (2") bar with 50 mm (2") distance between the supports. During the test, the force exerted by the applied load and the amount of the central deflection are measured.

Flexural modulus is the ratio of flexural stress to flexural strain before permanent deformation sets in. Modulus has units of KPa.

A high value of modulus indicates a stiff material and vice versa. Tables 3.5 and 3.6 show the flexural strengths of some industrially important thermoplastic and thermosetting polymers respectively.

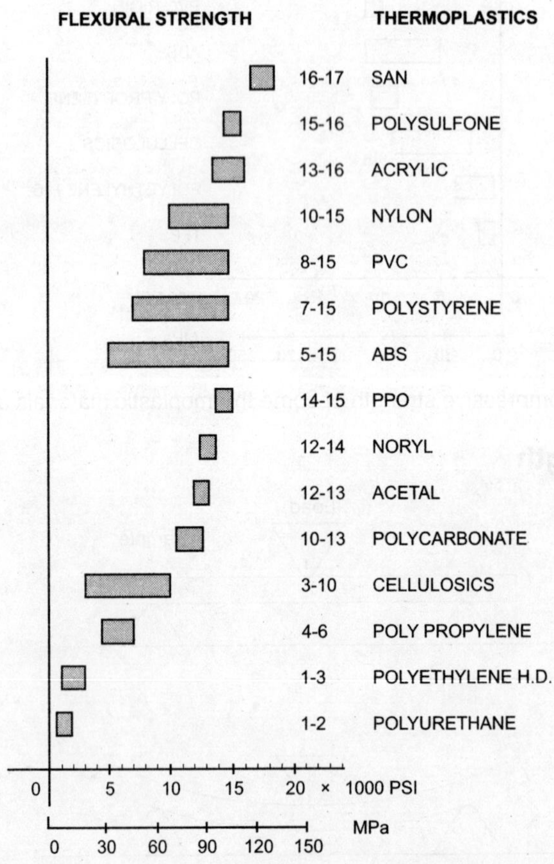

Table 3.5: Average flexural strength of some thermoplastic materials at room temperature.

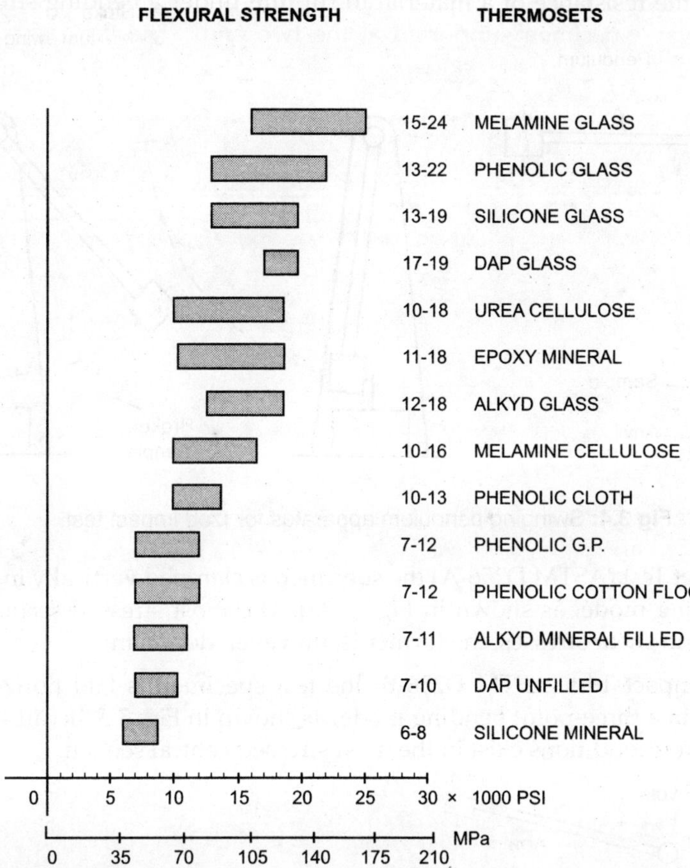

Table 3.6: Average flexural strength of some thermosetting plastic materials at room temperature.

3.4 Impact Strength

Impact strength is a measure of the energy required to fracture a material by a sharp blow or fall of a weight. There are a number of impact tests, viz.,

1. Pendulum Flexural Impact Tests
   (a) Izod Impact Test
   (b) Charpy Impact Test
2. Pendulum Tensile Impact Test
3. Falling Weight Impact Test
4. Perforation Impact Tests.

Two widely used impact tests, the Izod and Charpy tests, make use of a swing pendulum apparatus, Fig. 3.4, with test specimens loaded in bending mode. The energy consumed in fracturing the test specimen is normally the only quantity measured. It is expressed in joules (N–m) or ft. lb.f (1J = 0.737 ft-lbf). The striking speed of the pendulum is typically around 2.5 m/sec (≈8 ft/sec) and its energy can be in the range of 1.4 to 21.7J (1-16 ft.lbf).

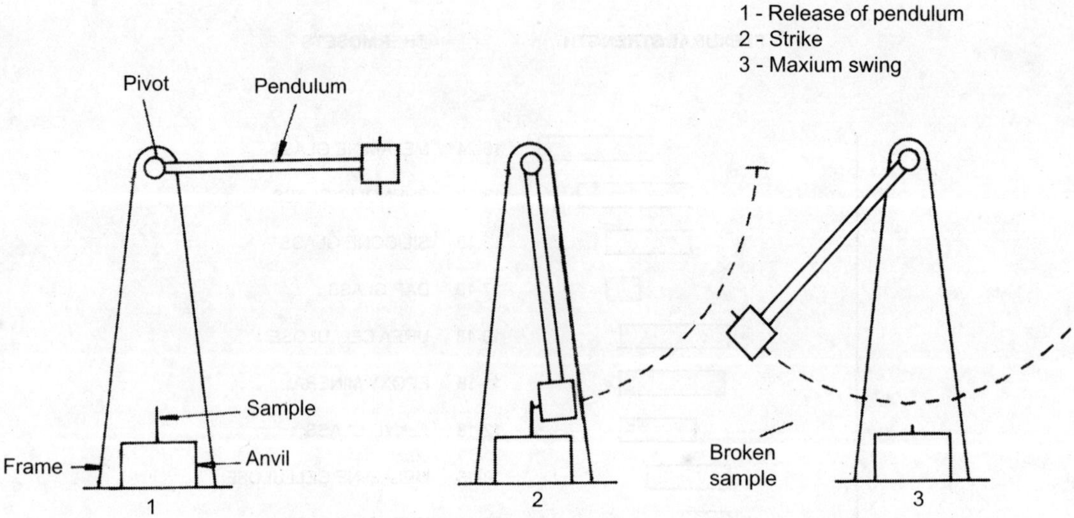

Fig 3.4: Swinging pendulum apparatus for Izod impact test.

In the Izod Impact Test (ASTM D256-A) the specimen is clamped vertically in a vise and struck in a cantilever bending mode, as shown in Fig. 3.5(a). The most stressed section is subjected to combined bending and shear stresses, the former is, however, dominant.

In the Charpy Impact Test (ASTM D256-B) the test specimen is laid horizontally over two supports and struck in a three-point bending mode, as shown in Fig. 3.5. Because of the geometry of loading true bending conditions exist in the most stressed central section.

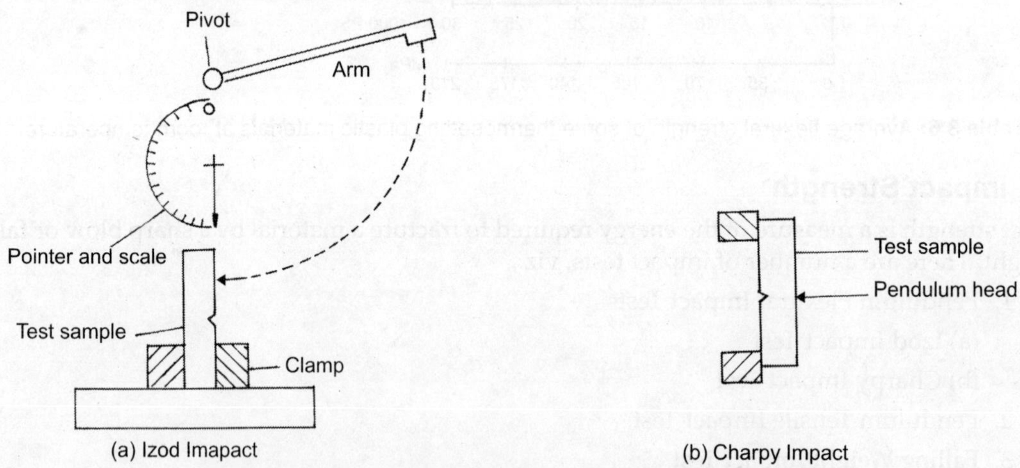

Fig 3.5: Impact testing methods; (a) Izod test; (b) Charpy test.

The test specimens used in both Izod and Charpy tests are ½″×½″×2½″ (12.7 mm×12.7 mm×63.5 mm) and is notched approximately with a notch depth of 0.1″ (2.54 mm). Tables 3.7 and 3.8 show Izod impact strength of some industrially important thermoplastic and thermosetting polymers respectively.

It is to be noted that the impact strength is usually quoted as energy per unit dimension (D′) for Izod test (J/m or ft.lbf/ in with 1KJ/m=18.7 ft.lbf) or energy per unit cross-section (D×D′) for charpy test (J/m^2, or ft.lbf/in^2). The dimension D and D′ are shown in Fig. 3.6.

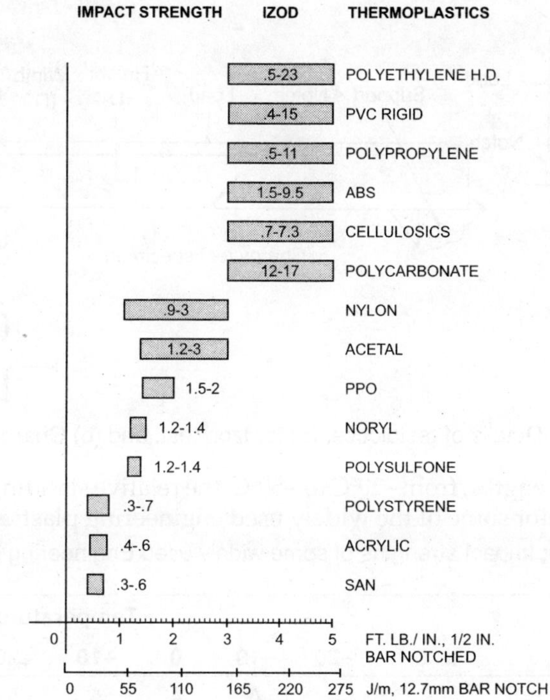

Table 3.7: Average impact strength of some thermoplastic materials at room temperature.

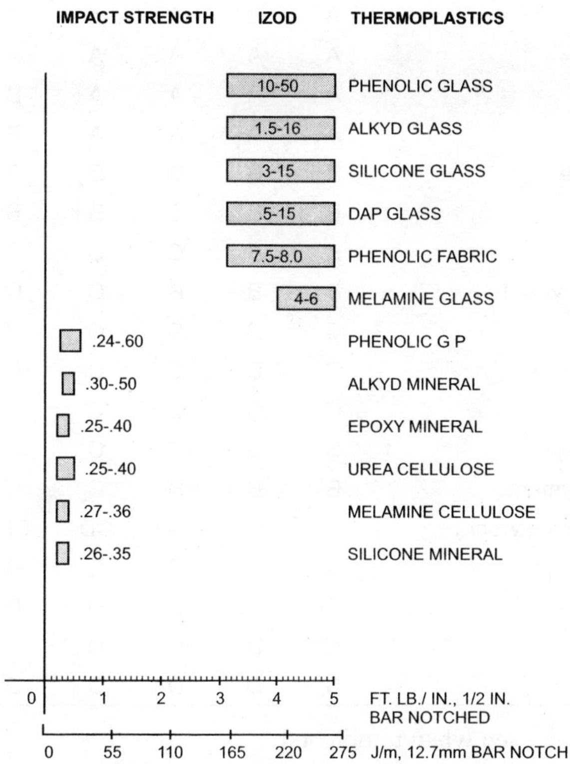

Table 3.8: Average impact strength of some thermosetting plastic materials at room temperature.

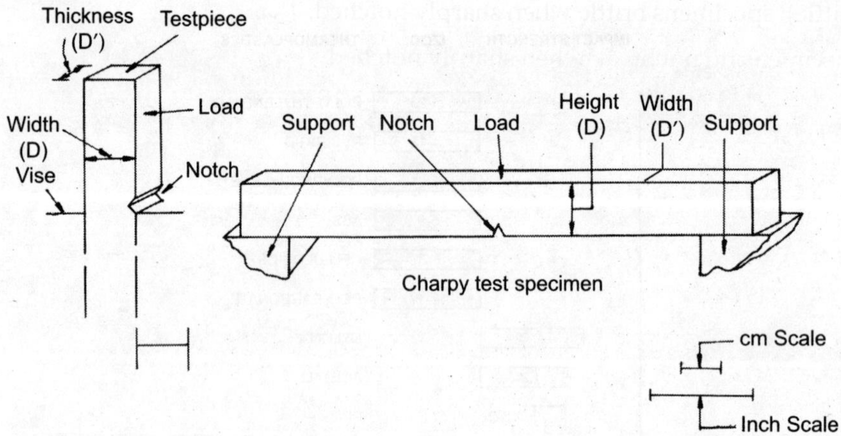

Fig 3.6: Details of testpieces; (a) for Izod test, and (b) Charpy test.

Based on the impact strengths, from −20°C to +50°C, the relative standings in order from the most brittle to the most ductile, for some of the widely used engineering plastics are given in Table 3.9.

Table 3.9: Impact strengths of some widely used Engineering Plastics

Material	Temperature(°C)							
	−20	−10	0	+10	+20	+30	+40	+50
Polystyrene	A	A	A	A	A	A	A	A
Acrylic	A	A	A	A	A	A	A	A
Glass-filled nylon(dry)	A	A	A	A	A	A	A	B
Polymethylpentene	A	A	A	A	A	A	A	AB
Polypropylene	A	A	A	A	B	B	B	B
Craze resistant acrylic	A	A	A	A	B	B	B	B
Polyethylene terephthalate	B	B	B	B	B	B	B	B
Polyacetal	B	B	B	B	B	B	C	C
Rigid PVC	B	B	C	C	C	C	D	D
CAB (Cellulose acetatebutyrate)	B	B	B	C	C	C	C	C
Nylon (dry)	C	C	C	C	C	C	C	C
Polysulphone	C	C	C	C	C	C	C	C
High density polyethylene	C	C	C	C	C	C	C	C
PPO (Polyphenyleneoxide)	C	C	C	C	C	CD	D	D
Propylene-ethylene copolymers	B	B	B	C	D	D	D	D
ABS (Acrylonitrile butadiene styrene)	B	D	D	CD	CD	CD	CD	D
Polycarbonate	C	C	C	C	D	D	D	D
Nylon (wet)	C	C	C	D	D	D	D	D
PTFE	BC	D	D	D	D	D	D	D
Low density polyethylene	D	D	D	D	D	D	D	D

A. Brittle: specimens break even when unnotched

B. (Notch brittle): specimens brittle when bluntly notched but do not break when unnotched.

C. (Notch brittle): specimens brittle when sharply notched.

D. Tough: specimens do not break when sharply notched.

3.4.1 Tensile Impact Test

Tensile impact test is generally carried out with a modified charpy impact machine, as per ASTM D1822. One end of the specimen, Fig. 3.7, is clamped in the pendulum head, H, and the other end in the T-shaped cross-head clamp. After release of the pendulum the two arms of the crosshead clamp hit the stationary double anvil at the lowest point of the swing, producing a sudden tensile load in the specimen causing fracture; and the energy absorbed is recorded in ft.lbf/in^2 orJ/m^2. Such a tensile impact test is particularly convenient when the material is available only in relatively thin form, or when orientation effects are to be assessed.

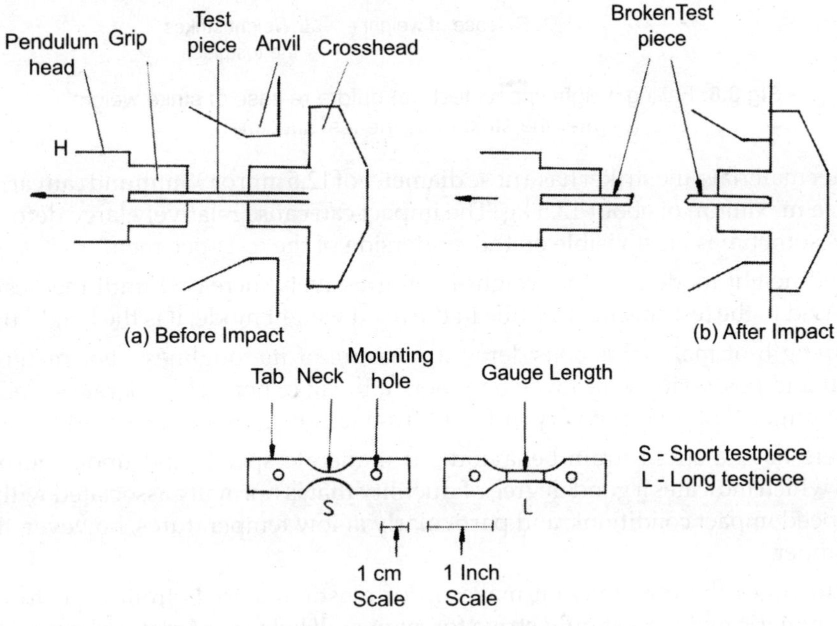

Fig 3.7: Tensile impact technique; (a) Before impact; and (b) After impact; and two types of tensile impact test specimen.

3.4.2 Falling Weight Impact Test

This test, based on ASTM 3029, assesses the impact resistance of rigid plastics through the effect of striker in the form of a ball or a dart, etc. of weight, W, dropped from a vertical height, H, on a horizontal test specimen, as shown in Fig. 3.8. The test specimen is generally a sheet or plate of thickness, t, held on an annular support of diameter, D. The values of t, D, and the nose radius r (6.25, 12.5, 25 mm), of the striker are important parameters in this test.

For very brittle materials, the striker can simply be a steel ball whose weight is a direct function of its radius. Damage to the test specimen in the form of cracks, local shattering, or total body shattering may occur.

The minimum height causing an arbitrarily defined level of damage is reported as a measure of the impact resistance which is determined from the weight of the striker and the height from which it is dropped to give the desired level of damage.

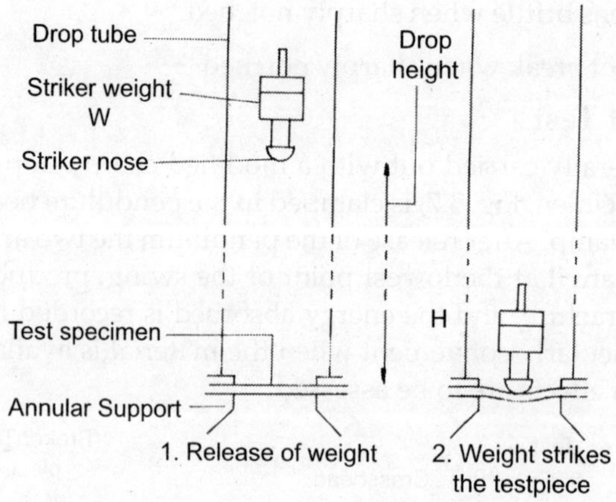

Fig 3.8: Falling weight impact test, (a) before release of strike weight, (b) after the striker hits the test specimen.

For tougher materials, the striker has a nose diameter of 12.5 mm or 38 mm and can carry adjustable weights, with a maximum of about 13.5 kg. The impact can cause relatively large deformation, and the damage is sometimes most visible on the underside of the test specimen.

In the fixed height mode test, the weight is progressively increased until the desired level of damage is caused to the test specimen, while in the fixed weight mode, it is the height that is varied.

Impact strength of material is considered a measure of its toughness. For materials that are brittle i.e., stiff and possessing very low elongation at break in normal moderate speed testing, the mechanism of impact failure is not very different from failure at moderate speed.

Many polymers have a stress-strain behaviour, at moderate speeds and under normal ambient temperatures which indicates a good degree of ductility that is normally associated with toughness. Under high speed impact conditions, and particularly at low temperatures, however, they fracture in a brittle manner.

Fig. 3.9 illustrates the fundamental make up of plastic materials from rigid to flexible. The designer of polymeric materials should strive for an overall balance of material properties. In most of the cases a tough plastic material will be better than either a rigid or flexible material.

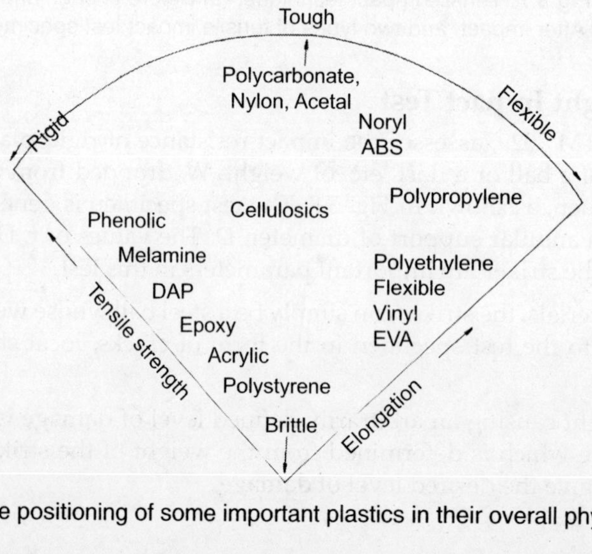

Fig 3.9: Relative positioning of some important plastics in their overall physical make up.

3.5 Shear Strength

Sheet plastics are tested for shear strength vide ASTM D732 using a disk-shaped test specimen clamped in a stationary block over a circular hole as shown in Fig. 3.10(a). A movable punch is used to shear off the central position of the test specimen. The force required is recorded and, when divided by the sheared area (thickness×circumference), gives the shear strength. Reported shear strength values for engineering plastics are in the range of 40-50 MPa(6-8 Kpsi).

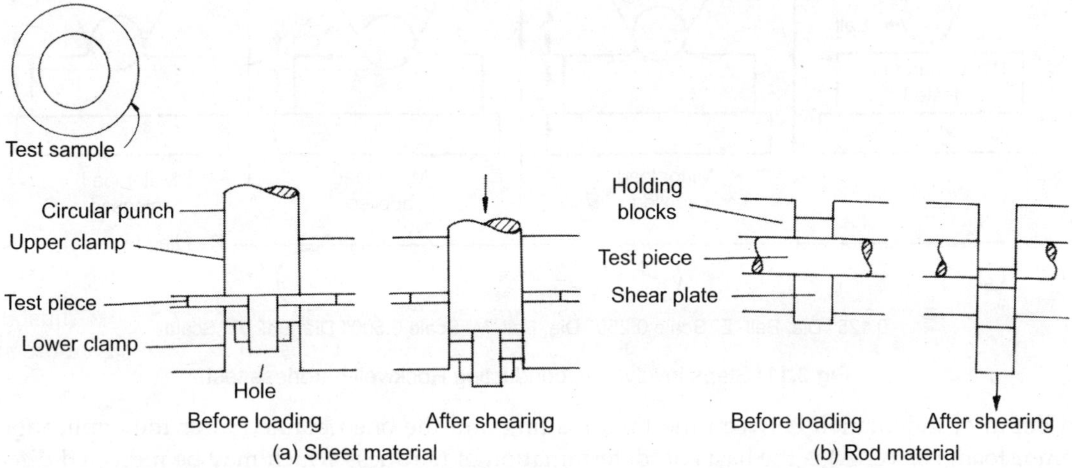

Fig 3.10: Shear testing of plastics; (a) Sheet material; and (b) Rod material.

Shear strength of rod type plastic can be obtained by using a cylindrical bar as illustrated in Fig. 3.10(b). This test is sometimes referred to as Double Shear Test.

3.6 Hardness Tests

There are a number of Hardness Tests used for determining hardness of polymers. The more often used among them are:

1. Rockwell Hardness Test
2. Shore Hardness Test
3. Barcol Hardness Test.

Brief description of these tests follows.

3.6.1 Rockwell Hardness Test

The Rockwell hardness test (ASTM D785) is a measure of the resistance of a material to indentation by a hardened steel ball under load. The hardness is expressed as a number derived from the net increase in depth of impression as the load on the indenter is increased from a fixed amount to a major load, and then returned to the minor load. Fig. 3.11 illustrates the steps involved to obtain a Rockwell hardness reading on a polymer material.

In *Step 1* is shown the initial position before the test is commenced. *Step 2* involves setting the dial at zero and the application of minor load (usually 10 kg) to ensure good contact between the indenter and the test piece with a depth of penetration A. *Step 3* involves the application of major load which is then held for 15 seconds. The surface is indented to depth B. In *Step 4* the major load is removed resulting in partial recovery of indentation, leaving behind an indentation of depth C. After another 15 seconds (total of about 45 seconds), the hardness is read on the dial. The difference

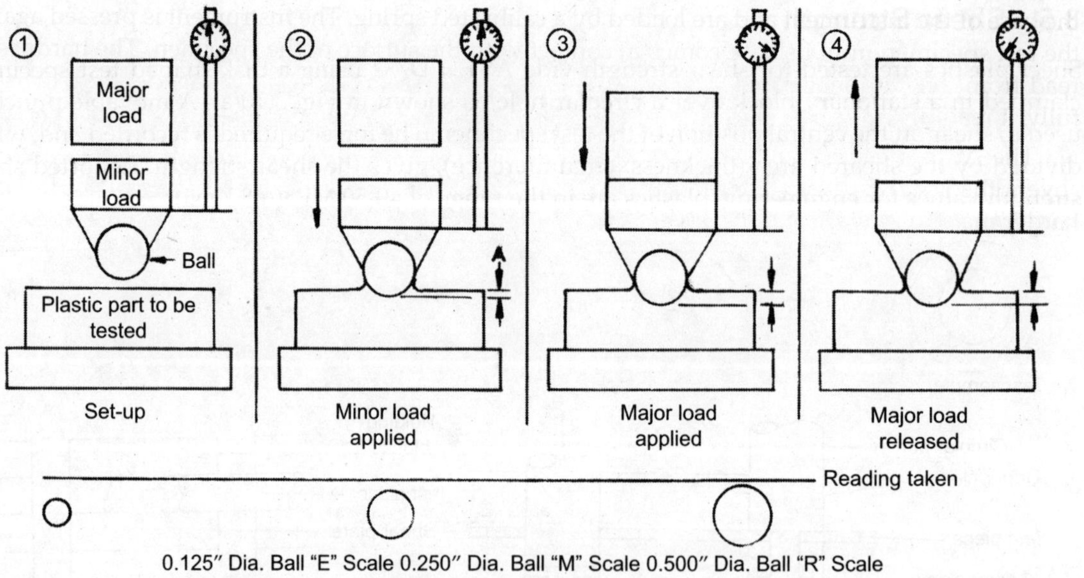

Fig 3.11: Steps involved in conducting Rockwell hardness test.

in depth of indentation between the final reading and the one obtained after the application of minor load, i.e., (C–A) is the basis of determination of hardness which may be indicated directly by the meter. The minor load is fixed at 10 kg but the major load varies with the different sized balls used as listed in Table 3.10.

Table 3.10: Indenter Shape and Load Application Method for Testing Hardness by Different Techniques

S.No.	Method	Indenter	Force	Load	Reading
1.	Rockwell-E	Ball 3.15 mm	100 kg	15 sec.	15 sec.
	Rockwell-M	Ball 6.25 mm	100 kg	15 sec.	15 sec.
	Rockwell-R	Ball 12.50 mm	60 kg	15 sec.	15 sec.
2.	Durometer	Dull Needle	56 to 822 grams (By spring)	Rapidly	At once
3.	Barcol	Sharp Needle	By Spring	Rapidly	At once

For polymeric materials, generally two conditions are used viz., Rockwell-R for softer materials, corresponding to d=½″ (12.5 mm) and the major load=10 kg for a range of R20 to R120; scale M for stiffer materials corresponding to d=¼ (6.25 mm) and major load of 100 kg for a range of Rockwell–M 20 to 140. The overlap in the two scales covers approximately the range R100–R120 or M20–M30. Scale E is used for reinforced plastics like DAP (Dially phthalate) glass filled.

Because of the relatively large size of the indenter, a minimum test specimen thickness of 6 mm and an area over 6 cm^2 are recommended; stacking of suitable thin specimens is possible.

Rigid thermosetting plastics often give Rockwell hardness in the range of M100-M120, while the engineering thermoplastics are often in the range of R110-R120. Table 3.11 lists the Rockwell hardnesses of some important industrial plastics on all the three scales, i.e., R, M, and E scales.

3.6.2 Shore Hardness

The shore hardness test (ASTM D2240) employs a small hand instrument called *Durometer.* It features pin-shaped indenters of two types (A or D), as shown in Fig. 3.12, which protrude from the base of the instrument and are loaded by a calibrated spring. The instrument is pressed against the test specimen until its base comes in contact with the surface of the specimen. The hardness is read from the Durometer dial. The scale readings range from 0 to 100. If the indenter penetrates fully, it reads 0 and in case it does not penetrate at all it reads 100.

The A scale, using the blunt indenter, is suitable for soft materials, such as elastomers, and covers the range between 20A and 95A. The D scale, with the pointed indenter, is intended for harder materials, and generally covers a range of 40D-90D; corresponds approximately to 90A.

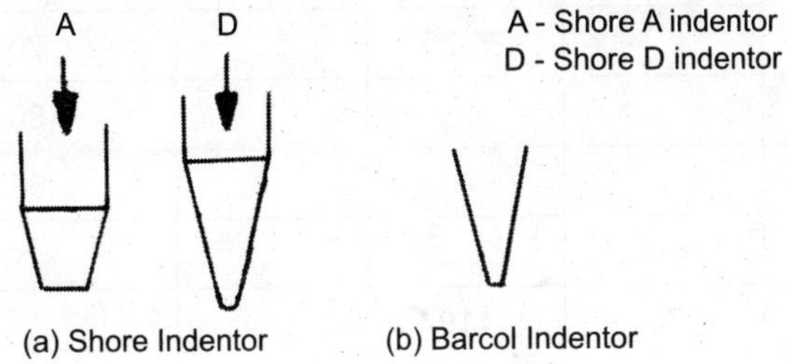

Fig 3.12: Hardness test indentors for; (a) Shore Hardness, and (b) Barcol Hardness.

Soft rubbers are in the range 20A-30A, and include items like rubber bands and soft erasers. The tyre treads are in the range 55A-70A. Typewriter rolls corresponds to about 90A. Soft plastics, such as low density polyethylene (LDPE) correspond to about 45D, while high density polyethylene (HDPE) and other medium stiffness plastics correspond to about 65 D. Bowling balls correspond to about 90D. Shore hardness of some other plastics are listed in Table 3.11.

3.6.3 Barcol Hardness

The hardness of rigid plastics particularly that of reinforced thermosetting plastics, is often reported in Barcol hardness unit, as per ASTM 2583. The Barcol (Barber and Coleman) tester, uses an indenter similar to that used in the shore D method, except for a flat tip, with a diameter of 0.006″ (≈0.15 mm), instead of rounded one as shown in Fig. 3.12(b). The minimum recommended test specimen thickness is 1.5 mm (≈1/16″).

Like in Shore hardness testing methods, the indenter protudes from a flat surface that is pressed against the test specimen. The indenter is loaded with a calibrated spring and its penetration is related to a Barcol hardness number that ranges from 0 to 100.

The Barcol tester is used in the reinforced plastics industry as a simple means of determining the degree of cure of polyester resins. The usual hardness range is about 50B to 90 B. Barcol 60B hardness is roughly equivalent to shore 80D and to Rockwell M100 hardness.

Relative hardnesses of different plastics in Rockwell M, R, E scales, Durometer shore D scale and Barcol scale for a number of industrially important plastics are given in Table 3.11.

Table 3.11: Hardness of various Plastic Materials

PLASTIC MATERIALS	ROCKWELL			DUROMETER	BARCOL
	M	R	E	SHORE D	
ABS HIGH IMPACT		85-109			
ACETAL	94	120			
ACRYLIC	85-105				49
CELLULOSICS		30-125			

DAP GLASS FILLED			80-87		52-65
EPOXY G. P.	75-108				
EPOXY GLASS FILLED	100				
PTFE FLUOROPLASTIC				50-65	
PFEP FLUOROPLASTIC		25		50-60	
PCTFE FLUOROPLASTIC		75-95		76	
VF_2 FLUOROPLASTICS				80	
SURLYN IONOMER				60	
MELAMINE CELLULOSE	116-124				
PPO MODIFIED	78	119			
NYLON 66		108-120			
NYLON 6		120			
PHENOLIC G. P.	45-60				
POLYCARBONATE	70	116			
POLYESTER RIGID	65-115				30-50
POLYETHYLENE H. D.				60-70	
POLYETHYLENE M. D.				50-60	
POLYETHYLENE L. D.				41-46	
POLYPROPYLENE		90-110		75-85	
POLYSTYRENE G. P.	68-80				
VINYL RIGID		117		65-85	
VINYL PLASTICIZED				40-100 SHORE A	
UREA CELLULOSE	110				
SILICONE	84				
POLYSULFONE	69	120			

3.7 Heat Deflection Temperature

The heat deflection temperature or Deflection temperature underload of a plastic material indicates the temperature at which it begins to yield. For this test (ASTM D648), a bar-shaped test specimen,S, is loaded in the three-point pending mode, as shown in Fig. 3.13.

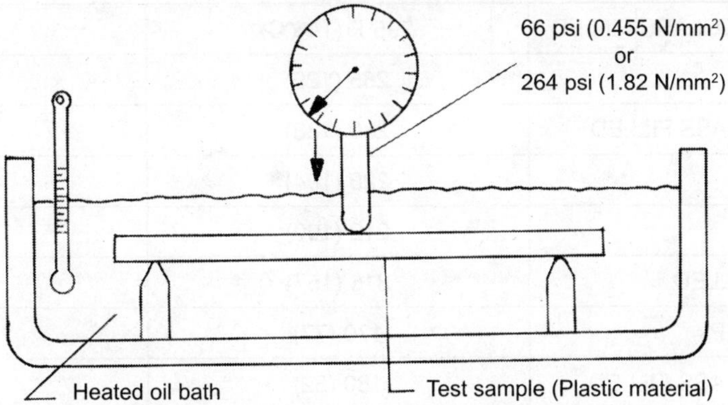

Fig 3.13: A schematic showing the method used to obtain the heat deflection point of plastics.

The constant load is chosen to produce a maximum surface stress in the test specimen, of either 0.455 MPa ($\simeq$ 66 psi) or 1.82 MPa ($\simeq$ 264 psi), as calculated from basic strength of materials considerations. The temperature is then raised at a specified rate of 120°C/h (216°F/h). The temperature at which the central deflection increases by 0.254 (0.010") is reported as the Heat Deflection Temperature (HDT) or Deflection Temperature Under Load (DTUL). The temperature obtained depends upon the applied stress level. DTUL for some industrially important plastics are listed in Table 3.12.

Table 3.12: Approximate range of deflection temperature under load of some plastic materials.

MATERIAL	264 PSI ($\approx$1820 × 10^3 Nm^{-2})	66 PSI ($\approx$455 × 10^3 Nm^{-2})
SILICONE GLASS FILLED	500-900° F (260 – 482°C)	
ALKYD GLASS FILLED	330-540 (166 – 282°C)	
ALKYD MINERAL FILLED	320-540 (160 – 282°C)	
MELAMINE CELLULOSE FILLED	350 (177°C)	
MELAMINE GLASS FILLED	400 (204°C)	
DAP UNFILLED	310 (154)	
DAP GLASS FILLED	330 (166)	
DAP ACRYLIC FIBER FILLED	225-266 (107 – 130)	
EPOXY MINERAL FILLED	250 (121)	
EPOXY GLASS FILLED	250 (121)	
PHENOLIC G. P.	260 (127)	
PHENOLIC GLASS FILLED	300 (149)	
POLYESTER GLASS FILLED	150-450 (66 – 232°C)	
THERMOPLASTICS		
POLYSULFONE	345 (174)	
POLYCARBONATE	265 (129)	

POLYCARBONATE GLASS FILLED	295°F (146°C)	
PPO MODIFIED	265 (129)	
PPO MODIFIED GLASS FILLED	280 (138)	
ABS	220 (104)	
ACETAL	212 (100)	
ACETAL GLASS FILLED	315 (157)	
POLYSTYRENE G. P.	170 (77)	
POLYSTYRENE GLASS FILLED	180 (82)	
POLYSTYRENE HIGH HEAT	190 (88)	
CELLULOSICS	100-190 (38-88)	
POLYETHYLENE H. D.		140-190°F (66 – 88°C)
POLYPROPYLENE	125 (52)	200-230 (93 – 110)
POLYPROPYLENE GLASS FILLED	230 (110)	305-310 (152 – 154)
NYLON 6	150 (66)	
NYLON 6 GLASS FILLED	320 (160)	
NYLON 6/6	150 (66)	
NYLON 6/6 GLASS FILLED	320 (160)	
PTFE FLUOROPLASTIC		258 (126)
PCTFE FLUOROPLASTIC		250 (121)
PVF_2 FLUOROPLASTIC	195 (91)	300 (149)

3.8 Heat Resistance

The effect of heat on the properties of polymers is of great importance. The maximum safe continuous operating temperature of a polymer is that temperature which the material can withstand over a wide variety of service conditions without blistering, shrinkage, distortion, or failing to function as required.

In general, plastics are low in heat resistance. The fluoroplastic PTFE can withstand a temperature of around 288°C (550°F), but will not stand up under much stress at that point. Some reinforced thermosetting plastics (the silicones, phenolics, and epoxies) can withstand temperatures between 150-250°C (300-400°F). Most thermosetting plastics can be safely used at temperatures up to around 120°C (250°F).

Usually all thermoplastics materials are ruled out where high heat resistance is needed. Polyethylene oxide, poly-sulfone, and polycarbonate are the best for temperatures between 120 and 150°C (250 to 300°F).

The great majority of thermoplastic materials are not servicable above 93°C (200°F). Table 3.13 lists the ranges of recommended continuous heat resistance temperatures for some of the more widely used plastics

Table 3.13: Approximate Recommended Continuous Heat Resistant Temperature for Some Plastic Materials

CHARING/BURNING TEMPERATURE °C	°F	MATERIAL
316	600	ALL PLASTIC MATERIALS CHAR OR BURN
302	575	
288	550	PTFE FLUOROPLASTIC
274	525	
260	500	POLYIMIDE, SILICONE GLASS
246	475	
232	450	PHENOLIC ASBESTOS FILLED
218	425	SILICONE MINERAL FILLED
204	400	PFEP FLUOROPLASTIC, DAP GLASS FILLED
191	375	PCTFE FLUOROPLASTIC
177	350	
163	325	POLYSULFONE
149	300	ALKYDS, PHENOLICS, EPOXY GLASS, PVF_2 – CHLORINATED POLYETHER
135	275	
121	250	MELAMINE, NYLON, POLYPROPYLENE, PPO, POLYCARBONATE
107	225	ABS/POLYCARBONATE ALLOY
93	200	ACETAL, ABS, SURLYN, NORYL, POLYURETHANE
79	175	190 ... ACRYLICS, CELLULOSICS, POLYETHYLENE,
66	150	POLYSTYRENE, SAN, VINYLS, UREA 130 ...
52	125	
38	100	
24	75	AMBIENT ROOM TEMPERATURE

3.9 Vicat Softening Point of Thermoplastics

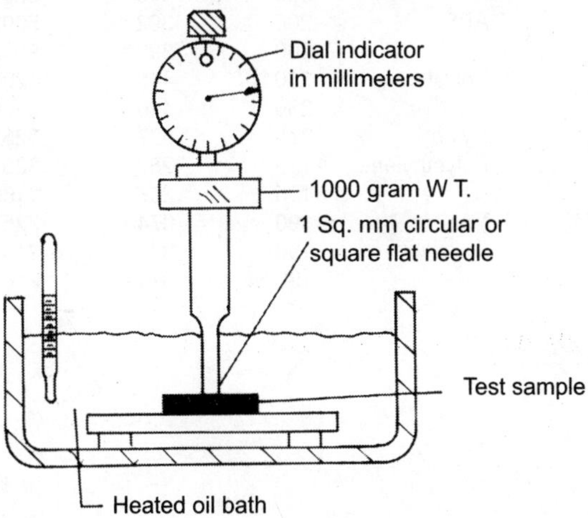

Softening point of some thermoplastics

Acetals	323-343°F (162 - 173°C)
Nylon 6/6	480-498 (249 - 259°C)
Acrylic	185-248 (85 - 120°C)
Polyethylene	
Low density	179-210(82 - 33°C)
Med. density	221-237(105 - 114°C)
High density	255-260(124 - 127°C)
Polypropylene	291-302(144 - 150°C)
Polystyrene	
Gen. purpose	213-302 (101-112°C)

Fig 3.14: Set-up for determining softening point of thermoplastic polymers by Vicat Method.

The vicat softening point is the temperature at which a flat-ended needle of one square mm circular or square cross-section will penetrate a thermoplastic test specimen to a depth of one mm. For this test (ASTM D1525), a small flat test specimen thicker than 3.2 mm ($\frac{1}{8}$") is used. The set up for the test is shown in Fig. 3.14 and the Vicat Softening Points of some of the well known thermoplastics are listed in Table 3.14.

Table 3.14: Softening point of some thermoplastic.

ACETALS	323-343° F	162-173° C
NYLON 6/6	480-498	249-259
ACRYLIC	185-248	85-120
POLYETHYLENE		
... LOW DENSITY	179-210	82-99
... MED. DENSITY	221-237	105-114
... HIGH DENSITY	255-260	124-127
POLYPROPYLENE	291-302	144-150
POLYSTYRENE		
GEN. PURPOSE	213-233	101-112

3.10 Melt Flow Index

The melt flow index test (ASTM D-1238) measures the rate of extrusion of a thermoplastic polymer through an orifice of a specified length and diameter, under prescribed conditions of temperature and pressure. Fig. 3.15 shows the essential features of the test equipment.

The test is useful for thermoplastic polymers having relatively low melt viscosities. It indicates the uniformity of the flow rate of a thermoplastic polymer. The property measured by this test is basically the melt viscosity. The melt flow index test is primarily useful to manufacturers of raw material, as a method of controlling material uniformity.

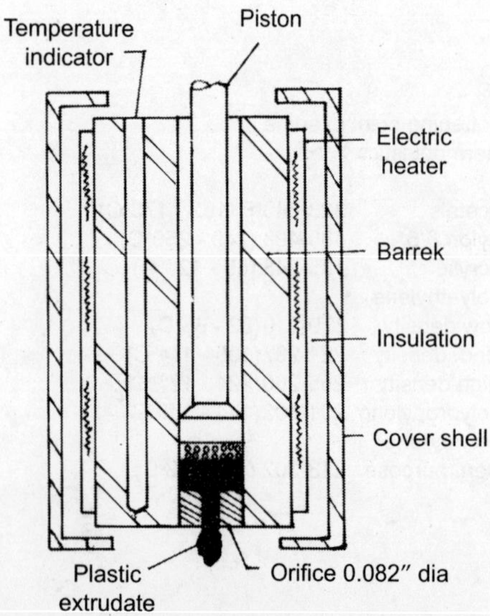

Material	Temp°C	Temp°F	Load GMS
Acrylics	230	446	1200
	230	446	3800
ABS	200	392	5000
Polystyrene	200	392	5000
	230	446	1200
	230	446	3800
Nylon	275	527	325
Polyethylene	125	257	325
	125	257	2160
	190	374	325
	190	374	2160
	190	374	2160

Fig 3.15: Essential elements of the set-up for determining the Melt Flow Index of thermoplastic Polymers.

The material to be tested is placed in the barrel of the plastometer. A weighted piston is applied on the top of the material, and heat is applied to the barrel. The heat melts the grannular thermoplastic material that is forced out through the orifice. After a predetermined time, the extruded portion is cut off and weighed. *The flow rate is reported as the rate of extrusion in grams per ten minutes.* The melt flow indices of some of the well known polymers are listed in Table 3.15.

Table 3.15: Melt Flow Indices of some well known thermoplastic polymers.

MATERIAL	TEMP °F	°C	LOAD GMS.
ACRYLICS	446	230	1200
	446	230	3800
ABS	392	200	5000
POLYSTYRENE	392	200	5000
	446	230	1200
	446	230	3800
NYLON	527	275	325
POLYETHYLENE	257	125	325
	257	125	2160
	374	190	325
	374	190	2160
	374	190	21,600

3.11 Specific Gravity/Density

Specific gravity is defined as the ratio of mass of a given volume of material to that of the mass of same volume of water both measured at 23°C. It, being a ratio, is a dimensionless quantity. *Density*, on the other hand, is mass per unit volume (g/cm^3 or g/ml ≈ 62.4lb/fl^3 or 0.036 lb/in^3). The density of water at 23°C is about 1 g/cm^3.

The conventional determination of density or specific gravity can be made by weight and volume measurements as is done in grade school laboratory practicals; alternatively ASTM D792 may be followed.

The determination of specific gravity or density of polymeric materials is important for a number of purposes including design considerations when weight is a factor, the identification of unknown meterials, and the assessment of the structure like crystalline, cellular, etc., of the material in a product.

For pure polymers, the specific gravity ranges from about 0.83 for that of PMP (polymethylpente) to about 2.15 for PTFE (polytetrafluoroethylene). The range 0.83-1.0, besides PMP, also includes PP (polypropylene) and polyethylene (PE). The range 1.00-1.25 includes PS (polystryrene) and other styrenics, polyurethanes (PU), polyamides (PA), acrylics (PMMA), and polycarbonate (PC). The range 1.25-1.50 includes many vinyls (PVC), polyesters (PET, PBT), acetal (POM), polyimides, phenolics, etc. Above 1.5, are the highly chlorinated or fluorinated polymers (PVDC, PVDF, PTFE, etc.) and silicone polymers.

Table 3.16 lists specific gravities and densities of a few specific polymers.

Table 3.16: Specific Gravities and Densities of some well known Polymers

S.No.	Polymer	Specific Gravity	Density		
			g/cu.in	lbs/in³	kg/m³
1.	Water	1	16.39	0.036	9.97×10^2
2.	Polyethylene, HD	0.942	15.42	0.034	9.42×10^2
3.	Nylon	1.14	18.70	0.041	11.35×10^2
4.	Phenolic G.P.	1.36	22.20	0.048	13.29×10^2
5.	PTFE Fluoroplastic	2.15	35.23	0.078	21.59×10^2

A more sophisticated test method (ASTM 1505) is widely used for polymers, particularly to assess very small differences in density.

3.12 Coefficient of Thermal Expansion

All materials change in dimensions as the temperature changes. This property represents the increase in size, length, area or volume of a polymer per unit of temperature rise. The thermal expansion values for most plastics are relatively high. *They expand and contact, with changes in temperatures about five times as much as mild steel.* The units used are (m/m)/K in SI units and (in/in)/°F in British Unit (BU) system. However, as those values are very small so the units normally employed are (μm/m)/K or (μin/in)/°F, i.e., μSI and μBU; where, 1 SI unit=1.8 BU.

A standard method for the measurement of the coefficient of thermal expansion (ASTM D696) uses a quartz dilatometer as shown in Fig. 3.16. The apparatus features a long, low expansion vitreous silica tube and rod of the same material. A test specimen, at least 5 cm long, is placed between the tube base and the bottom of the rod. The top of the rod contacts a dial gauge for measuring the linear expansion. A suitable length of the tube is immersed in a temperature-controlled bath and the temperature is slowly raised.

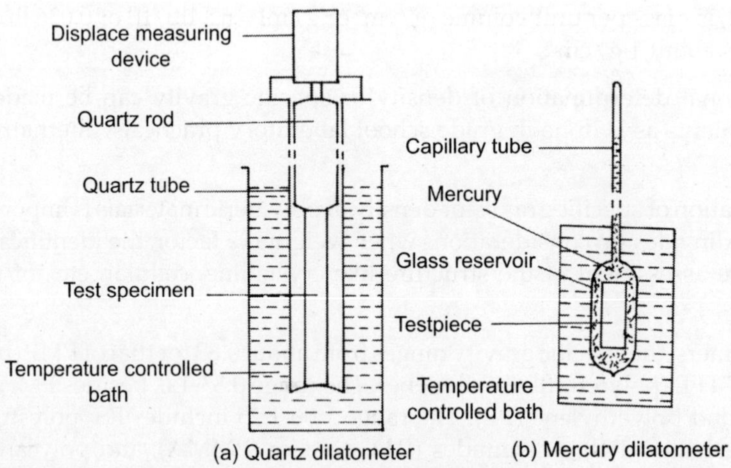

Fig 3.16: Set-up for measurement of coefficient of, (a) Linear thermal expansion, and (b) Cubic thermal expansion.

The coefficient of cubic expansion, as per ASTM, D864 can be measured with a mercury dilatometer; which is shown in Fig. 3.16(b). The test specimen is immersed in the mercury within the reservoir, and the expansion is derived from the rise of the mercury in the capillary tube, when the temperature of the bath is raised. Table 3.17 shows the coefficient of linear thermal; expansion of some well known plastics used in the industry.

Table 3.17: Approximate Range of the Coefficient of Thermal Expansion of some Plastics

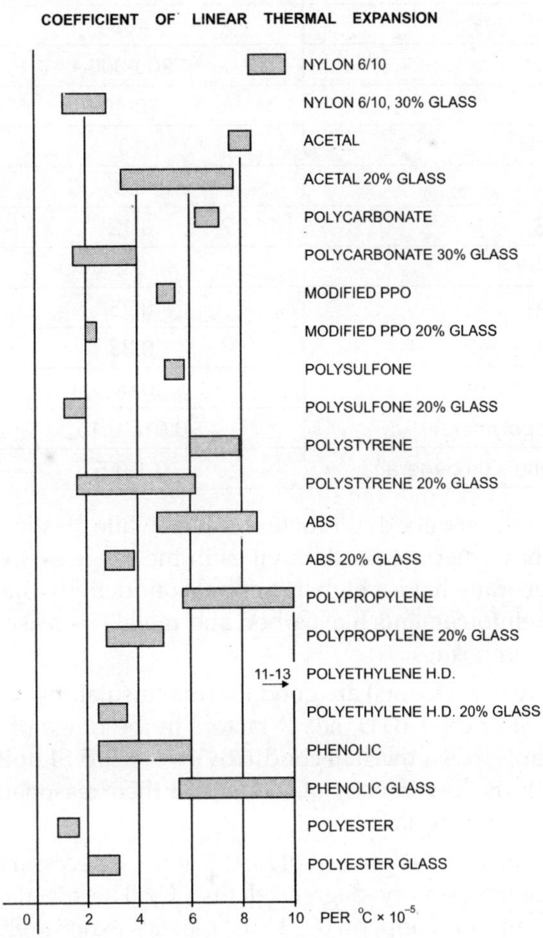

Note that with the addition of fiber glass the thermal expansion in all cases is reduced.

3.13 Thermal Conductivity and Specific Heat

Thermal conductivity is the ability of a given substance to conduct heat. Plastic materials have low thermal conductivity and are, in fact, excellent thermal insulators. Metals like silver and copper transmit over 2000 times as much heat as most plastics, asbestos four times as much, and wood and paper about twice as much.

The thermal conductivity (K), in SI system, is expressed in W/(m.k), though in more popularly used metric system of units K is expressed in Cal/(cm-S.°C); while in the British Thermal units it is expressed as BTU.in/(ft^2.h.°F), the area being in square feet and the thickness in inches. One BTU=0.144 SI unit. The numerical value of the thermal conductivity expressed in BTU is sometimes referred to as *K factor* and inverse i.e., (1/k), as the *R factor*.

Thermal conductivity of polymeric materials is determined as pex ASTM C177 and ASTM C518. Thermal conductivity of some metals and popular plastics are listed in Table 3.18.

Table 3.18: Thermal Conductivities of Some Common Engineering Materials

S.No.	Material	Thermal Conductivity SI(Cal/cm.S.°C)	(BTU.in)/(ft².h.°F)
1.	Copper	400	2800
2.	Aluminium	200-300	1400-2100
3.	Steel	20-40	140-480
4.	Cement/Concrete	10	70
5.	Glass	0.8	5.5
6.	Polystyrene (PS)	0.15	1
7.	Polycarbonate (PC)	0.2	1.4
8.	Polyamides (PA)	0.25	1.75
9.	Polyethylene (PE)	0.33	2.3
10.	Air without convection)	0.04	0.28
11.	Foams (cellular polymers)	0.015-0.15	0.1-1
12.	Polymeric Materials (in general)	0.1-0.5	0.7-3.5

The metals are, in general, very good conductors of heat while plastics are good insulators. Most *pure* polymeric materials have thermal conductivities in the range of 0.1-.5 SI units (0.7-3.5 BTU). Semicrystalline plastics generally have a higher thermal conductivity than do amosphous plastics. The presence of fillers or reinforcements like carbon and metals increase the thermal conductivity of reinforced plastics and composites.

Closed-cell cellular polymers (foams) are good thermal insulators, with thermal conductivities in the range of 0.015-0.15 SI or 0.1-1 BTU, i.e., R factors in the range of 1-10. A typical expanded polystrene (EPS), for example, has a thermal conductivity of 0.035 SI units (0.24BTU), i.e., R factor = 4.15. This combined with the low density of foams and their reasonable mechanical properties make them popular insulating material.

Specific Heat : Specific heat is the amount of heat (calories) necessary to raise the temperature of one gram of a substance through one degree celcius (°C). The specific heat of plastic materials is between 0.025 and 0.50; and for copper 0.09. Thus, plastics exhibit relatively high specific heat, approximately five times that of copper. Because the plastics are so much lighter than most metals, it takes less heat to raise the temperature of a given volume of plastic materials than an equivalent volume of metal. The lower the specific heat of a plastic, the lower will be its heating cost when it is subjected to a moulding process. This information is important to the designers and manufacturers of plastic goods.

3.14 Optical Properties

Most polymeric materials are capable of transmitting electromagnetic radiations in the wavelength range corresponding to visible light with wavelengths of about 3900-8000A° (390-800 nm). Light sources (sunlight, incandescent, vapour-arc, and arc lamps) have a characteristic radiation spectrum, and filters can be used to enhance certain wavelengths such as violet ($\simeq$ 410 nm), yellow ($\simeq$ 580 nm), or red (650 nm).

Amorphous or non-crystallizing pure polymers often allow high levels of light transmission and are therefore referred to as *transparent polymers.* They include many styrenics (PS, SAN, etc.), acrylics (PMMA, HEMA, etc.), cellulosics (CA, CAB, etc.), as well as engineering plastics such as polycarbonate (PC), polysulfone (PSU), also thermosets, such as unsaturated polyester (UP), epoxy

(EP) and urethane resins (UR), and unfilled, specially compounded elastomers like polyisoprene (PI), silicone, urethane, etc.

The presence of crystalline regions in crystallizing polymers tends to reduce the level of light transmission, and pure semi-crystalline polymers in moderate thicknesses are generally translucent. They include olefinics (PE, PP, etc), polyamides (PA6, PA66, etc.), and thermoplastic polyesters (PET, PBT, etc.). There are, however, some crystallizing polymers that can be made into highly transparent, relatively thick objects. They include polymethyl pentene (PMP) and polyethylene terephthalate (PET). Films of many crystallizing polymers, particularly oriented films, can also be transparent.

The presence of additives, particularly colourants and fillers, can reduce or suppress light transmission through any polymer.

Polymers that have high-light-transmission and low haze properties are used in optical applications. The advantages that polymers have over glass in optical uses include low density, high impact strength and high light transmission. Their low density makes it possible to produce binoculors that are 30% less in weight than equivalent binoculars with glass prisms.

Transparent protective and decorative applications include large items such as flat glazing, corrugated panels, skylights, domes, canopies, cockpit covers, blisters or bubbles for aeroplanes and helicopters, solar collectors and mirrorized sheets. Smaller parts include windshields, face shields, and safety glasses. Large translucent panels are used for internally lighted signs.

For applications requiring excellent transmission quality such as automotive lenses for tail lights and head lights, spectacles and sunglass lenses, and even camera lenses.

Thin polymeric coatings of acrylic, silicone or urethane are used on polymeric or inorganic transparent parts to protect them against chemicals or abrasion, or to alter their surface characteristic by using them as anti-glare coatings.

Transparent polymers are also widely used for applications involving fluid sighting including fluid level indicators, syringes, filters, bowls, flowmeters, pressure gauge lenses, and see-through tubing and piping.

Applications in the electrical fields include transparent cable connectors, fuse carriers and covers, and relay casings and covers.

Transparent polymers are now widely used for applications concerning optical fibres used for light transmission or communication.

3.15 Flammability

Most polymers being organic chemicals contain carbon and hydrogen atoms so are prone to burning under certain conditions. Because polymers are now widely used where their combustion could have serious consequences; for example electrical insulation, wall-, ceiling-, and floor panelling and insulation, upholstery, bedding, children's sleepwear, and draperies; tanks and containers for flammable liquids, and interior elements in transportation.

A number of tests have been developed by different organisations involved with safety, to determine the flammability of polymers. One of the most widely used such test is called *Oxygen Index Test* and is described in the following section.

3.15.1 Oxygen Index Test

Oxygen Index Test (ASTM 2863), also known as *Fenimore-Martin Method,* determines the minimum concentration of oxygen in an oxygen-nitrogen mixture, which supports flame burning of a polymer under test. The schematic of the test set-up is shown in Fig. 3.17.

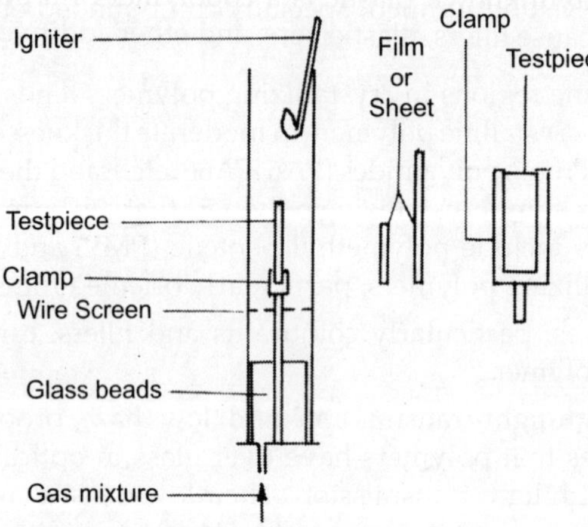

Fig 3.17: Set-up for Flammability Test of Polylmers.

A heat resistant glass tube is partially filled with glass beads to homogenise the upward flow of O_2–N_2 mixture fed at the base. The linear gas flow rate is specified at 4 cm/sec (1.6"/sec) and the volumetric gas flow, depending upon O_2–N_2 mixture, can be adjusted to achieve that. Selfsuppporting polymer test specimen, 2-3 mm thick or a sheet or a film material held in a frame (clamp) is placed in an upright position in the tube. A flame is ignited, with a gas flame igniter, at the upper end of the tube as shown. The test criterion of 3 minute duration or burning of 50 mm length is used to determine critical volumetric O_2 concentration above which the test specimen continues to burn and below which it extinguishes. The concentration, on a percent basis, is referred to as the *Oxygen Index* (OI) or *Limiting Oxygen Index* (LOI). Since the volumetric oxygen concentration of standard air is about 21%, materials with an Oxygen Index above 21 should, in principal be self-extinguishing, and materials with OI lesser than 21 should burn in air. However, combustion being a complex mechanism, it may depend upon the specific conditions in a given situation. LOI for some of the widely used polymers are listed in Table 3.19.

Table 3.19: Limiting Oxygen Index of Some Widely Used Polymers

Polymer	Limiting Oxygen Index Range
Polytetra Fluoroethylene (PTFE) Polyvinylidene Chloride (PVDC) Polyaryl Sulfone (PASU)	Above 50
Polyvinyl Chloride (PVC) Polyvinylidene Fluoride (PVDF) Polyethylene Sulfide (PPS) Polyether Sulfone (PESU) Polyimides (PI) Polyphenylene Oxide (PPO) Polycarbonate (PC) Polyamides (PA)	50 (In decreasing order) 21
Polystyrene (PS) Polyethylene (PE) Polypropylene (PP) Polymethylene Oxide (PMO) or Acetal Cellulose Acetate (CA)	< 21

It may, however, be kept in mind that the addition of fire retardants or some other modifications can markedly modify the LOI rating of polymeric materials.

3.16 Identification of Plastics

It is not easy, even by experience, to identify more than a few plastics by visual inspection or a simple mechanical test. A systematic chemical analysis of a plastic material is perhaps the only reliable method of true identification. One of the best test methods is the infrared spectrophotometer, but that is quite expensive and cumbersome. However, some very simple, inexpensive tests have been evolved for quick tentative checks. The most reliable results are obtained when the unknown plastic is of simple type. If the unknown plastic is a copolymer or a polyalloy, the identification may be ambiguous. This is because fillers, plasticizers, and other additives all have one or the other effect upon the test results.

In order to identify an unknown polymer, it is always a good practice to use a standard known plastic for comparison, whose identification characteristics are known. Both are tested under similar conditions and their results are compared. The first step is to determine whether the unknown polymer is of thermoplastic or thermosetting type. This is done by heating a small piece from the sample. If that softens, it is most probably a thermoplastic while if it chars/darkens without softening then that is a thermoset. This test can as well be carried out with the help of soldering iron, if the hot iron indents the plastic surface it is a thermoplastic resin, but will not soften a thermosetting resin. Once the unknown plastic has been categorised as thermosetting or thermoplastic, the following tests may be carried out to identify it more definitely.

1. Burning Test
2. Melting Point Test
3. Specific Gravity Test
4. Solubility Test
5. Copper Wire Test

3.16.1 Burning Test

Almost all plastics burn when exposed to a flame. The burning characteristics to be noted include odour, colour of the flame from the burning plastic, and melt behaviour. A gas burner or a small gas torch can be used to conduct the test. Hold the test piece of the unknown plastic to the edge of the non-luminous part of the flame until it ignites. Note the character of the flame and carefully smell any odour produced after the flame from the burning plastic has been extinguished.

3.16.2 Melting Point Test

The melting point of the polymer under test can be determined by using Fisher-Johns apparatus shown in Fig. 3.18.

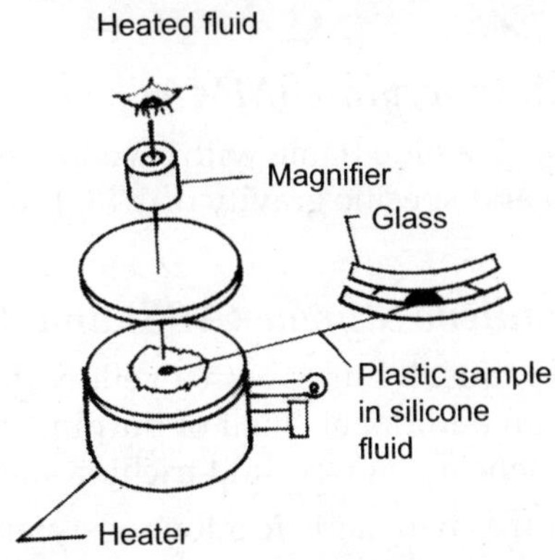

Fig 3.18: Fisher-Johns Method of determining melting point of a polymer.

In this test (ASTM D 789), the polymer sample in the form of a pellet and a few drops of silicone oil are placed together between two cover glass dishes. The dishes are then placed on an aluminium stage of the electric heater heated by a current passing through a variable controlled resistance. The oil forms a miniscus on the glass dish. The temperature is raised until the polymer pellet melts or softens enough to deform and the upper glass dish along with miniscus moves downwards slowly. The temperature at which the miniscus starts moving downwards is the melting point of the polymer under test.

A standard sample is helpful in this test because all polymeric materials do not have the same melt characteristics. Crystalline plastic materials have a sharp melting point, while amorphous polymers have a gradual melting point. The melting point of fluoroplastic resins are difficult to observe, because the melt does not become fluid. Thermosetting plastics do not melt but char or burn.

3.16.3 Specific Gravity Test

The specific gravity of an unknown plastic can be determined by using the standard apparatus having an analytical balance and a beaker of water. The test is conducted at 23°C or close to that.

3.16.4 Solubility Test

This test supplements the findings of the above three tests. Each thermoplastic polymer will be attacked or dissolved in a definite chemical solvent. Put the test piece in a test tube and pour in enough solvent to cover the sample. It may require a few minutes to an hour to dissolve some polymers.

3.16.5 Copper Wire Test

Polymers that contain chlorine may be identified by using a copper wire and a flame. Touch the heated copper wire to the polymer under test in such a way as to retain a small amount of it on the wire. Then return the wire with polymer to the flame. A green flame indicates the presence of chlorine in the polymer. Chlorinated polyether, vinyl chloride, vinyl chloride-vinyl acetate, vinylidene chloride polymers and fluorocarbon polymers give positive results by this test.

3.16.6 Identification of some specific thermoplastics

Test results of some of the well known thermoplastic polymers are included alphabatically in this section.

3.16.6.1 *Acetals/Polymethylene Oxide (PMO)*

This is a rigid plastic that burns with a pale blue flame and gives off an odour of formaldehyde. The dripping may also burn. The crystalline melting point is around 175°C (347°F). Its specific gravity is 1.42.

3.16.6.2 *Acrylic/Methyl Methacrylate (MMA)*

This polymer will burn slowly with a blue flame with a yellow tip. It has a fruit-like odour. It has a melting point of 190°C (374° F) and specific gravity of 1.18. It is soluble in acetone, benzene, and ethyl acetate.

3.16.6.3 *Acrylonitrile Butadiene Styrene (ABS) and Styrene Acrylonitrile (SAN)*

ABC is a rigid plastic which is opaque. It burns with a yellow flame, black smoke, and it drips. It has an acrid (bitter) odour with an additional smell of burning rubber. It has a specific gravity of 1.04. It is soluble in acetone, ethylene dichloride, and trichloromethane.

When SAN is burnt, it gives off a characteristic odour of styrene with an additional smell which is bitter.

3.16.6.4 *Cellulose Acetate (CA)*

This plastic is extremely slow burning with a yellow flame and has the odour of acetic acid or vinegar. It has a wide range of melting points, depending upon its composition. Its melting point is around 230°C (446°F) and the specific gravity between 1.25 and 1.35. It is soluble in acetone, acetic acid, and trichloromethane.

3.16.6.5 *Cellulose Acetate Butyrate (CAB)*

This polymer of cellulosic material burns with a blue flame and gives off an odour of rancid (stale) butter. It has a wide range of melting points, depending upon its composition. However, the average melting point is taken as 180°C (356°F), and it has a specific gravity of 1.24. It is soluble in acetone and trichloromethane.

3.16.6.6 *Cellulose Acetate Propinate (CAP)*

CAP burns with a blue flame with a yellow tip. It has an agreeable sweet odour. Although CAP has a wide range of melting points but the average melting point is taken as 235°C (455°F). It has a specific gravity of 1.20 to 1.24. It is soluble in carbon tetrachloride, and trichloromethane, and acetone.

3.16.6.7 *Cellulose Nitrate (CN)*

This polymer burns very rapidly with white flame, and decomposes. It has a very pungent odour. An odour of camphor is given off when this plastic is scratched or abraded with knife. It has a specific gravity of 1.34 to 1.40.

3.16.6.8 *Ethyl Cellulose (EC)*

This polymer burns with a yellow flame with a blue tip. It has an odour of burnt sugar. It has a melting point of 160°C (554°F). It is soluble in acetone, warm ethyl alcohol, and trichloromethane.

3.16.6.9 *Fluoroplastics/Fluoropolymers/Fluorocarbons*

The members of this family of polymers do not burn in flame, but they get deformed. The fluoroplastics are extremely inert, and the best method of their identification is by fluorine detection. PTFE, PFEP, PCTFE, and PVDF are important members of this family of plastics.

PTFE (Polytetra fluoroethylene: It has a melting point of 327°C (621°F) and is the only fluoroplastic that will not flow above its melting point. It has a specific gravity of 2.14 to 2.20 and volatises at about 399°C (750°F). This material has a milky, waxy appearance and a *slippery* feel.

PFEP (Poly Fluorinated Ethylene-Propylene): It has a melting point of 290°C (554°F) and a specific gravity of 2.12 to 2.17.

PCTFE (Polychloro trifluoro ethylene): It has a melting point of 210°C (410°F) and a specific gravity of 2.1 to 2.2.

PVDF (Polyvinylidene fluoride): This fluoroplastic is also known as PVF_2 and even PVdF. It has a crystalline melting point of 171°C (340°F) and a specific gravity of 1.75 to 1.78. When exposed to a flame, it degrades essentially to hydrogen fluoride gas and a carbon residue, without dripping. It is attacked by fuming sulphuric acid at room temperature. It will swell in acetone.

3.16.6.10 *Polymethyl pentene (PMP)*

This polymer has a crystalline melting point of 240°C (464°F). It has a specific gravity of 0.83, thus will float on water. It is attacked by chlorinated solvents.

3.16.6.11 *Nylons/Polyamides (PA)*

Polyamides or Nylons burn with a blue flame with a yellow tip. They give off an odour of burnt wool, burnt hair, or burning vegetation. The melting point ranges between 145-254°C (293-489°F), and the specific gravity ranges between 1.04 and 1.17. Nylons are unaffected by common solvents and are attacked by strong acids only. They are soluble in phenol and in formic acid.

3.16.6.12 *Polycarbonate (PC)*

This thermoplastic polymer burns with difficulty, has a yellow flame, and is self-extinguishing. It has a week odour of phenol. The melting point of polycarbonate is 221°C (430°F), and its specific gravity is 1.2. It is soluble both in methylene chloride and ethylene dichloride.

3.16.6.13 *Polyimide (PI)*

This opaque polymer is classified as thermoplastic, but it has all the characteristics of a thermoset as well. It has a specific gravity of 1.43. It is non-burning and is attacked by strong alkalies, nitrogen tetraoxide, and hydrazine. It gives off a faint odour of benzene when removed from a flame.

3.16.6.13 *Polyvinyl Acetate (PVAC)*

Polyvinyl acetate burns with a yellow flame and has an odour that of acetic acid. Its melting point ranges from 60 to 88°C, and specific gravity lies between 1.18 and 1.20.

3.16.6.14 *Polyvinyl Chloride (PVC)*

PVC burns with a yellow flame with green at the edges. It gives off a bright green flame with the copper wire test. It gives off a pungent, irritating odour of hydrochloric acid. This polymer has a melting point of about 150°C (302°F) and a specific gravity of 1.16 to 1.72. It is soluble in acetone and cyclohexanone. PVC has an ignition temperature of 391°C (734°F).

3.16.6.15 *Polyvinyl Chloride Acetate [P(VC–VAC)]*

This co-polymer burns with difficulty and is self-extinguishing. The flame is yellow with green at the bottom edges. It gives off a bright green flame with the copper wire test. Its melting point is around 127°C (261°F) and a specific gravity of 1.16 to 1.37. It is soluble in acetone and trichloromethane.

3.16.6.16 *Polyvinylidene Chloride (PVDC)*

PVDC is slow to ignite and burns with a yellow flame. It has an odour of hydrochloric acid. It gives off a bright green flame with the copper wire test. Its melting point is around 177°C (350°F) and specific gravity between 1.63 and 1.73. It is soluble in hot toluene or ether. It has an ignition temperature of around 532°C (990°F).

3.16.6.17 *Polythylene (PE)*

This polyolefin polymer burns at a fairly rapid rate and melts and drip like wax. The flame is blue with a yellow tip. It has an odour of burning paraffin wax. Low density polyethylene (LDPE) has a melting point 105°C (221°F), while the high density polyethylene (HDPE) has a melting point of 120°C (248°F). PE has a density of less than one, it therefore will float on water. It is soluble in hot benzene or hot toluene, but insoluble in other commonly used solvents. It is easy to identify by its waxy feel and natural milky-white appearance. PE has an ignition temperature of about 341°C (645°F).

3.16.6.18 *Ionomer (IO)*

This copolymeric material burns with a yellow flame and has an odour of paraffin. It has a melting point of about 107°C (225°F) and a specific gravity of 0.93 to 0.96; thus floats on water.

3.16.6.19 *Polypropylene (PP)*

Polypropylene burns slowly with a yellow tipped blue flame. Its odour is similar to that of diesel fumes. It has a melting point of about 168°C (334°F) and a specific gravity of 0.906; thus will float on water. It is soluble in hot toluene. It is stiffer than polyethylene.

3.16.6.20 *Polystyrene (PS)*

Polystyrene burns with an orange-yellow flame that produces dense smoke with clumps of carbon in the air. It smells like marigold. It has a melting point of 190°C (374°F) and a density of 1.09. It is soluble in acetone, benzene, ether, and trichloromethane. PS gives a metallic ring when dropped or tapped with a hard object. It has an ignition temperature of 360°C (680°F).

3.16.6.21 *Polyphenylene Oxide (PPO)*

This polymer has a melting point of about 262°C (504°F) and a specific gravity of 1.06. It is self-extinguishing and non-dripping. PPO has an odour of illuminating gas when burnt. It is soluble in toluene and dichloroethylene. Modified PPO is opaque and gives off a slight odour of rubber when it is burnt. Modified PPO has a melting point of around 260°C (500°F).

3.16.6.22 *Polysulfone (PSU)*

Polysulfone has a melting point of about 260°C (500°F) and a specific gravity of 1.24. It burns with a white flame and gives off an odour of sulphur. It gives a metallic ring when dropped or struck with a hard object.

3.16.7 Identification of Specific Thermosetting Plastics

Identification of some of the well known thermosetting plastics, considered alphabetically, can be carried out as follows.

3.16.7.1 *Diallyl Phthalate (DAP)*

The DAP filled moulding compounds have a specific gravity of 1.34 to 1.78. DAP is self-extinguishing to non-burning. (See also test for DAP, Alkyd, Phenolic, or Epoxy section 3.16.8).

3.16.7.2 *Epoxy (E)*

This thermosetting polymer burns with a yellow flame and in most cases will give off a sharp irritating amine odour. It has a specific gravity of 1.11 to 2.10. It is affected slightly by chlorinated hydrocarbons and ketones (see also section 3.16.8).

3.16.7.3 *Phenol Formaldehyde (FM resins)*

This thermosetting resin burns with difficulty and is often self-extinguishing, depending on the filler. The material cracks when burnt and gives off phenol and the decomposed filler. Phenolic materials are generally black and brown; however sometimes they may also be of dark green colour.

3.16.7.4 *Urea Formaldehyde (UF)*

Urea formaldehyde gives a strong odour of formaldehyde and also a musty odour (stalely smell) when it is burnt. Urea moulding materials have a specific gravity of 1.47 to 1.52.

3.16.7.5 *Melamine formaldehyde (MF)*

This thermosetting polymer has a specific gravity of 1.5 with cellulose filler, 1.8 with mineral filling, and 2.0 with glass filling. It gives off a fishlike odour when burnt. It is self-extinguishing and cracks when burnt.

3.16.7.6 *Polyesters (Alkyds)*

Unsaturated polyester resins (UPR) or Reinforced Polyester moulded plastics have specific gravity of 1.2 to 1.9. They burn with a yellow flame with black soot and have a sour cinnamon odour. Saturated polyesters are widely used in films and fibres. They burn with a yellow flame with black soot and have a mild sweet odour. *Polyesters modified with fatty acids are called alkyds.* The moulded alkyds are difficult to identify, because the burnt odour is generally that of the filler used. The alkyds are moulded in many colours and will show a white mark when scratched with a sharp instrument. (see also section 3.16.8)

3.16.7.7 *Silicones*

The flexible silicone polymers have a specific gravity of 1.05 to 1.23, while the filled silicones have a specific gravity of 1.18 to 2.82. When a silicone is burnt, it shows a white ash. A silicone polymer will show the SiO_2 bond in an infrared spectrometer.

3.16.8 Test for DAP, Alkyds, Phenolic and Epoxy Polymers

The test procedure for determining an unknown thermosetting polymer that may be DAP, alkyd, phenolic, or epoxy is as follows.

The test sample of moulded unknown polymer is pulverized into small pieces. Place some of the pulverized sample on a hot plate that has a temperature between 260-315°C (500-600°F). Hold a piece of aluminium foil at one end in a metallic clamp and position it about 13 mm above the polymer sample getting heated on the hot plate. A white thin cloudy film will form on the bright surface of the aluminium foil if it is DAP or alkyd. If nothing forms on the aluminium foil, it is an epoxy or phenolic.

In case a white cloudy film has formed on the aluminium foil, a part of the pulverized polymer sample should be boiled in a closed vessel in a solution of 20 to 30% sodium hydroxide for one to two hours. If the sample degrades, it is an alkyd, otherwise it is DAP.

In case no white cloudy film forms on the aluminium foil, the unknown polymer is a phenolic or an epoxy. Boil a part of the pulverized polymer sample in a 20 to 30% solution of sodium hydroxide for one to two hours. If the sample of unknown polymer degrades, it is phenolic, otherwise it is an epoxy.

3.16.9 Detection of Fibreglass and Asbestos Fillers

The fibreglass and asbestos fillers in moulded plastic articles can be detected by burning the material to ash in a flame. The moulded polymer should be placed on a fine wire mesh screen over a large gas burner placed under a hood, as a large amount of burnt carbon in the form of soot will be given off. After the polymer sample has burnt out, the remains in the form of ash are examined for detecting the fibre glass or asbestos left in that.

3.17 Relative Costs of Engineering Plastics

For selecting a polymer for making a component to serve in a given situation it is imperative to know the relative costs of the suitable contenders. When two or more polymers are found suitable for the same purpose, the choice will obviously be based on economic considerations. The final cost of an article will, however, depend upon the cost of the material and the cost of manufacturing; and those need to be analysed before the final selection. Table 3.20 lists the relative costs of some well known engineering plastics taking the cost of polypropylene as unity. These costs are prone to change with time as newer methods of production are developed.

Table 3.20: Relative Costs of Engineering Plastics

S.No.		Polymer	Abbreviation	Relative Cost (mass)
1. Thermoplastics				
	(A) Amorphous Thermoplastics			
	1.	Polystyrene	PS	1.15
	2.	High impact Polystyrene	HIPS	1.20
	3.	Styrene-acrylonitrile (Copolymer)	SAN	1.65
	4.	Acrylonitrile Butadiene styrene	ABS	3.4
	5.	Polystyrene modified Polyphenylene Oxide	MPPO	3.2
	6.	Polymethyl methacrylate	PMMA	2.4
	7.	Cellulose Acetate	CA	2.55
	8.	Polycarbonate	PC	3.8
	9.	Unmodified Polyvinyl Chloride (Rigid PVC)	UPVC/PVC	1.0
	10.	Plasticized Polyvinyl Chloride (Flexible PVC)	PPVC	1.1
	(B) Crystallizing Thermoplastics			

	1.	Polypropylene	PP	1.00
	2.	Glass Fibre Reinforced Polypropylene	GRPP	1.45
	3.	Impact modified Polypropylene (Block copolymer of propylene and ethylene)	IPP	1.10
	4.	Low Density Polyethylene	LDPE	1.00
	5.	High Density Polyethylene	HDPE	1.25
	6.	Poly methyl pentene/Poly 4-methyl pentene	PMP/P4MP	3.5
	7.	Polyamide 6/Nylon 6	PA6	3.75
	8.	Polyamide 66/Nylon 66	PA66	3.85
	9.	Polyamide with 66.33% glass fibre	Impact Modified PA66	3.85
	10.	Polyamide 6, 10/Nylon 610	PA610	5.5
	11.	Polyamide 11/Nylon 11	PA11	7.05
	12.	Polyethylene terephthalate	PET	3.5
	13.	Polytetra methylene terephthalate/Polybutylene terephthalate	PTMT/PBT	4.0
	14.	Polytetra fluoroethylene	PTFE	16
	15.	Polyvinylidene fluoride	PVDF/PVdF/ PVF_2	16
	16.	Polyether sulfone	PES	17
	17.	Polyphenylene sulphide	PPS	6
2. Thermosetting Polymers				
	1.	Phenol Formaldehyde	PF	1.3
	2.	Urea Formaldehyde	UF	1.2
	3.	Melamine Formaldehyde	MF	1.75
	4.	Unsaturated Polyester Resin	UPR	2.9
		(i) Sheet Moulding Compound	SMC	2.1
		(ii) Dough Moulding Compound	DMC	1.9
	5.	Epoxy Resins	E	3.8
	6.	Polyurethane	PU	2.3
3. Elastomers				
	1.	Ethylene Vinyl Acetate	EVA	1.9
	2.	Polybutene	PB	2.0

4 PROCESSING OF POLYMERS

4.0 General

Polymers are processed by various methods which may involve moulding/casting, forming, welding and machining. Apart from these major processing methods, there are other methods which are used for assembling, embedding, and decorating the plastics. It is not possible to detail all these processes in this text, so only those processes are described which are widely used in the polymer industry. To create a sort of similarity with processing of metals, the processing of plastics is classified as shown in Fig. 4.1.

Manufacturing Processes for Plastics

- Moulding/Casting Processes
  - Compression Moulding
  - Transfer Moulding
  - Injection Moulding
  - Reaction-Injection Moulding
  - Rotational Moulding
  - Shell Moulding
  - Two Colour Moulding
  - Casting
  - Centrifugal Casting
  - Casting of Plastisols
  - Hand Layups
- Forming Processes
  - Extrusion
  - Extrusion Coating
  - Blow Moulding
  - Film and Sheet Forming
  - Tubular Forming
  - Thermoforming
- Welding and Adhesive Bonding
  - Hot Gas Welding
  - Extruded Filler Welding
  - Hot Tool Welding
  - High Frequency Induction Welding
  - Ultrasonic Welding
  - Friction Welding
  - Infared Welding
  - Nuclear Welding
  - Adhesive Welding
- Machining Processes
  - Turning
  - Drilling
  - Milling
  - Grinding
- Miscellaneous Processes
  - Punching
  - Staking
  - Broaching
  - Riveting
  - Snap-in and Snap-on
  - Silk Screening
  - Pultrusion
  - Thread Cutting

Fig 4.1: Classification of Manufacturing Processes for Polymers.

Brief descriptions of these processes follows.

4.1 Moulding and Casting Processes

Moulding and casting of polymers is different from that of metals; while the metals can be heated to a molten state, the polymers can be heated to only the viscous- fluidic state. This is because of the very high molecular weight of polymers. Thus, the plastics cannot flow freely so have to be forced into dies or heated in the form of slugs and then pressed into the desired shape. So, this class of processes covers both the casting and working (forming and forging) processes as applicable to metals. Brief descriptions of the processes as listed under moulding/casting Processes in Fig. 4.1 follows in this section.

Compounding and Preforming

The polymeric material used in most manufacturing processes is in a powder or granular form, although some are preformed to an appropriate shape before use.

Although a few polymers are moulded without additives, most industrial products demand a compromise of properties that necessitates mixing the ingredients before moulding. Most *thermoplastic polymers* are purchased in granular form and are compounded dry. *Thermosetting polymers,* on the other hand, are purchased as a liquid or as a partially polymerized compound.

Compounding of ingredients like resins, stabilizers, colour pigments, plasticizers, and fillers, is carried out in a muller. Compounded materials that have been mixed and sometimes melted are placed in the feed hoppers of injection, extrusion, or calendering machines.

Some thermoplastic materials are preformed into small pellets/slugs of the proper size and shape for a given mould cavity. The use of pellets/slugs avoids waste of material, speeds up production due to rapid feeding, and avoids overloading of the machine. In preforming thermosetting resins, the material is cold-moulded, and no curing takes place. *Preforms are used only in compression, and transfer moulding processes.*

4.1.1 Compression Moulding

In compression moulding a given amount of material is placed into a heated metallic mould and, as the mould closes, pressure is applied causing the softened material to flow and conform to the shape of the mould, as shown in Fig. 4.2.

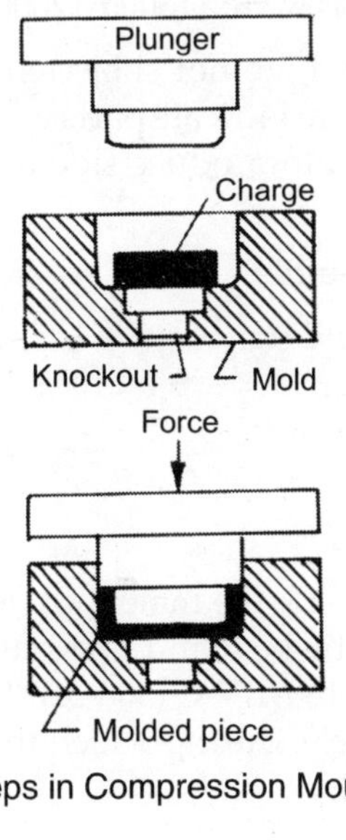

Fig 4.2: Two steps in Compression Moulding of plastics.

Compression moulding temperatures are generally in the range of 140-200°C(284-392°F); mould pressure can vary from 3.5 MPa to70 MPa (500-10,000 psi), depending upon the material used. The setting time can vary widely from about 1 minute to over 1 hour for very thick rubber components.

A large variety of hydraulic presses ranging from hand-operated to completely automatic, are used for compression moulding. The process not only applies the necessary pressure but also the required heat to plasticize properly to cure the polymeric material. Heat may be transferred from heated platens or applied directly to the metal mould. It is supplied by steam, heated liquids, electrical resistance, or ultra-high frquency electric current.

Compression moulding is used almost exclusively for thermosetting polymers; however it is occasionally used for processing thermoplastic polymers also. Thermosets most commonly used for compression moulding include phenolics, amino and alkyd resins, and DAP (diallyl phthalate) as well as unsaturated polyesters (UP) and epoxy resins (EP).

Vulcanizable rubbers are most commonly processed by compression moulding. In this case the material charge being deformable so preheating is not essential.

Compression moulding of thermoplastics requires cooling of the mould after shaping the part necessitating a longer production cycle, so the process is not commercially attractive. However, exceptions include the moulding of extremely viscous thermoplastics like PTFE, and UHMWPE (Ultra-high molecular weight polyethylene), or special products such as phonograph records made from vinyl slugs.

Compression moulded rubber products are innumerable. Many products involve metal inserts that form strong bonds with rubber and are used to attach the products to structures. Automobile tyres are also moulded to their final shape by compression moulding.

4.1.2 Transfer Moulding

Transfer moulding like compression moulding is mainly used for thermosetting plastics and vulcanizable rubbers.

In this proces, a softened thermosetting plastic or a vulcanizable rubber is transferred through a narrow gate into the closed cavity of a heated mould, where it cures to a solid state. Hydraulic presses similar to those used for compression moulding are used. Pressure used are controlled accurately and set at high levels of 42 to 85 MPa (6000-12000 psi).

Delicate inserts can be moulded and are not subjected to movement, unlike in compression moulding. It is also possible to mould more intricate parts with side cores, uneven wall section, small holes, etc. Longer core pins can be used to produce side holes and very deep holes unobtainable with compression moulding.

There are three types of transfer moulding, viz.,

(i) Pot or sprue type transfer moulding;

(ii) Plunger transfer moulding; and

(iii) Screw transfer moulding.

1. Pot Transfer Moulding

In pot transfer moulding, shown in Fig. 4.3, the material is loaded into the pot (chamber), where it is plasticized into viscous fluidic mass by heat and pressure and is then forced into mould cavity by pressure. Moulding temperatures of 150-177°C (300-350°F) and pressure up to 6000 psi (42 MPa) are employed. Following the curing cycle, during which the material is polymerized, the press is opened and the moulded part is ejected.

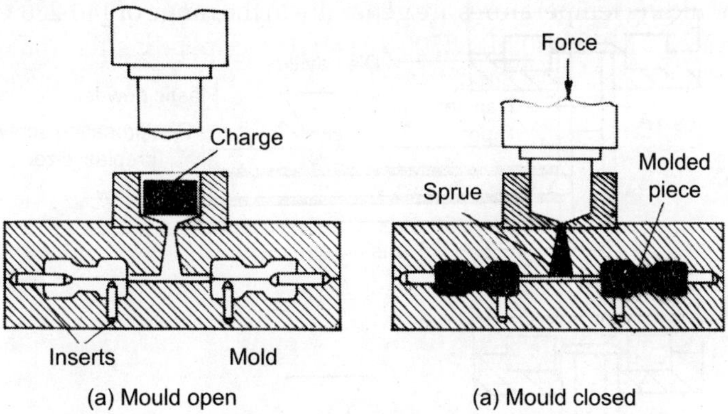

Fig 4.3: Two steps in pot Transfer Moulding of plastics.

2. Plunger Transfer Moulding

In plunger transfer moulding, shown in Fig. 4.4, the material charge is placed in the loading well, that extends to the mould parting line. The plunger then transfers the material to the mould cavaties through runners and gates. The material left at the bottom of the loading well, known as the *transfer pad*, is essential for pressure control but should be minimal.

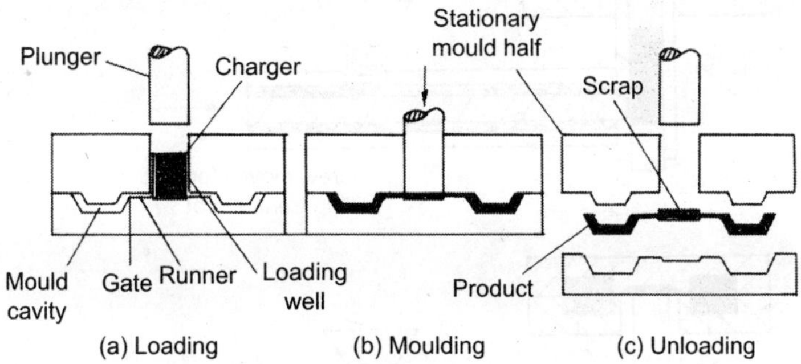

Fig 4.4: Plunger Transfer Moulding of plastics.

The plunger then backs off, the mould opens, and the knockout pins eject the part(s) from the mould. Plunger transfer moulding develops more frictional heat than the pot-type transfer moulding and thus reduces the moulding cycle time. This process, however, is not normally suitable for single part moulding.

3. Screw Transfer Moulding

In this process the moulding compound is hopper-fed to the reciprocating screw barrel. As the screw rotates, material travels forward along the flights and is thoroughly pre-plastisized by the mechanical shearing action. The heated pre-plasticized material is forced from the barrel into the transfer pot by the screw. The transfer ram, then advances to inject the material through the runner system. The mould is opened and the cycle is completed as the part is ejected. The whole process is shown in five steps in Fig. 4.5.

Transfer moulding process is used for general-purpose and reinforced plastics and reinforced phenolics, melamines, filled alkyds, diallyl phthalate (DAP), epoxies, and glass-reinforced polyesters. With the development of mouldable, soft, flowing material such as the epoxies, DAP, and silicones, it is now possible to transfer mould or encapsulate extremely fragile insert components and assemblies.

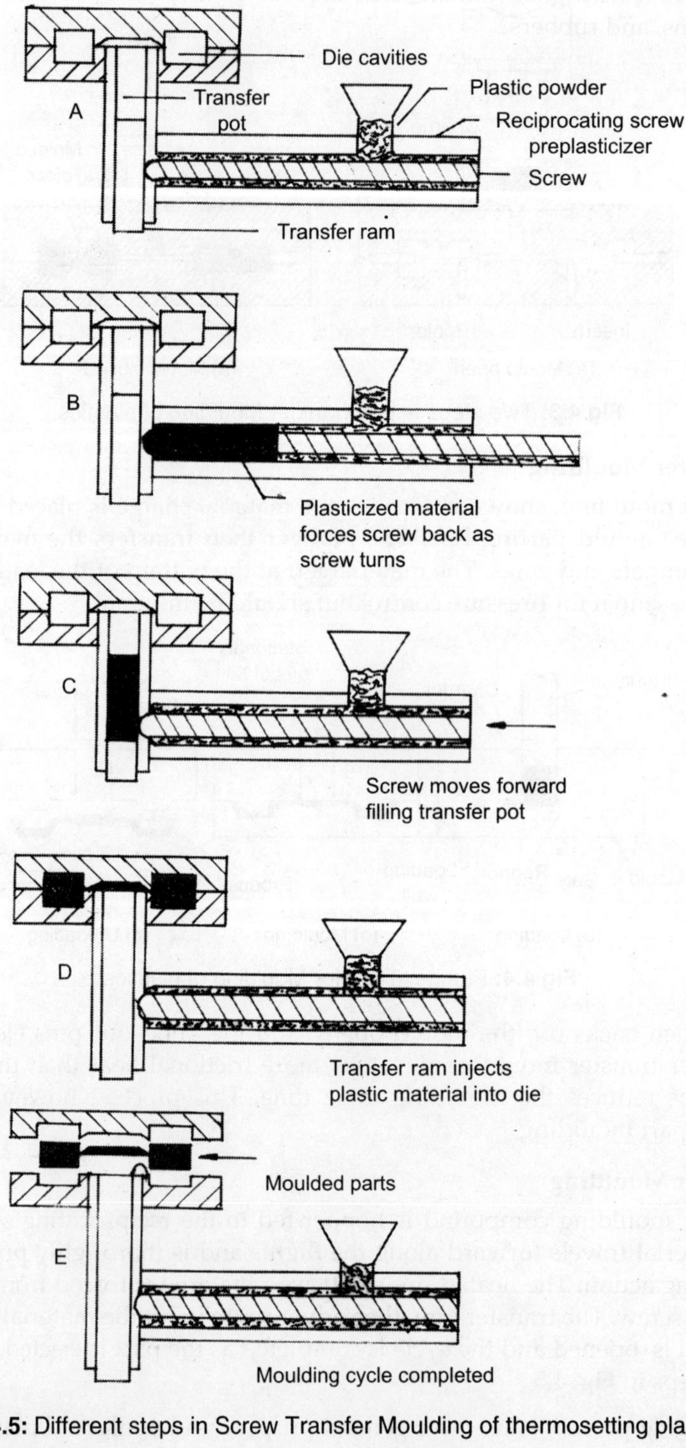

Fig 4.5: Different steps in Screw Transfer Moulding of thermosetting plastics.

4.1.3 Injection Moulding

The injection moulding process is one of the most versatile production methods in the plastics manufacturing industry. It is a process that is capable of producing moulded parts of relatively intricate configuration with good dimensional accuracy. The process is used for thermoplastics, thermosetting resins, and rubbers.

The granular or powedered plastic is fed from a hopper to a heated chamber where it is heated to fluidic state and is then injected into the mould by a ram or screw-ram. Pressure is held on the molten plastic material until the mass has hardened enough for removal from the mould.

Injection moulding machines are designated by the maximum amount of general-purpose protystyrene material that the machine can process. This refers to the capacity of the heating cylinder and is generally stated as the number of ounces (gms.) of uniformly heated polystyrene that can be shot with one complete stroke of the injection ram or screw-ram. Moulding machine capacities range from a fraction of an ounce* (a few grammes) to about 300 ounce (9 kg.). Another way of designating injection moulding machines is by the tonnage with which the dies may be clamped. Most machines of this type have a 50 to 2500 ton (0.4-22 MN) clamping force.

The heating chamber of an injection moulding machine is cylindrical in shape with a torpedo-shaped spreader in the centre for a layer thin enough to be heated both uniformly and rapidly. The heating chamber temperature ranges from 120 to 260°C (250 to 500°F) depending upon the material being charged and the size of the mould. Heat is supplied by a series of electrical resistance coils. These chambers are of rigid construction to withstand pressure that may go up to 200 MPa (30,000 sprue psi). The shot size, including the sprue and runner system should not exceed two-thirds the rated shot capacity of the press.

The mould temperature must be maintained below the softening point of the material being moulded. It must also be high enough to prevent shrinkage of the plastic, due to the hardening of outer layers of the mass too quickly. Mould temperatures usually run from 65 to 150°C (125 to 300°F). Cold circulating water is used to maintain proper temperature of the mould.

The average injection moulding pressure ranges from 65-200 MPa (10,000 to 30,000 psi) depending on the moulding material and the mould design.

The injection moulding machines are of three types, viz.,

1. Conventional Single-stage Plunger type,
2. Single-stage Reciprocating Screw type, and
3. Two-stage Plunger or Screw Plasticizer type.

1. Conventional Single-Stage Injection Moulding Machine

In this system the plastic material in the form of granules is fed into the machine hopper wherefrom it flows into the chamber of heating cylinder. A plunger then compresses the material, forcing it through progressively hotter zones of the cylinder, where it is spread thin by a torpedo-shaped spreader. The torpedo-spreader may also be heated so that the plastic material is heated both from inside and outside. The set-up for this moulding system is shown in Fig. 4.6.

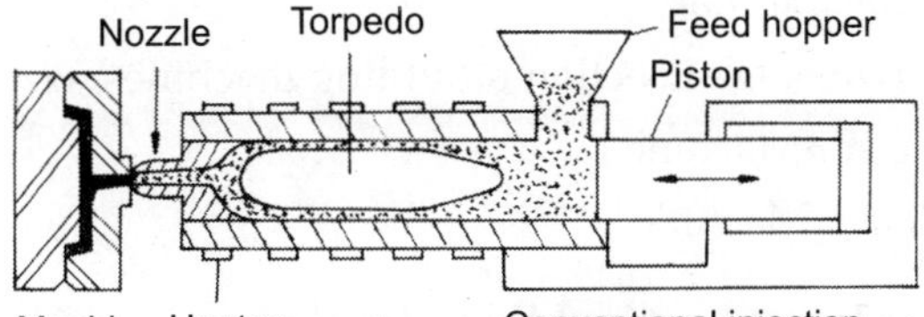

Fig 4.6: Conventional single stage Injection Molding Machine.

---

* 1 Ounce (oz) = 28.35 grammes (gm)

The material flows from the heating cylinder through a nozzle into the mould. The nozzle acts as a seal between the cylinder and the mould. It prevents leaking of material by the pressure used. The mould is held shut by the clamp end of the machine. For polystyrene, the pressure required on the clamp end of the machine is generally 2 to 3 tons/in^2 (30-45 MN/m^2) of projected area of the part and runner system.

The conventional single-stage plunger-type injection moulding machine is the only type of machine that can produce a mottled-coloured moulded part. The other types of injection moulding machines mix the plastic material so thoroughly that only one colour will be produced.

2. Single-Stage Reciprocating-Screw Injection Moulding Machine

This type of injection moulding machine employs a horizontal extruder instead of the heating chamber/cylinder of conventional injection moulding machine. The plastic material is moved forward through the extruder barrel by the rotation of the screw conveyor. As the material progresses through the heated barrel with the screw, it changes from the granular to the viscous-fluidic state. In the reciprocating screw the heat conveyed to the moulding compound is caused both by conduction and friction between the screw and the walls of the extruder. As the material moves forward, the screw retracts backwards to a limit switch that determines the volume of the material in the front of the extruder barrel.

For injection of the viscous fluidic polymer into the die, the screw moves forward to displace the material in the barrel. In this type of machine the screw performs both as a plunger and a conveyor. As the material freezes in the gate section of the mould, it prevents the backflow and the screw begins to rotate and moves backwards for the next cycle. Fig. 4.7 shows the essential features of a reciprocating screw injection machine.

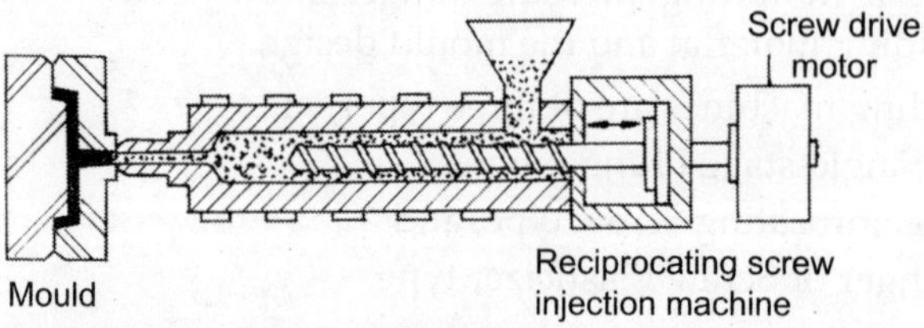

Fig 4.7: Schematic of Reciprocating-Screw Injection Machine.

There are several advantages of this method of injection moulding. Due to the mixing action of the screw it plastisizes the polymeric material more efficiently and blends colours more rapidly. Higher heat generation results in shorter overall cycle time. A breakdown of the injection moulding cycle time is illustrated in Fig. 4.8.

3. *Two-stage Injection Moulding Machine*

There are two types of two-stage injection moulding machines, viz.,

(i) Piston-type Plasticizor Machine

(ii) Screw-type Plasticizer Machine

(i) Piston-Type Plasticizer Injection Moulding Machine

This machine, shown in Fig. 4.9, employs a combination of conventional injection moulding machine and a holding chmber to injection-mould a component. The material is plasticized in the conventional-type chamber and a definite amount is transferred by a plunger into the holding chamber from which the plunger injects it into the mould cavity.

This type of machine produces pieces faster than the conventional type machine because the moulding chamber is filled to shot capacity during the cooling time of the part. Due to the fact that the injection plunger acts on the fluidized material no pressure loss is encountered in compacting the granules. This allows for the production of large parts with more projected area.

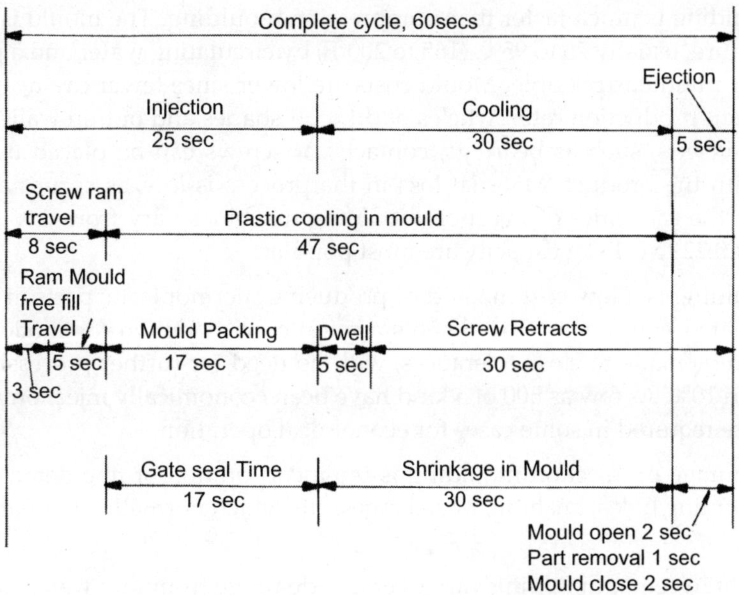

Fig 4.8: Different elements of Injection Moulding cycle for Reciprocating-Screw Injection Moulding Machine.

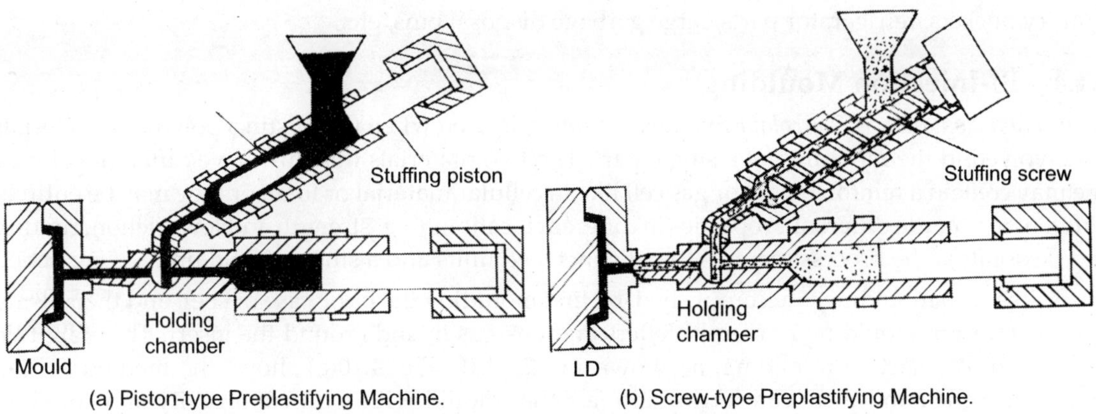

Fig 4.9: Two types of two-stage Plasticisor Injection Moulding Machines, (a) Piston type, and (b) Screw type.

(ii) Screw-Type Two-Stage Plasticizor Injection Moulding Machine

When the conventional-type injection moulding cylinder is replaced by the Reciprocating-Screw type cylinder, as shown in Fig. 4.9(b), then it is called the Screw-type two-stage plasticizor injection moulding machine.

Apart from the fact that this machine can produce larger components, the use of screw conveyor results in the following advantages.

1. Better mixing and shearing action of the plastic melt,

2. a broader range of stiffer flow and heat-sensitive materials can be run,
3. colour changes can be handled in a shorter time, and
4. lower stresses are developed in the moulded part.

Injection moulding is much faster than compression moulding. The mould is maintained at a constant temperature, usually 75 to 95°C (165 to 200°F) by circulating water and a production cycle of 2 to 6 shots per minute is possible. Mould costs are lower since fewer cavities are necessary to maintain equivalent production rate. Articles of difficult shapes and of thin walls are successfully produced. Metal inserts, such as bearings, contacts, or screws can be placed in the mould and cast integrally with the product. Material loss in the process is low, as sprues and gates can be reused. Although the capacities of injection moulding machines vary from 0.06 to 3.6 kg, small-parts machines of 0.22 to 0.45 kg capacity are most popular.

Injection moulding is a low cost method of producing thermoplastic parts in large quantities. With fully automated equipment, 3 to 6 shots a minute are common for moderate-sized parts. Intricate parts can be made to close tolerances, with no need for further processing. Scrap loss is generally less than 10%. As few as 500 of a kind have been economically injection moulded, but as many as 10,000 are required in some cases for economical operation.

A major advantage of injection moulding is the incorporation of fine details such as bosses, locating pins, mounting holes, bushing, ribs, flanges, etc. which normally eliminates assembly and finishing operations.

Applications of injection moulding vary over a wide range from tiny watch gears to medium-sized parts like buckets, bowls, bottle crates, telephone handsets, large products including automotive dashboard frames, garden furniture, small boat hulls, etc. Other large products include laundry buckets, refrigerator parts, large garbage disposal bins, etc.

4.1.4 Bi-Injection Moulding

Bi-Injection moulding or *Two colour Injection Moulding* is used when two distinct polymeric materials are involved in the moulding of a single part. The two materials may just have different colours, one may contain a reinforcement or gas cell like in cellular material or foam or they may be entirely different to provide different properties in different locations (e.g. strength and low friction, rigidity and flexibility). The process requires two distinct feed units and a single mould unit.

Bi-Injection moulding is accomplished by first moulding the basic case or shell and then, using that as an insert, moulding numerals, letters and designs in and around the insert. There are two methods of Bi-Injection moulding, as shown in Fig. 4.10. Fig. 4.10(a) shows the method of first moulding a shell and then transferring it to another mould with a different cavity and core pin. The second colour is injected into the shell and around the second core pin. In the second method, shown in Fig. 4.10(b), the inner plug is moulded first and then transferred to another cavity where the second colour is injected around the plug.

In this type of moulding, the mould design is very critical and gating is specialized. Typical parts made by this process include caps for typewriter and business machine keyboards, automotive taillight lenses, push buttons, and telephone dial components.

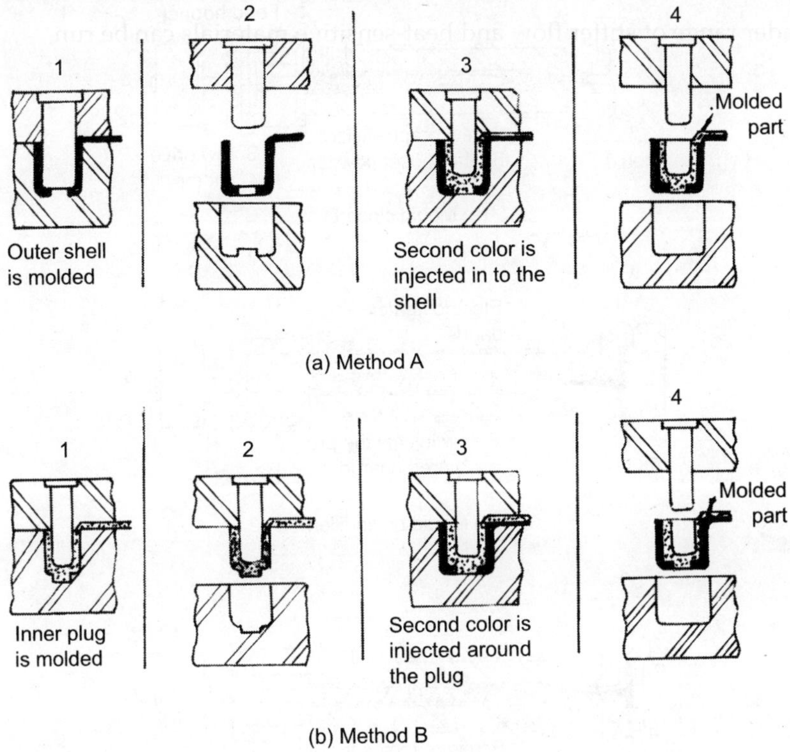

Fig 4.10: Steps involved in Bi-injector or Two-Colour Moulding by two different methods.

4.1.5 Injection Moulding of Thermosetting Resins

Thermosetting resin systems, such as phenolics (PF) or unsaturated polyesters (UP), often used with fillers or reinforcements are increasingly injection moulded at relatively high speeds. Two types of machines are used for the purpose. These are:

1. Screw Injection Moulding Machine
2. Ram Injection Moulding Machine

Brief description of the two processes follows.

1. Screw Injection Moulding

In this process, loose thermosetting powder is heated, softened, and compacted by a reciprocating screw. Heat is generated by the electric heater bands around the barrel as well as by friction caused by the rotation of the screw conveyor. The barrel temperature ranges from 55 to 115°C (130 to 240°F). The mould temperature varies from 120-175°C (250 to 350°F). The injection pressure ranges from 80-135MPa (12,000 to 20,000 psi). As the screw rotates, the material travels forward along the flights of the screw and is thoroughly plasticized by a mechanical shearing action. The material builds up infront of the screw and pushes it backwards. The screw is then used as ram to inject the material into the mould. After the material in the mould has polymerized, the mould is opened and the part is ejected. Fig. 4.11 shows the three steps of the process.

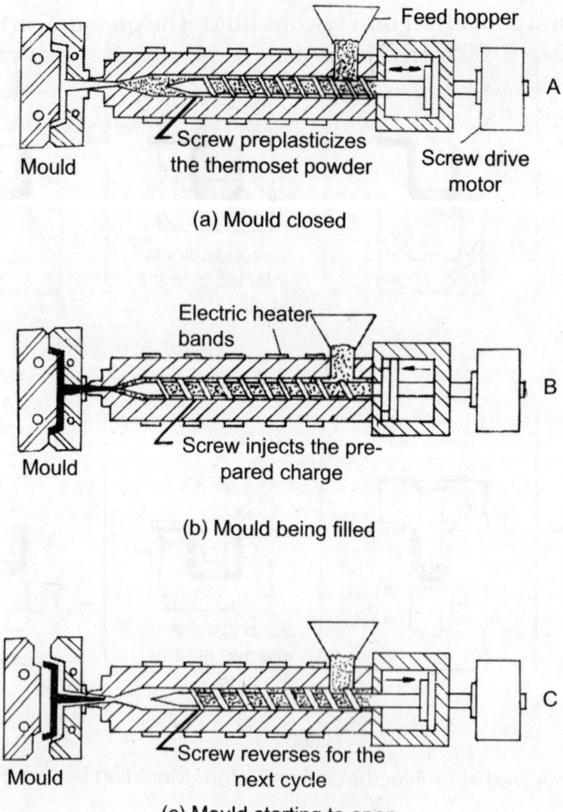

Fig 4.11: Three stages in Screw Injection Moulding of thermosetting resin.

The sprue length should be kept as short as possible. The *Sprue break* in regular injection moulding of thermoplastics is at the end of the nozzle while for thermosets it is at the beginning of the nozzle as is shown in Fig. 4.12. The tappered nozzle and sprue openings must be cleared after each shot.

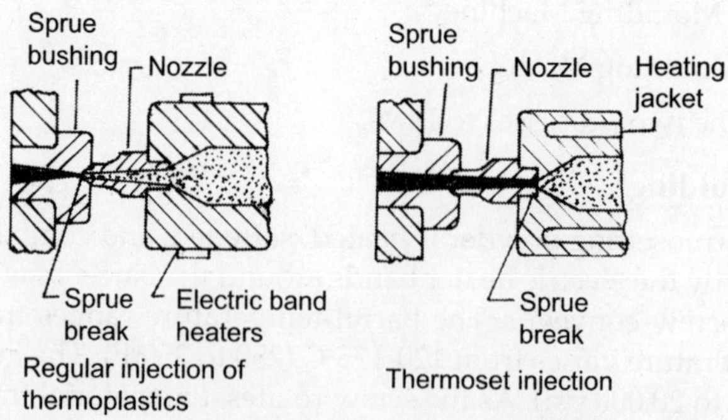

Fig 4.12: The position of sprue break vis-a-vis injection nozzle for injection moulding of; (a) thermoplastics, and (b) thermosets.

2. Ram Injection Moulding

In Ram injection moulding the thermoset moulding compound, in granular form, is pre-weighed and preheated before it is fed into the heated injection barrel/cylinder. The barrel temperature ranges between 80-120°C (175 and 250°F). As the ram moves forward, it compresses the preheated thermosetting compound and converts it into viscous fluid. The pressure on the material may range from 140 to 180 MPa (20,000 to 40,000 psi). The viscous fluid is injected into a mould that is heated to 120-150°C (250-350°F). After the material is plasticized or set, the sprue is automatically removed, and the mould is opened and the part is ejected. Fig. 4.13 shows the different steps of operation for such an injection moulding machine, which is also sometimes referred to as *horizontal transfer moulding*.

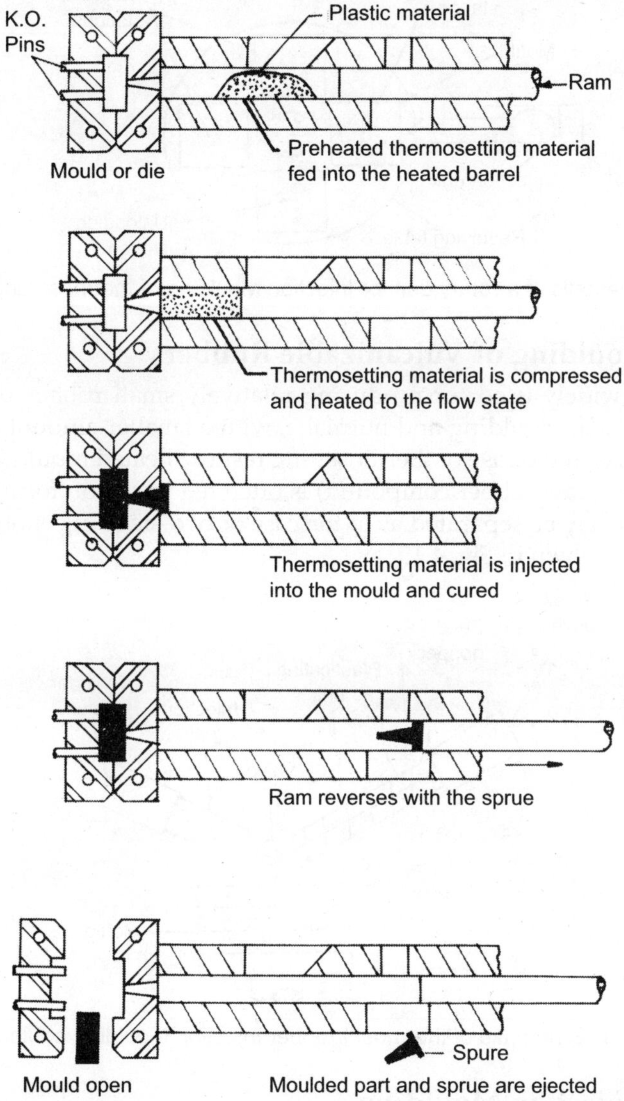

Fig 4.13: Different steps in Ram-Injection Moulding of thermosetting resins.

The size and location of the sprue, runners, gates and cavities must be determined for each type of thermosetting resin that is to be injection moulded. This moulding process is much faster than *standard transfer moulding* and it shows a decrease in material loss due to spillage, preform damage, and dirt getting into the powder/moulding compound.

Thermosetting polymers cure or polymerize in the mould and are discharged hot with no cooling required, while thermoplastic parts must be cooled in the mould to become rigid enough to be removed without distortion. However, because curing of thermosets involves chemical reactions so generally takes much longer than the injection operation. Multi-mode machines are thus often used with shuttle or rotary systems to achieve the desired result. Fig. 4.14 shows a schematic of a rotary system having four stations, each featuring a clamping mechanism; injection is followed by curing and ejection.

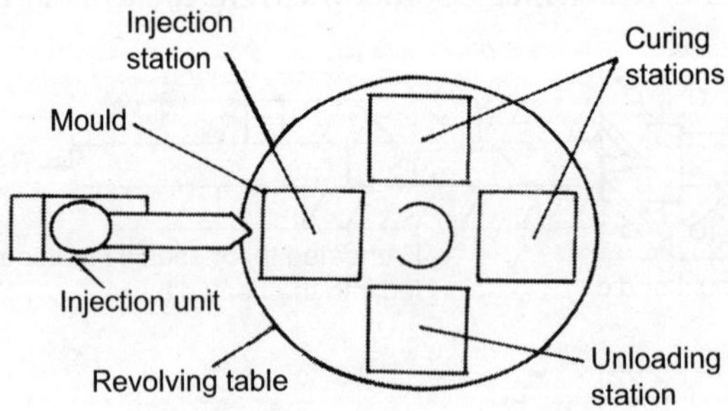

Fig 4.14: Schematic of a rotary unit for Injection Moulding of thermosetting polymers.

4.1.6 Injection Moulding of Vulcanizable Rubbers

Injection moulding is widely used for producing relatively small rubber products significantly faster than by compression moulding and normally, with a smaller amount of scrap and a better dimensional accuracy. As in the case of thermosetting resins a heated mould is needed for curing/vulcanization. The stock (raw rubber compound) is often fed in ribbon form, and the plasticating and injection functions may be separated as in the case of two-stage injection moulding machines used for thermoplastics shown in Fig. 4.15.

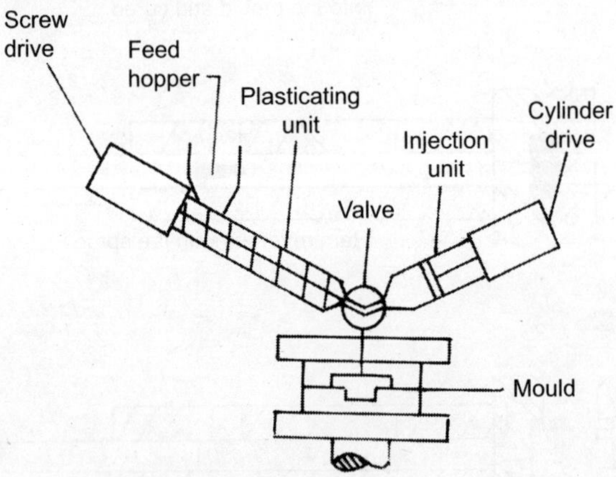

Fig 4.15: Schematic of thermoset/rubber Injection Moulding equipment.

4.1.7 Reaction - Injection Moulding

Reaction-Injection Moulding (RIM) is a later development in injection moulding that involves rapid mixing, in precise proportions, of two or more highly reactive low molecular weight liquid components and the immediate injection of the mixture in the closed mould. Polymerization and molecular network formation take place in the mould in a very short time (typically around 30 seconds) yielding a solid product in a total cycle time of the order of 1 to 2 minutes. The RIM process was developed for urethane resins and is still used primarily with these polymers. This process is particularly suited to the production of large, relatively thin parts because the capital investment and operating costs are much less than those for injection moulding of thermoplastics. The process is energy efficient but requires good control of complex reactions.

Equipment: RIM process equipment involves two or even more components, each one corresponding to a stream up to the mixing heat. Each stream features several major elements, as shown schematically in Fig. 4.16, for a two component unit.

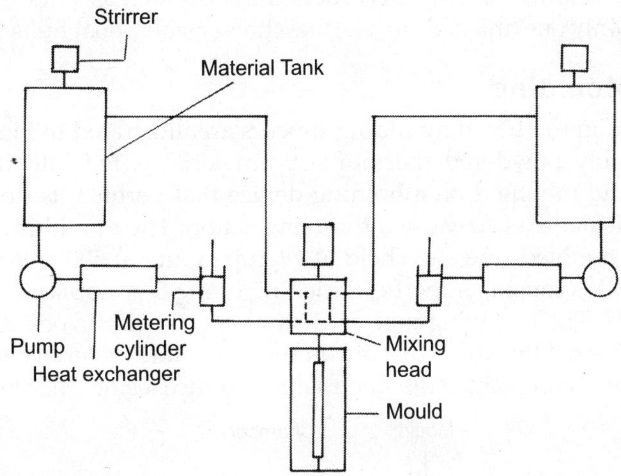

Fig 4.16: Schematic of double component Reaction Injection Moulding unit.

Suitable amount of each material is stored in tanks of 25 litre capacity which is kept under a pressure of about 110 psi (0.75 MPa). The tanks are given a shield of nitrogen gas to protect the chemicals from oxygen or moisture, and a vortex-free stirrer particularly for materials containing filler(s).

A high pressure 30 MPa (4400 psi) pump is used for circulating the liquid component. A heat exchanger is incorporated after the pump to control the temperature of each liquid component.

A cylinder-piston unit of 2 to 40 litre capacity is used for the rapid and accurate measured dispensing of the liquid material, through a servo-hydraulic drive system.

The mixing head of self-cleaning type accomplishes rapid and thorough mixing of liquids coming from different streams before injecting the mixture into the mould.

The two halves of the mould are mounted on the moulding press platens which often allow rotation or tilting or angle/alligator opening of the mould. Platen sizes can be 4′×9′($\simeq$1.20 m×2.74 m). Because of low in-mould pressure of 50-100 psi (0.4-0.8 MPa), a moderate clamping force of 300 tons (3000 KN) is found to be adequate. Some systems incorporate shuttle or rotory-type mould carriers to increase the production rate.

Polyurethane resins are processed at low temperatures around 65°C (150°F), but some resin systems like polyamides (nylons) require much higher processing temperatures necessitating the use of jacketed tanks and pipelines to conserve heat.

Tooling: Because of the low temperatures and pressures involved in processing polyurethanes, the mould material used may be epoxy, nickel-coated epoxy, kirksite, aluminium, and steel, depending upon the number of parts required. The overall injection rates can be as high as 5 kg/sec.

Product Materials: *Urethane* systems are most widely used. They are normally processed at near ambient temperatures.

Urea resins with superior temperature-resistance are moulded for parts like auto-motive body components.

Polyamides (nylons) are processed at temperatures above 85°C (185°F) and the mould temperature must be over 140°C (285°F).

Epoxy systems are generally used as an impregnating matrix for continuous glass reinforcements.

Unsaturated Polyester and *acrylic* systems are also reportedly processed to a limited extent.

Applications: Because of the low viscosity fluids and low mould pressures involved, the RIM process is particularly well suited to the fabrication of large thin parts, which include exterior automotive body panels such as front-end instrument panel and fenders, as well as furniture, electronic cabinets, and window frames. This process has also been used for the rapid fabrication of large parts that require only one finished surface like shower stalls, boat hulls, camper canopies, etc.

4.1.8 Rotational Moulding

Different steps involved in rotational moulding process are illustrated in Fig. 4.17. At the loading stage a charge of suitably powdered thermoplastic material is fed into the open mould. The mould is then closed and mounted on a holding device that permits its double rotation around two mutually perpendicular axes to cause a tumbling action. The mould is then heated in special chambers, while being tumbled, and then held at a temperature of 500 to 900°F (260-480°C). Heat penetrates the mould walls causing superficial melting of the thermoplastic material causing it to stick to the mould walls. The revolving motion distributes the plastic powder uniformly to form a hollow object of the desired thickness by consuming the whole material charge. Further heating may help consolidate and homogenize the material to produce a smoother inner surface.

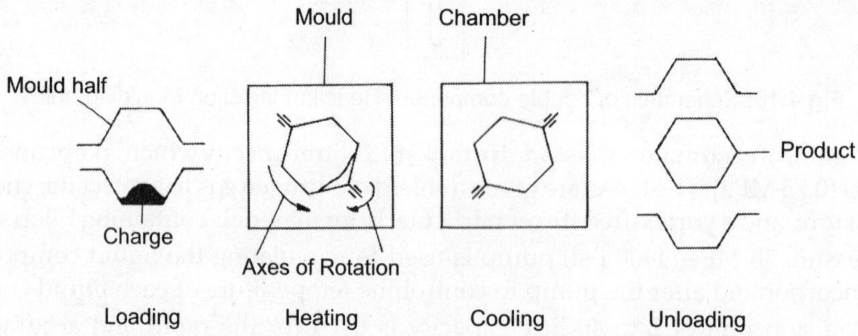

Fig 4.17: Different steps involved in Rotational Moulding Process.

Then follows the cooling stage which results in rapid cooling of the mould walls to cause solidification of polymeric material allowing it to be unloaded. The process is characterized by relatively long thermal cycles.

For rotational moulding of small parts cast aluminium moulds of about 6 mm thickness are used. However, large moulds are generally fabricated from steel or aluminium sheet. Electroformed copper-nickel, Beryllium-copper and epoxy-based nickel moulds are also used.

Materials: The thermoplastic polymers used for moulding are required to be in fine powder form in the range of 200 to 500 μm, with 250-400 μm range being more popular.

The thermoplastic materials, in powder form, used in rotational moulding include polyethylene particularly in its high density form (HDPE). Polyproplylene (PP) and acetal (PMO) are among other crystallizing thermoplastics occasionally used.Amorphous thermoplastics moulded by this process include HIPS (High Impact Strength Polystyrene), ABS, PVC, polycarbonate (PC), cellulosics (CAB), and thermoplastic polyurethane (PU). EVA (ethylene vinyl acetate) copolymers are used for flexible items. Thermosetting resins such as epoxies (EP) which melt before curing are widely used not so much to make parts, but for the internal coating of metallic vessels, pipes and other structures which serve as moulds.

Advantages claimed for rotational moulding include low initial investment, a flexibility that allows a variety of parts to be made on the same equipment, low tooling costs, totally enclosed and open-end pieces, fine details, excellent surface finish, and low cost. The same equipment can be used for either thermoplastic powder or plastisol moulding.

Applications: Some typical parts made by rotational moulding include drums and tanks (chemical storage, fuel, septic tanks, etc.), pipes and ducts, bins, road markers and signs, automotive body components for trucks, tractors, trailers, furniture (children's chairs), luggage shells, boat hulls, fishing floats, rocket case liners, squeeze toys, fashion models (mannequins), hobby horses, machinery guards, garbage containers, etc. In fact, there is hardly any shape or size that cannot be rotationally moulded.

4.1.9 Casting

In general casting of polymers involves pouring the liquid polymeric resin into a mould under normal atmospheric pressure, also called *gravity pouring*. Thus, plastics that are cast are resins available in liquid form such as phenolics, polyesters, silicones, and acrylics; and thermoplastics that can be softened enough to pour, such as some cellulosics. Casting of polymers generally involve minimal stresses due to gravity pouring, slow chemical reactions, moderate temperature, etc. Because of the low stresses involved, light built open moulds or two-piece moulds may be used, as are shown in Fig. 4.18. The mould material used is lead, rubber, glass, plaster, etc.

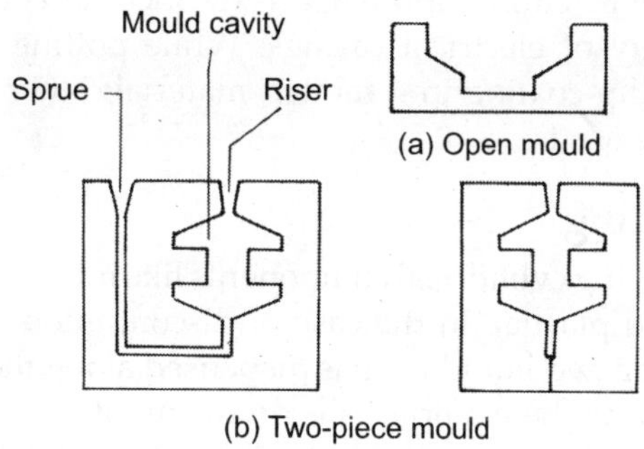

Fig 4.18: Types of casting moulds for plastics, (a) Open mould, and (b) two-piece mould.

Plastics can be cast in many shapes. A model is made of wood, plaster, porcelain, metal, etc. and a mould is formed around it. This may be a simple draw, split, cored, or flexible mould depending on the size and shape of the part and the kind of plastic. The equipment is not expensive, but the process is slow.

A catalyst is added to unpolymerized resins, and they are heated in an oven while in the mould for hours or days at about 150 to 250°F (65-120°C) to harden. Acrylic tubes are normally made by centrifugal casting in spinning cylindrical moulds with tolerances held to ±0.125 mm; the free inner surface allows unhindered shrinkage.

In some cases, the chemical reaction taking place during a casting process converts a low molecular weight monomer into a high molecular weight thermoplastic, the most common examples are acrylics such as PMMA (poly methyl methacrylate) and polyamides or nylons. In other cases, polymerization and cross-linking take place simultaneously in the casting process, leading to themosets; examples include polyurethane resins (UR), unsaturated polyester resins (UP), epoxy resins (EP), and silicone resins (SI).

Plastics are cast when the number of parts required is not enough to justify the making of expensive dies. Frequently open moulds of lead are formed by dipping a steel mandrel of special shape into molten lead and stripping the shell from the sides of the mandrel after it solidifies. Cores of lead, plaster, or rubber may be introduced if desired. Solid objects may be made from moulds of plaster, glass, wood, or metal. When parts have numerous undercuts, the moulds are made of synthetic rubber.

Applications: Casting is commonly done to make ornaments, prototypes, and dies, and to pot or encapsulate electrical parts.

Many polyurethane products are made by casting. They include ski boots or hockey skate shells, as well as roller skate or skate board wheels.

Marble-like plastic furniture components like sinks, vanity tops, etc. are conveniently made by casting a viscous resin system, to which a liquid colourant is added to form appropriate streaks.

Polyamide (nylon) cast products are generally limited to prototypes, limited runs or large, bulky parts; they include journal and slide bearings, gears etc.

Allyl resins are especially useful for optical lenses and other applications requiring excellent clarity. These resins have a wider use in casting than the thermoplastics as they have greater fluidity in pouring. Ethyl cellulose and CAB (cellulose acetate butyrate), needed for drop hammer and stretch dies. Acrylics are used in the casting of transparent articles and flat sheets.

Potting and Embedding

It is sometimes desirable to protect electrical or mechanical parts in blocks of plastics. *Potting* refers to the insulation and positioning of such parts, while embedding (encapsulation) is the enclosing of parts in a transparent plastic for presentation and display. Encapsulated products are often associated with the electrical-electronic industry. They include coils, transformers, connectors, terminal blocks, and a variety of electrical casings. While potting and embedding are usually accomplished by casting or dip-coating in a molten material; other processing methods such as transfer moulding can also be used.

4.1.10 Centrifugal Casting

This process is used for producing cylindrical components like tubing, pipes and tank shells, usually made of FRP (fibre-reinforced plastic). In the case of discontinuous fibre reinforcement a mix of chopped fibres and pre-crystalyzed liquid resin is dispensed along the axis of a rotating cylindrical mould, as shown in Fig. 4.19. As the material falls on the mould surface it is spread uniformly by centrifugal forces, also keeping it in place during cooling. Desired thickness is achieved by feeding the needed amount along the length of the cylinder. The final inner surface produced is reasonably smooth though not as good as the outer surface. Shaped features like flanges, threads, ribs, etc. can be formed provided the mould is made of suitable sections to facilitate its extraction. The process can be modified to allow continuous pipe production.

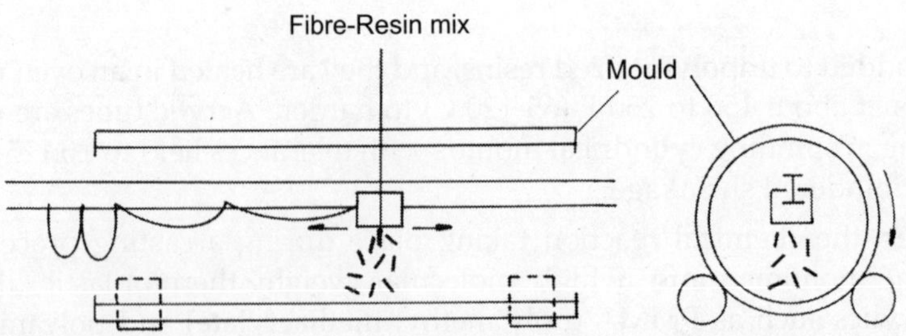

Fig 4.19: Set-up for centrifugal casting of fibre-reinforced plastic (FRP).

In a variant of the process, large pipes have been made that involve the successive formation of an external FRP layer, a rigid foam core layer, and an internal FRP layer. Such pipes have outstanding insulation and crushing properties, in addition to pressure resistance.

4.1.11 Hand Lay-up and Spray Process

The oldest and the simplest method of processing polymers involves the mannual placement of mat or fabric reinforcement and batch-mixed resin on the open mould surface; and is referred to as *hand lay-up* or *hand laminating*.

It is often necessary to place several successive layers of reinforcement, each layer being compacted with the help of brushes, rollers (flat or finned), and squeezes (i.e. rubber blades). The method is well suited to the incorporation of other elements like foam cores, and fasteners; but it is very labour intensive. Moulds for hand lay-up are often made of reinforced plastic when room temperature cure is used. Prepregs (pre-impregnated reinforcements) can also be placed manually, but heat cure is then required.

In *spray process,* chopped glass fibres are incorporated in a spray of catalysed resin applied to the mould by spray gun. Faster production rates are achieved as compared to hand lay-up method. Compaction is often required between successive passes. Spraying and compaction can be automated for relatively rapid production rates, particularly in the case of simple convex shapes such as cylindrical tanks.

4.1.12 Shell Moulding

Powder moulding of hollow parts is done by placing powdered resin in a heated mould until a layer of desired thickness gets deposited on the mould walls. The unfused resin inside the mould is dumped out of the mould and the hollow moulding removed. Polyethylene, polypropylene, and vinyls may be used. Powder shell moulding produces thin-walled parts like toys, tanks, and other containers. Walls are stress-free and crack-resistant because there is no moulding pressure and no flow of material. Wall thickness is uniformly controllable within a tolerance limit of ±10%. Operation time is more but mould costs less than for injection moulding. The break-even point is reported at around 10,000 parts.

4.1.13 Casting of Plastisols

The methods popularly used for casting plastisols are:

1. Slush Casting
2. Dip Casting

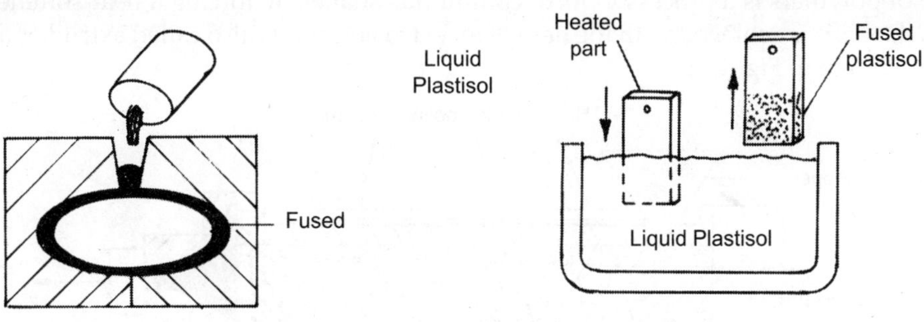

(a) Plastisol slush casting (b) Plastisol dipping

Fig 4.20: Casting of plastisol by, (a) Slush Casting, and (b) Dip Casting.

1. Slush Casting

This process uses a liquid material charge called a *plastisol* which is a suspension of finely divided particles of a thermoplastic resin, usually PVC (polyvinyl chloride) in a plasticizer such as DOP (di-2-ethyl hexyl (octyl) phthalate). The liquid plastisol is poured into the heated, 175°C (350°F), metal split-cavity mould as shown in Fig. 4.20(a). The mould cavity produces the outside of the object to be made. The heating of plastisol causes gelling that is, the irreversible combination of plasticizer and resin to form a homogeneous layer of plasticized PVC (PPVC). As in the case of powder shell moulding heat transfer through the forming PPVC layer determines the rate of retention of plastisol and thus the product thickness. The excess plastisol, still in the liquid form, is poured out from the mould. The cast plastic part and the mould are cured in an oven at 175°C ($\simeq$ 350°F). Upon cooling, the mould is opened and the casting removed.

2. Dip Casting

In this process a pattern, usually made of metal, is heated to 175°C ($\simeq$ 350°F) and then dipped or submerged into a container of liquid plastisol. The plastisol in immediate contact with the heated metal pattern gets deposited on it by fusing as shown in Fig. 4.20(b). The pattern with the solidified plastisol is removed from the container and cured in an oven at about 175°C ($\simeq$ 350°F). After curing, the fused casting is stripped from the mould.

4.2 Forming Processes

In these processes the polymer is heated to give it shape by one method or the other. Though no melting is involved, still the term mould is often used for some processes. In fact, in the processing of polymers no clear demarcation is made in categorising the processes used for shaping them. In this text an attempt is made to categorise the processing of polymers on the basis similar to those adopted for categorising the processes used for shaping metals.

In this class are included the following processes:

1. Extrusion
2. Extrusion Coatings
3. Blow Moulding
4. Film and Sheet forming
5. Thermoforming

Brief description of these processes follows in this section.

4.2.1 Extrusion

Extrusion of polymers is a process to form continuous shapes by forcing a heat-softened material through a die. This is achieved with the help of an extrusion press, also called extruder; a schematic of which is shown in Fig. 4.21.

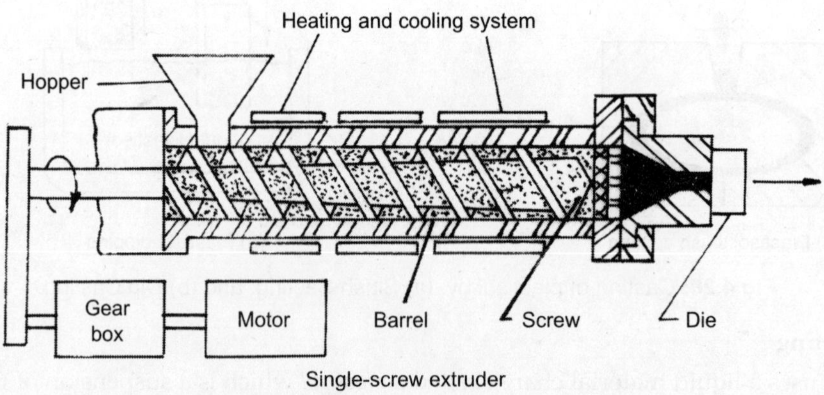

Fig 4.21: Set-up for Single-Screw Extrusion of plastics.

The core of the screw unit is the screw conveyor that consists of three sections, viz., the feed and compression zone, the plasticizing zone and the metering section.

Powder or granular material is fed from the hopper, through the throat, and on to the spiral screw. Rotation of the screw transports the polymer along the cylinder through the compression zone, where it is compacted, and the trapped air is expelled. Progressing down the cylinder, it is heated further by shearing action between the screw and the cylinder walls as also by conduction of heat from the cylinder walls so that by the time it reaches the third zone it becomes a softened polymer in the form of thick viscous mass in which form it is forced through the die which produces the required profiled product. As the shaped material leaves the die, it is cooled by air, water, or by contact with chilled surface and gradually hardens as it rests on the conveyor.

It is usual to include a *sizing die* as a part of the cooling train to rectify the profile of the extrudate after it leaves the die. The size of the die will, almost always, be larger than the size of the desired finished product. The extrudate is pulled away faster than it is extruded, thus causing it to *draw down* or get smaller as it leaves the die. The draw down keeps the extrusion straight and permits size adjustments. The amount of draw down is defined as *draw down ratio* and is the ratio of the size of the die opening to the final cross-sectional area of the extruded polymer.

There are a number of different kinds of extruders in use, but the *single-screw extruder* is the most common type. Many adaptions are used with the single-screw extruder, such as dual extrusion, metal embedment, and cross-head die extrusion; as shown in Fig, 4.22.

Dual Extrusion uses two extruders, as shown in Fig. 4.22(a). In this process, two polymners are combined into an integral profile. *Metal embedment* provides a means of incorporating metal stripes, wires, or roll-formed shapes in a plastic profile extrusion, Fig. 4.22(b). A *cross-head die extrusion*, Fig. 4.22(c), is used mostly for wire or cable covering. Here in the die is placed at right angles to the extruder.

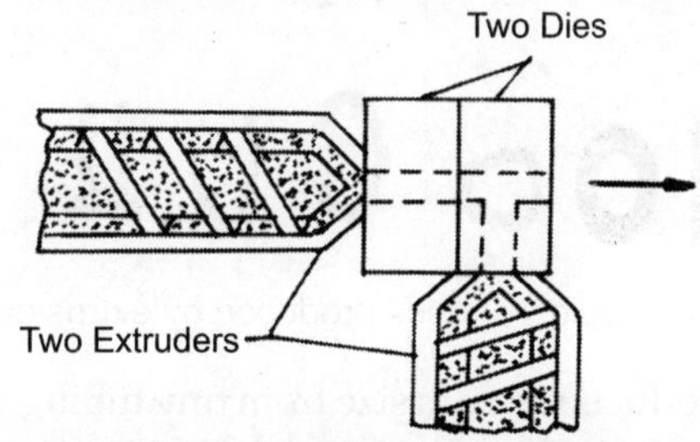

(i) Dual extrusion

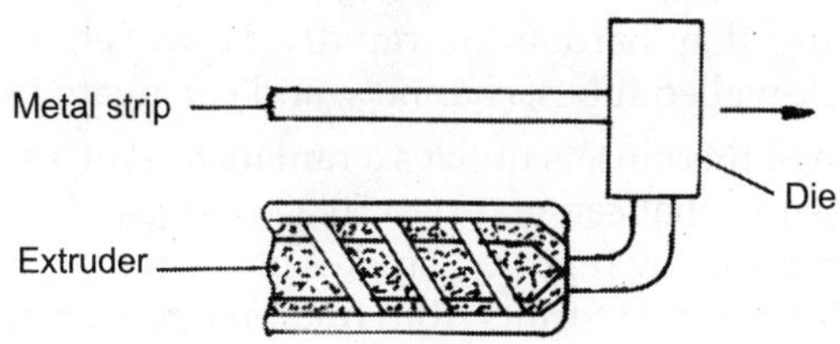

(ii) Metal embedment

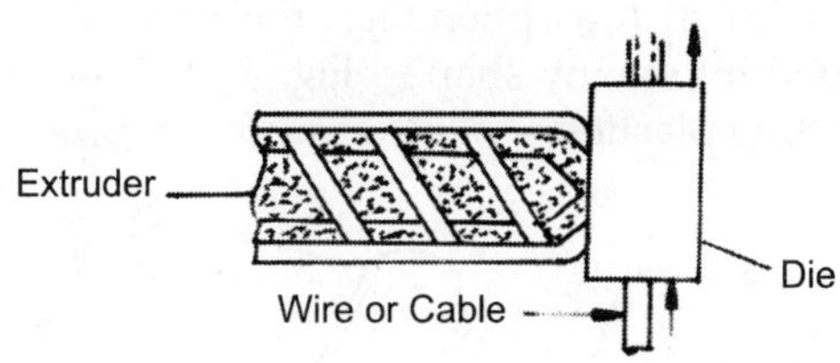

(iii) Cross-head Die Extrusion

Fig 4.22: Three adaptions used with single-Screw extruders for plastics.

The size of the extruder is given by the diameter of its screw conveyor and that may vary between 40 mm and 160 mm.

Materials: Practically all thermoplastics can be processed by extrusion, but the more popularly used polymers are polyethylene (PE), polypropylene (PP), polyvinyl chloride (PVC), polytetra fluoroethylene (PTFE), UHMWPE (ultra high molecular weight polyethylene), polyamides (nylons), acrylics, polyesters, vinylics, styrenic and olefine resins.

Applications: Common products manufactured by extrusion process are pipes, curtain rails, flat and tubular films, sheet, wire coverings and fibres. Other applications include production of hand rail coverings filaments, film and sections with a wide variety of profiles, as shown in Fig. 4.23.

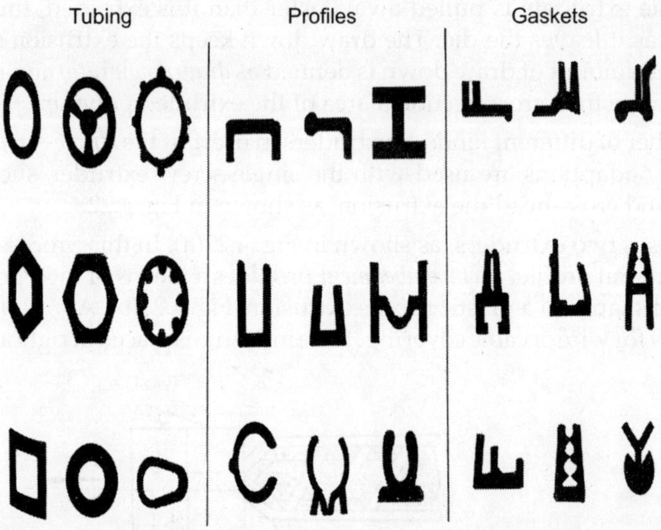

Fig 4.23: Some typical shapes produced by extrusion of thermoplastic polymers.

Extruded tubular products range in size from tiny tubing, used in artificial kidneys and drinking straw to pipes as large as 1.6 m in diameter and 2 cm thick made of HDPE. Polyvinyl chloride and polyethylene are widely used as tubing or pipe resins.

Extrusion of Thermosetting Plastics: Thermosetting compounds are not well adapted to extrusion by screw type extruders because they harden too rapidly. However, they are used to a limited extent in the production of thick-walled tubes, channels, and corrugated sheeting.

A machine for the extrusion of thermosets utilizes a ram instead of a screw to force the material through the die involving pressures that can go up to 50,000 psi (345 MPa). Material is fed from a hopper at the rear of the cylinder and, by repeated strokes of the ram, is forced into a long tapered die that has heated zones. Additional heat results from frictional resistance as the material is forced through the cylinder and die. Curing is complete as it reaches the forward end.

Applications: Thermosetting plastics such as phenolics, melamines, epoxies, diallyl phthalates, polyesters and alkyds are extruded into many shapes. Fig. 4.24 shows some of the typical cross-sections extruded from thermosetting plastics.

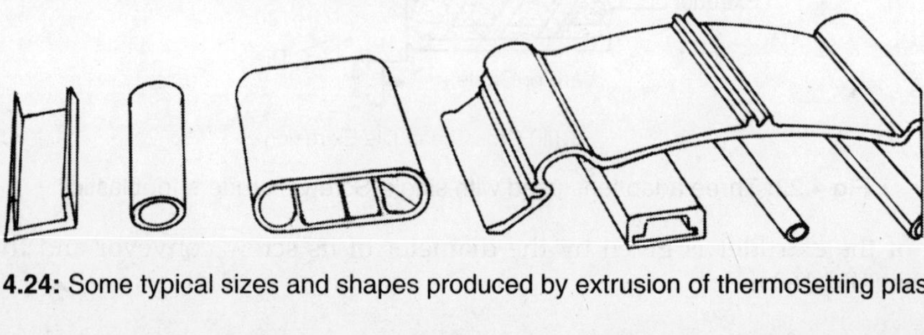

Fig 4.24: Some typical sizes and shapes produced by extrusion of thermosetting plastics.

Products include tubes, rods, mouldings, bearings, brake lining, and gears. Cross-sectional tolerances of 0.5% can be maintained. Fig. 4.24 shows some extruded shapes of plastics.

4.2.2 Extrusion Coatings/Covering

A process known as extrusion coating is used extensively for coating paper, fabrics, and metal foils. A thermoplastic material is extruded through a flat die on to sheet passing beneath it, as shown in Fig. 4.25. The extruded plastic (extrudate), while soft, blends onto the surface and is pressed by a rubber roll that holds it against the water-cooled steel roll at a desired pressure. The edges of the sheet are trimmed prior to its spooling in the wind up roll. Although any thermoplastic material can be extruded as a coating, but those used most are the vinyls, polyethylene and polypropylene.

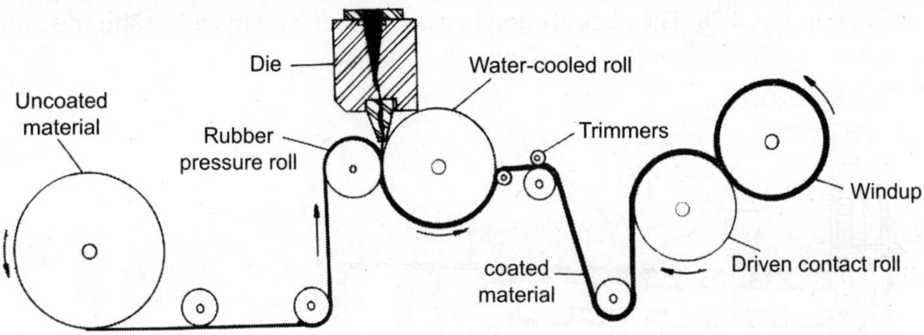

Fig 4.25: Schematic of Extrusion Coating process.

A variant of the process called *Extrusion Covering* is most commonly used for the formation of an insulating layer or a protective jacket on power or communication electrical wires and cables. A block diagram for the process is shown in Fig. 4.26. The polymer is heated typically to about 100°C (212°F) before it is extruded on to the wire/cable by passing through the die. Instruments for the continuous testing of dimensions or properties are often incorporated before the product is spooled on to a reel. Production speeds as high as 30,000 m/min (10,000 ft/min) are reported.

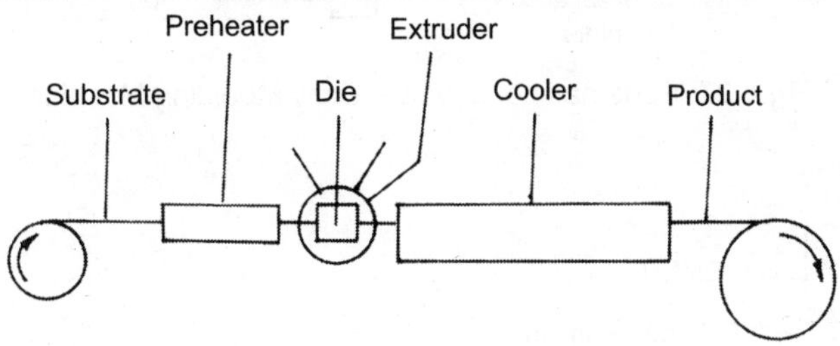

Fig 4.26: Block diagram for a set-up for Extrusion Covering Process.

Plasticized PVC (PPVC) and polyethylene (PE) are the most widely used thermoplastic covering materials, but other plastics such as polyamides, fluoropolymers, and cellulosics are used for demanding applications in wire insulation, or where transparency is desired. For conventional rubber covering an unvulcanized cover is applied which is cured through steam, motten salt, or high frequency heating.

Wire and cable coating, sheathing and jacketing are the most common applications of extrusion covering.

Foam can also be manufactured on standard extruders, when the major problem is to control the amount of foaming on leaving the extruder. This is usually achieved with several sizing and cooling rollers. PVC foam for replacement of wood is fast becoming an important market.

4.2.3 Blow Moulding

Blow moulding is a process of shaping a preform, called parison, of a polymer into a hollow article by the use of heat and air pressure. It involves stretching and then hardening the parison against the mould walls. The process is called *Extrusion Blow Moulding* (EBM) when the parison is produced by extrusion or Injection Blow Moulding (IBM) when the parison is produced by injection moulding.

In *Extrusion Blow Moulding* the parison is extruded as rapidly as possible and positioned between the jaws of a split mould as shown in Fig. 4.27. As the mould is closed, it pinches off the parison and air is blown into it forcing the material against the mould surface. As soon as the product is cooled sufficiently to prevent distortion, the mould opens and the product is ejected from it with the pinched-off parison material at the bottom of the article and the top of the neck section, as shown in Fig. 4.28. The neck section is trimmed off by a machine and the bottom part of the article, called a *tail*, is easily removed manually.

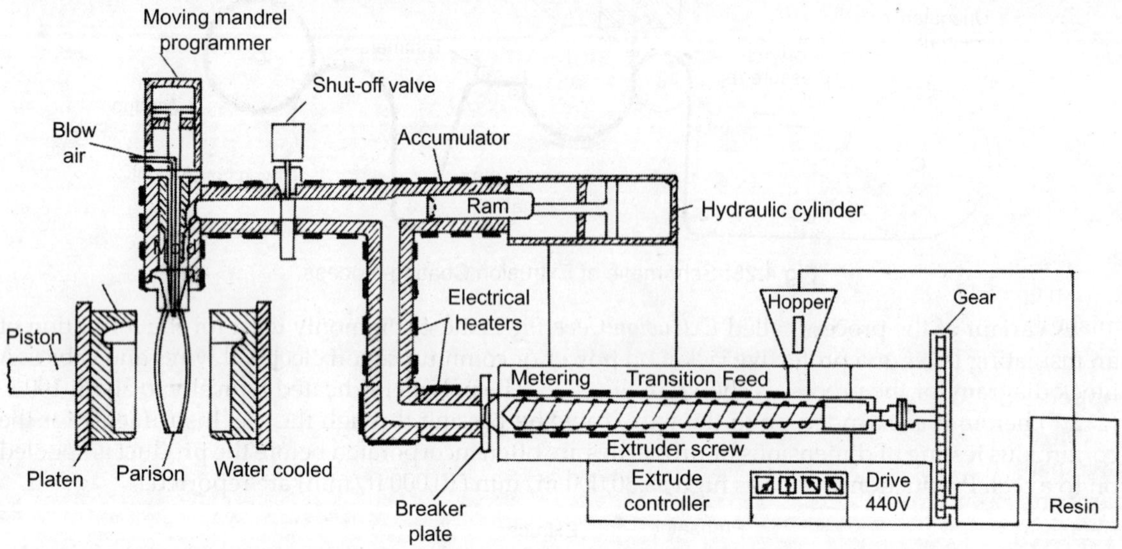

Fig 4.27: Schematic of a typical Blow Moulding Machine.

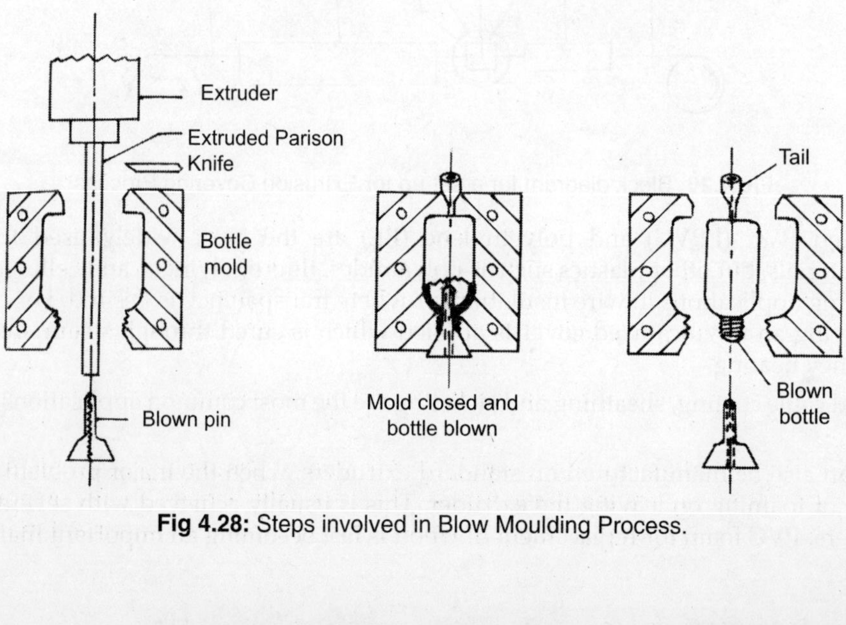

Fig 4.28: Steps involved in Blow Moulding Process.

In *Injection Blow Moulding* the parison is produced by injection moulding over a mandrel which is subsequently transferred to moulding station by indexing as is shown in Fig. 4.29. The blow mould closes, and internal air pressure through the *blow vent* forces the thermoplastic parison to the mould walls to produce the desired component.

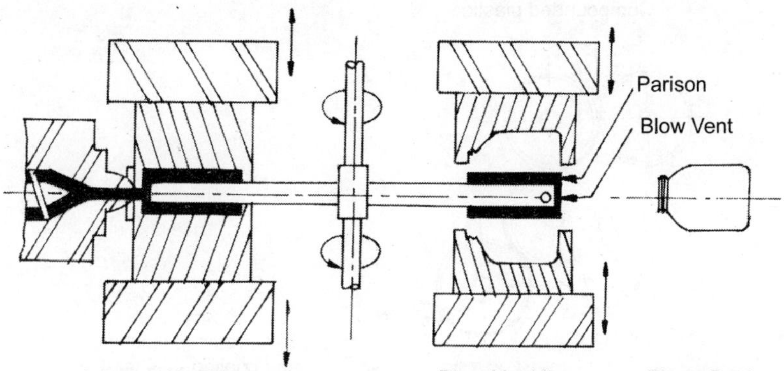

Fig 4.29: Schematic of Injection Blow Moulding Machine.

An advantage of injection blow moulding process is that it produces the part that is completely finished.

Blow moulds are made from cast aluminium alloys, zinc alloys, alloy steels, and beryllium-copper.

Most thermoplastics can be blow moulded, and objects as large as 50 gallon (≈200 litres) drums have been produced. A major drawback is that walls cannot be held uniformly because different parts are stretched by different amounts. Also, walls cannot be thickened for ribs or bosses. Blown surfaces do not have a high gloss because pressure in the mould does not exceed 100 psi (0.7 MPa).

High molecular weight polyethylene (HMWPE), polypropylene, PVC and thermoplastic polyester (PET) and cellulosic acetate are popularly used for blow moulding. Other thermoplastic resins that have been blow-moulded include PETG, Nitrile, SAN, PVDC, PPO, PC, and PA. It is estimated that the resin accounts for about 85% of the total cost of a mass produced beverage bottle.

Blow-moulded products include bottles, floats, automobile heater ducts, liquid detergent containers, hot water bottles, and cosmetic, toiletries, pharmaceutical and medical packaging. Other components produced by this process include shipping drums, stationary storage tanks, portable gas canisters, small boat hulls (produced in one piece that is split into two open hulls), toys like rattles and riding horces.

4.2.4 Film and Sheet Forming

There are four basic methods for producing film or thin sheet, viz., calendering, extrusion, blowing, and casting.

Calendering: It is the process of manufacturing sheets and films of uniform thickness by squeezing a thermoplastic polymeric material between rolls, as shown in Fig. 4.30. The material comprising resin, plasticizer, filler and colouring pigments is mixed and heated before being fed to the *calender**. Modern calenders usually consist of four rolls which are temperature controlled. Heat-softened

* A calender is a machine in which cloth, paper, plastic etc. is pressed between rolls to glaze or smooth it.

polymer is fed to the first pair of rolls from an internal mixer, a two-roll mill or an extruder; and the roll separating gap controls the feed rate. The thickness of the sheet/film produced depends upon the speed of the finishing rolls that stretch the plastic. Before the sheet/film is spooled, it passes through water-cooled rolls.

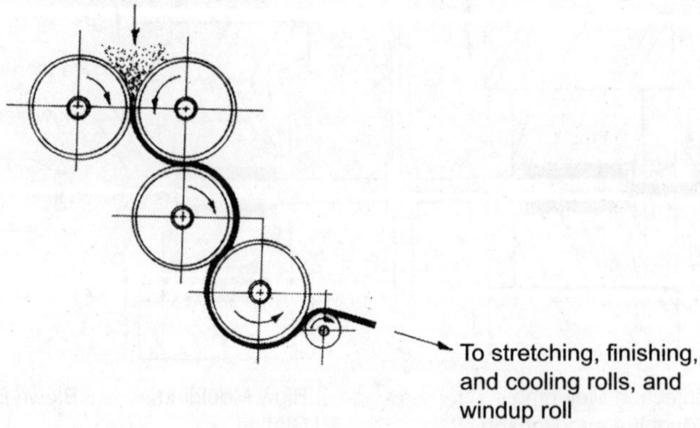

Fig 4.30: Schematic of sheet/film production by Calendering Process.

An outstanding feature of calendering is the very high output for a comparatively low energy consumption compared with extrusion; further it is particularly suitable for temperature-sensitive materials.

Calendering is used to produce film and sheet of vinyl, polyethylene and cellulose acetate as also for laminating plastic film to other substrates, including other plastic films, textiles, paper, and metal foil. Vinyl floor tiles are also produced by calendering. Also, this process is used for rolling out uncured rubber stock in tyre-retreading operations.

Extrusion Process is used for the production of sheet from polypropylene, polyethylene, polystyrene and ABS (Acrylonitrile Butadiene Styrene); Fig. 4.31 shows the essential elements of the process. After mixing the different ingredients of the material, it is placed in the feed hopper. The material is then heated to a maximum temperature of 315°C (600°F) and forced into the die area at a pressure of 2000-4000 psi (14-28MPa) by the screw conveyor. The thickness of the sheet is controlled by manipulating together the choker bar and the die opening. After extrusion the sheet passes through oil- or water-cooled, chromium-plated rolls before being cut to size. Oil cooling is recommended, since the temperature must be maintained at approximately120°C (250°F) for curing. Sheet material made in this manner can vary in thickness from 0.03 to 3.20 mm (0.001-0.125 inch).

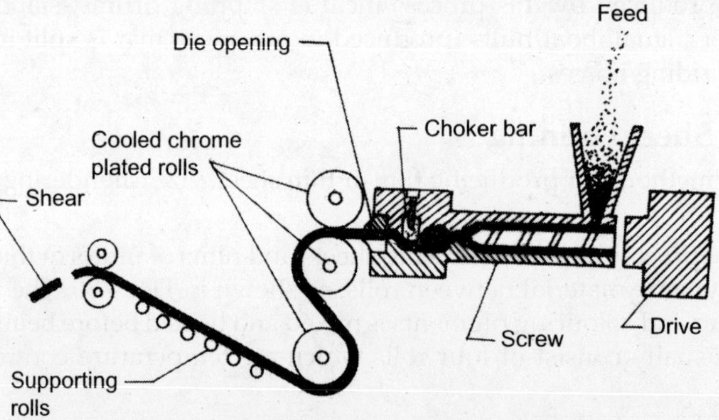

Fig 4.31: Set-up for extrusion of plastic sheet and film.

Blown Tubular Film: Most of the polymeric film is produced by this process. It involves extruding a tube vertically through a ring die and then blowing it with air into a large diameter cylinder, as shown in Fig. 4.32. It is then extended longitudinally by nip rolls running at a linear speed higher than the rate of extrusion. The nip-rolls also serve to prevent the immediate escape of the air. The flow of air into the tube is controlled to give a constant bubble size and thus a consistent film thickness. The blown cylinder is air-cooled as it rises vertically and is finally flattened by driven rolls (lay-flat) before it reaches the spooling/wind-up unit. The process permits the extrusion of thin film such as is used in trash bags and packaging materials.

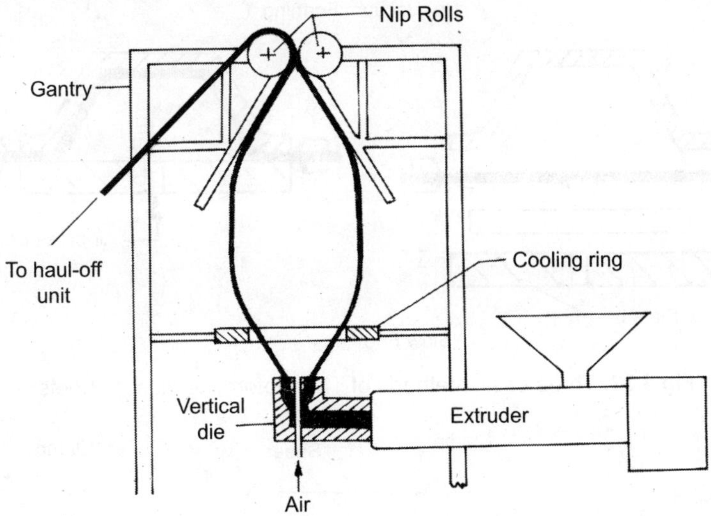

Fig 4.32: Set-up for Blown Film-Tubular Process.

In *Film Casting* the thermoplastic resins are dissolved in a solvent and spread on a polished continuous belt or a large drum, and conveyed through an oven where they are cured and the solvent removed. In *cell casting*, a cell is made up of two sheets of polished glass, separated according to the sheet thickness desired and gasketed about the edges to contain the liquid catalyzed monomer. The cell is then raised to the proper temperature in an oven, where it remains until curing takes place. Cell casting is used in the production of most acrylic transparent sheet.

4.2.5 Thermoforming

Thermoforming is a process of forming thin sheets of thermoplastic polymers into desired shape by heating and shaping using mechanical methods, vacuum, air pressure, or a combination of these, as shown in Fig. 4.33. There are several types or names given to the basic techniques for thermoforming of thermoplastic sheets. These include straight vacuum forming, free forming, drape forming, plug and ring forming, vacuum snap-back forming, plug-assist forming, and matched-die forming. Some of these methods are illustrated in Fig. 4.34.

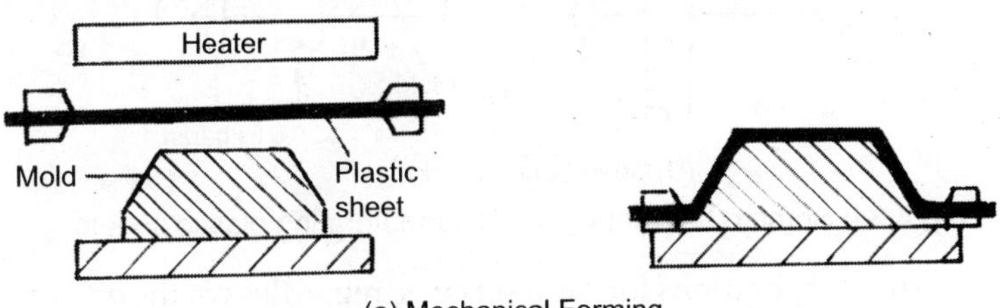

(a) Mechanical Forming

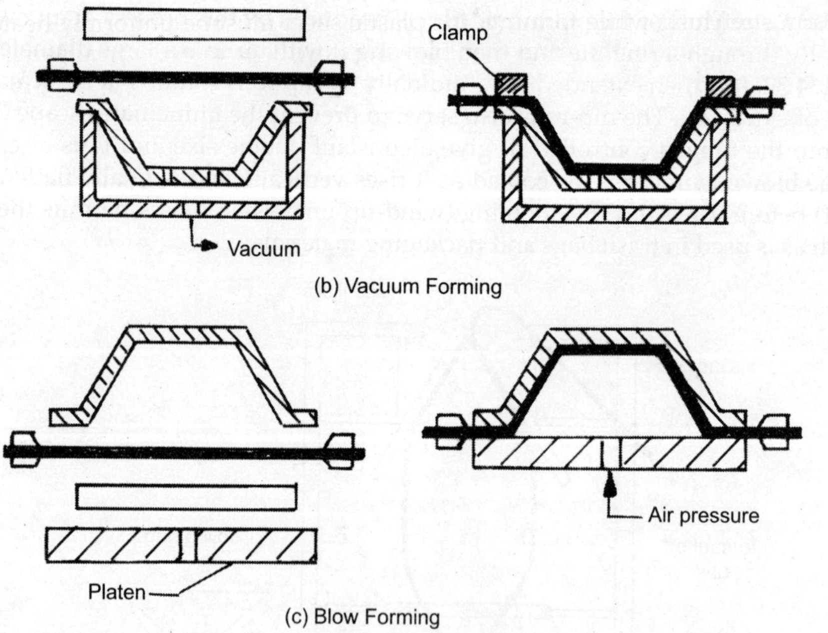

Fig 4.33: Three basic methods of Thermoforming plastic sheets.

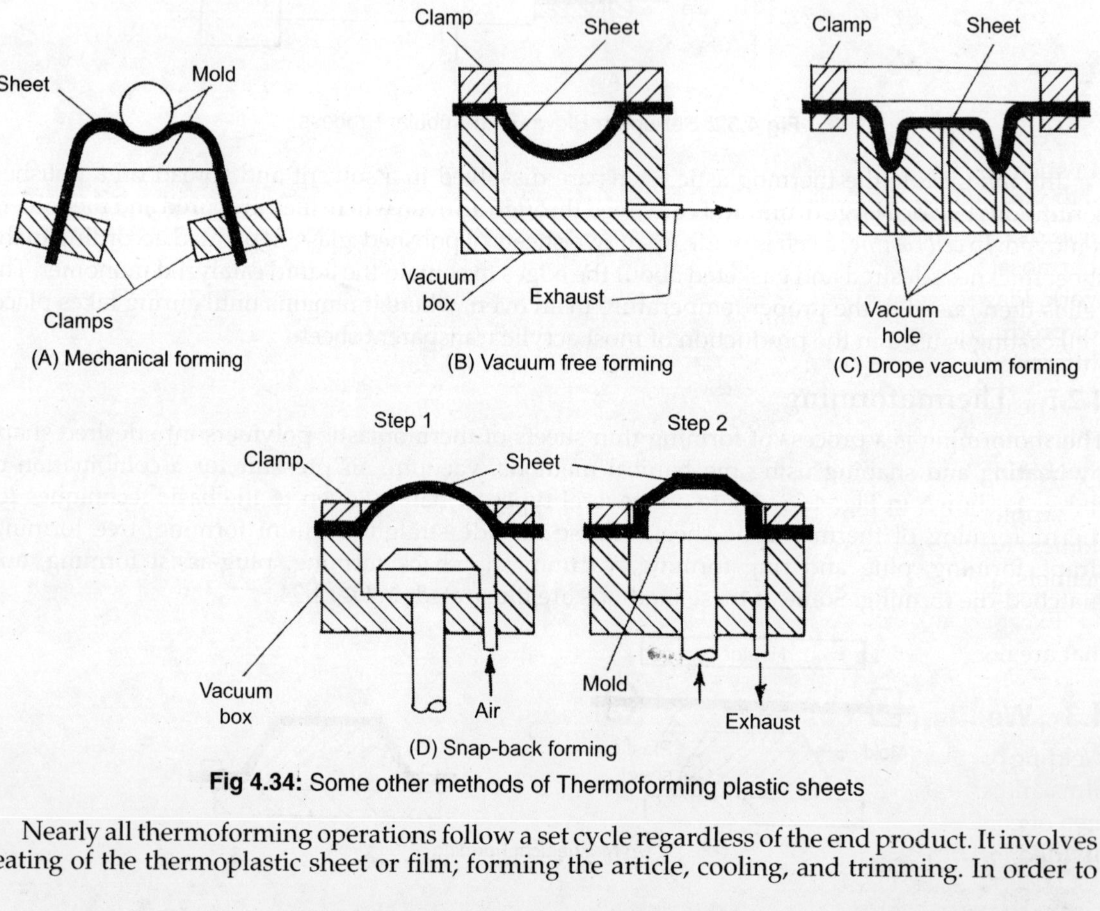

Fig 4.34: Some other methods of Thermoforming plastic sheets

Nearly all thermoforming operations follow a set cycle regardless of the end product. It involves heating of the thermoplastic sheet or film; forming the article, cooling, and trimming. In order to ensure uniform stretching while forming, the plastic sheet must be uniformly heated. The entire plastic sheet should be gradually and evenly brought up to the forming temperature. There are three modes of heating that can be used, viz., radiant heating, convective heating, and conductive heating.

The sheet is heated in an air oven or more often by infrared heaters. Single-side heating suffices for sheet up to 3 mm thickness, otherwise double-sided heating is employed. When the sheet has been softened, the component is shaped, using either a male or a female mould, with a plug frequently employed to assist drawing. In the female mould, vacuum is applied to draw the material into the cavity, as shown in Fig. 4.35(a). If the draw is deep, considerable thinning will occur at corners. However, thicker corners are obtained if a plug is used to stretch the material before the vacuum is applied, Fig. 4.35(b). Alternatively, a male mould can be used on deep drawn sections, especially if compressed air is used to expand the sheet initially, Fig. 4.35(c). The mould can then be brought into contact with stretched sheet and sheet collapsed to take up the mould shape.

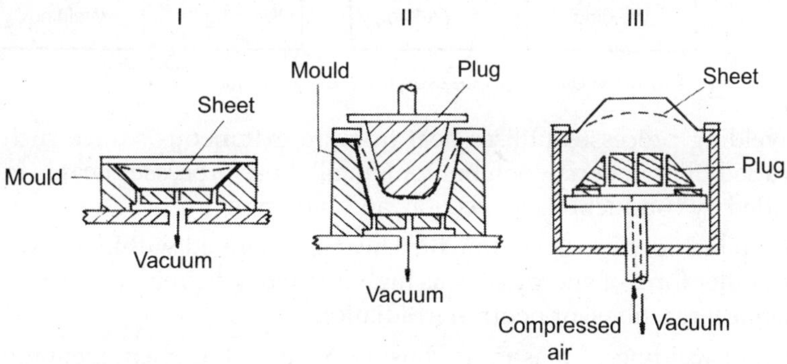

Fig 4.35: Three different techniques of Vacuum Thermoforming

A major advantage of sheet forming is that the sheets may be predecorated, coated, or preprinted for such effects as wood grains, lettered, or metallic finishes. The formed parts may have quite thin walls or large areas which are difficult to mould in other ways. Because sheet forming is not hard on moulds, a variety of materials can be used for making moulds. These include wood, plaster, epoxy, thermosetting plastics, and metals like cast iron. Wood and plastics are usually limited to prototype work. Plastic moulds are used for short runs and limited production; while metal moulds are used for production runs. The major difference between plastic and metal for production moulds is the difference in thermal conductivity of the two materials.

The thermoplastic sheet or film materials that are used in thermoforming include acrylics, ABS, polystyrene, vinyls, cellulose acetate, and cellulose acetate butyrate; however it is possible to form sheet/film of all thermoplastics.

Applications of thermoforming include door liners, signs and displays, contour maps, aircraft domes/windshield canopies, bowls and trays, refrigerator door liners, lighting fixtures, safety helmets, and luggage boxes.

Except for a limited number of cases,thermoforming of sheets is economical only for quantities that are not large, such as for prototypes or to get into production quickly.

4.3 Welding of Plastics

Welding of plastics is widely used in a number of industries particularly for joining ot thermoplastic films and sheets. All welding processes employed at present involve the application of heat to the area of contact. According to the source of heat employed the welding process for plastics may be divided into two broad classes as shown in Fig. 4.36.

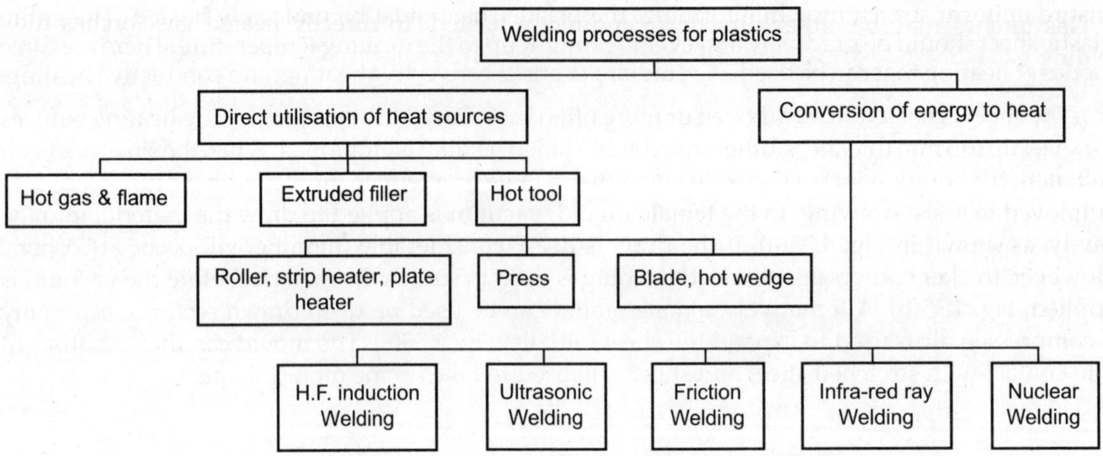

Fig 4.36: Welding processes for thermoplastic polymers.

One class of welding processes utilises heat from an extraneous source such as a stream of hot gas, a hot extruded filler material, or a hot tool. In all these processes heat is transferred to the surfaces being welded by conduction, convection, and radiation.

The second group includes processes in which heat is generated within the workpiece through conversion of some other form of energy such as high frequency current, ultrasonic waves, friction, infra-red light, chemical reactions, or neutron irradiation.

The mechanism of welding of plastics is considered to be the phenomenon of *auto-cohension* by which welding is accomplished by the diffusion of some molecular chains from one piece into another to form a strong macro-molecular bond between the two pieces.

Welding of plastics is done in the viscous fluidic state under the application of pressure. Better weldability is shown by thermoplastics which have a wider softening range rather than a sharp melting point. Because the coefficient of thermal expansion of plastics is several times that of metals, residual stresses may develop in the weldment resulting in reduced joint strength.

The factors affecting process selection for welding of plastics include workpiece thickness, physio-chemical properties of the plastic, design of the article, and the number of components to be produced. The filler material used in welding plastics should be as close in mechanical properties to the parent material as possible.

4.3.1 Hot Gas Welding

In this process a jet of hot gas which may be air, nitrogen, argon, products of combustion of some fuel gas (for example, acetylene, hydrogen, LPG) is played on the edges to be joind as shown in Fig. 4.37. Fuel gas cannot be used directly to weld plastics because the flame has a very high temperature. That is why special torches have been developed for the hot gas welding of plastics. The welding gas may be heated by electricity or by flame. Air temperature can be adjusted by varying its flow rate and the resistance of the electric element. The flow rate is set anywhere between 25 and 30 m/sec with a valve, and the resistance of the circuit with a rheostat. Air temperature can be checked by placing the tip of the torch within 5 mm of the bulb of a mercury thermometer. If the thermometer reads the required temperature in 10 to 15 second the operator may proceed with welding.

For operator safety, electrically heated torches operate on voltage not exceeding 36 volt. The efficiency of electrically heated torches is 60 percent. Such torches are simple to make and there is no open flame therefore these may be used in room holding inflammable materials. However, these torches are heavy and, thus, rather unwieldy for use in places difficult of access or in awkward positions.

Gas torches may be either directly or indirectly heated. In directly heated gas torches the welding gas is mixed with the products of combustion of fuel gas while in indirectly heated gas torches the products of combustion transfer their heat to the welding gas through the wall. The fuel gas (C_2H_2, H_2 LPG etc.) is used under a pressure of 0.5 to 10 N/cm^2. In comparison with electrically heated torches gas torches can weld at a higher rate, are lighter in weight and more durable. When used eight hours a day, the service life of a gas torch is 1.5 to 2 years. A major drawback of gas torches is that the gases used are inflammable and explosive.

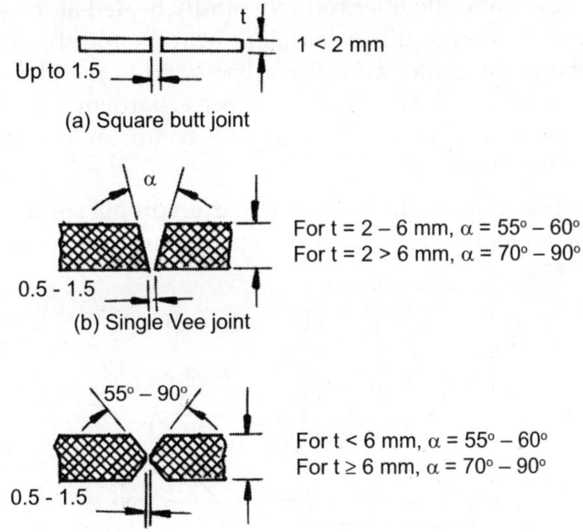

Fig 4.37: Edge preparation for butt joints in plastics.

Joint Design: Usually butt joints are preferred because lap, tee, and fillet joints are more difficult to make. Depending upon the work thickness, square edge, single vee, and double vee edge preparations are employed for butt joint preparation as shown in fig. 4.37. The standard edge preparation for butt welds requires a root gap but no root face.

Double vee joints are usually stronger than single vee joints and the groove angle has a decisive effect on joint strength. As a rule joint strength increases as the groove angle is increased because better penetration is obtained at the root; however the production rate is lowered.

Welding Procedure: the fusion faces are carefully cleaned and degreased, say with acetone; the glossy spots are removed with emery paper or scraper. Before welding torch is switched on or fired the welding gas is turned on and its flow rate adjusted. The gas is then fired in case of a gas torch or electric current switched on for an electric torch.

Filler rods used come in diameters of 2, 3, 4, ±0.5 mm and other shapes like triangular and trapzoidal of different sizes. The filler rods are fabricated from the same material as the work material but may be of different colour and usually contain higher percentage of plasticizer to lower down its softening point. The filler rods may either be cut to lengths of at least 0.5 m and tied up in bundles or uncut and supplied in coils of 3 to 4 kg. The size of filler rod is chosen to suit the work thickness, type of edge preparation and the strength desired. Thicker rods usually result in reduced joint strength.

The torch tip size is selected depending upon work thickness, and edge preparation. Tips with orifice diameter of 1.5-2 mm are used to weld sheets 3 to 5 mm thick while tips with diameter of 3.5-4 mm are used for sheets 16 to 20 mm thick. As a rule the orific diameter of a tip should be equal to the diameter of filler rod used. Otherwise the rod will not be heated adequately and the strength of the joint will be impaired.

The torch-to-work angle is chosen in relation to material thickness. For sheet under 5 mm this angle should preferably be 20°-25° and for sheets in the range of 10-20 mm, it should be 30°-45°. The torch tip-to-work distance should be kept constant between 5-8 mm. To produce a good bond between the filler and the work the rod should be heated and fused at the start of the weld so that its end extends 3-5 mm beyond the end of the work.

The hot gas stream must not be directed in any one position, instead it should be moved continuously over a short length of the welding rod and the surface being welded so as to heat both uniformly. The joint edges and the filler rod are rapidly heated at the surface because plastics are poor conductors of heat. However, it is essential to heat the rod throughout its bulk so that it can be thoroughly softened at the center and properly placed in the groove. That is why thicker rods cannot be used and welding is slow by hot gas process particularly with filler rod technique. If the pressure is not applied properly the softened rod is compressed in the direction opposite to its motion which produces waviness in it.

The filler rod should be fed square to the weld so as to exert proper control of pressure. As the filler rod is forced down by hand it welds to the softened edges and forms a weld, as shown in Fig. 4.38.

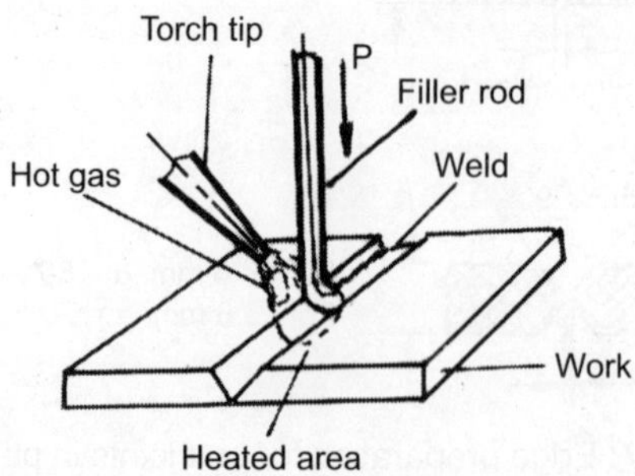

Fig 4.38: Manual hot gas welding using filler material.

Hot gas welding without filler rod accelerates the process and enhances the mechanical properties of the joint. A simple setup for this technique is shown in Fig. 4.39 (c). In this method of sheet edges are scarfed and fitted before being uniformly heated by hot gas. The hot gas jet is followed by cold rollers which exert the required pressure to complete the weld. Welding rate with this technique may be 12 to 20 m per hour, depending upon sheet thickness. The strength of the joint is 80 to 90 percent that of the parent material and the impact strength remains the same. Hot gas welding without filler material is most often applied to make lap joints in films.

For critical joints it is better to seal the weld root to improve the joint strength and quality.

After welding the joint is allowed to cool. Artificial cooloing particularly in material thicker than 10 mm may lead to cracking.

The strength of butt welds in plastics is 65% that of the parent material in shear, 75% in tension, 85% in compression, and 65% in bending while that of fillet weld is 65% in tension. The impact strength of the weld material is usually very low.

Apart from low strength of the joint hot gas welding also results in reduced plasticity in the weld and near-weld area, low production rate especially in thick sheets, danger of overheating and dependence on the operator skill. Inspite of these limitations hot gas welding is widely used for welding PVC, polyethylene, acrylics, and polyamide.

For welding PVC, hot gas welding process is most often used. PVC has no sharp melting point. At a temperature of over 80°C it softens. At 180°C it begins to flow, and at 200-220°C it passes into viscous fluidic state; if pressure is then applied it will weld. The welding temperature must be kept below the critical point at which the material begins to decompose.

To obtain an optimal temperature of 200-220°C for hot air in the welding zone, it should be heated to 230-270°C in the torch. The effect of air temperature on welding rate and joint strength are presented in table 4.1.

Table 4.1 Effect of Air Temperature and Welding Rate on Joint Strength in PVC Welds

Air temperature (°C)	Welding rate for single run (m/h)	U.T.S. for double vee joint (N/mm²)	Joint efficiency (%)	Remarks
210	4.8	14	25	Indirectly heated gas torch.
230	8.4	34	25	Air pressure 0.6 atmosphere.
250	11.4	32	58	Tip diameter, 2.5 mm
270	13.8	35	63	Rod diameter, 2.5 mm
300	15.0	17	30	
320	Material decomposes.			

If a correct welding temperature has been chosen, a dull spot appears on the PVC sheet 2 or 3 seconds after the jet of hot air has been played on it.

Weld quality in PVC depends upon the rate at which the filler rod is fed into the joint, the angle at which it is fed into the joint, the force applied to press the heated rod into the joint, the distance of the torch tip from the work surface, the position and direction of the torch during welding. A filler rod 3 mm in diameter should be fed to the joint at a rate of 12 to 15 m per hour.Welding made in PVC by hot gas filler rod technique show a low impact strength. PVC is sensitive to stress concentration to such an extent that even when a rod is welded to a tube the impact strength of the joint is just about 10% of the impact strength of the parent material.

The welding of PVC by hot gas welding is a slow process. For example to weld one metre of PVC, 18-20 mm thick, with V edge preparation it is necessary to lay 30 to 35 rods, 3 mm in diameter, requiring about 2 hours to accomplish the job. The welding speed can be increased by raising the gas temperature to 300°C and by preheating the filler rod but this requires careful monitoring of the process otherwise the higher temperature may lead to the decomposition of the material.

Acrylics are welded with a jet of air 200-220°C. Time taken for welding acrylic sheet is almost double that required for PVC sheet of the same thickness, and the welding rate is therefore nearly halved. The filler rods used are cut from acrylic sheet and have a cross-sectional area of 7-12 mm^2. Acrylics can also be welded satisfactorily by using PVC filler rods. To achieve quality welds in acrylics it is best to degrease the surfaces to be welded with acetone or dichloroethane prior to welding. The tensile strength of welding joints in acrylics is generally 30-45% that of the parent material.

Polyethylene should preferably be welded with N_2 or CO_2 gas heated to 200-220°C, although gas flame torches may also be used.

Hot gas welding is also frequently used to weld vynil plastics, polystyrene, and some other plastic materials.

The major use of hot gas welding is in the production of very large fabrications made from sheet materials, for example, ducting, pipe work and ventilator hoods for chemical plant installations. This method is normally not used for joining small parts.

4.3.2 Extruded-Filler Welding

In this method the filler in a viscous fluidic state is fed into the joint. The hot filler material melts the edges of the plastic being joined and a strong bond is formed between the filler and the parent material. In a way this process resembles the hot gas process with filler rod technique. Satisfactory welds can be accomplished by this process both in films and heavy guage sheets.

4.3.3 Hot Tool Welding

This process can be performed by several techniques depending upon the type of tool employed which may include hot blade, hot wedge, hot plate, strip heater or a press.

In *hot blade welding,* the heated blade is placed between the surfaces to be joined as shown in Fig. 4.39 (a). After the hot blade has softened the surfaces it is rapidly withdrawn and the surfaces are brought in contact under pressure to accomplish the weld. This process can be used to make butt and lap joints over the entire surface of contact at the same time.

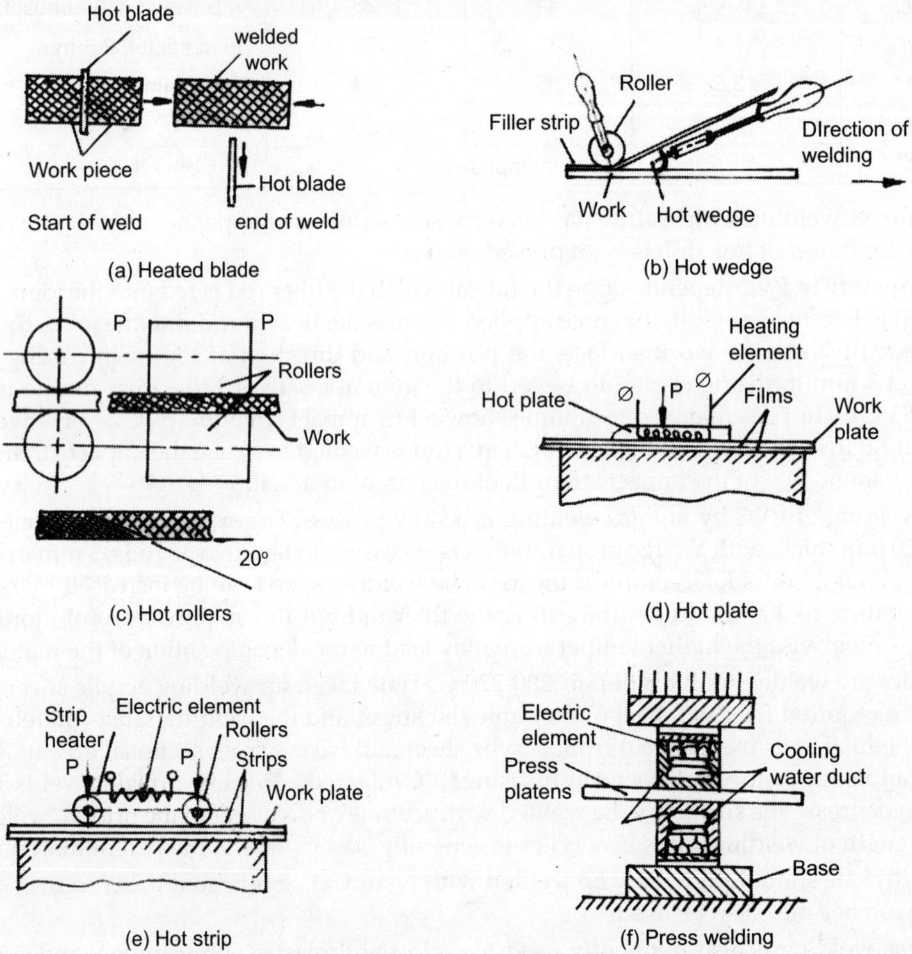

Fig 4.39: Typical method for hot tool welding of plastics.

In *hot wedge welding* shown in Fig. 4.39 (b) the heated wedge is placed between the surfaces to be joined and is moved along the line of welding as the edges are softened. Pressure is applied through a roller to the top strip to weld it to the bottom sheet. This process is used to weld elastic materials but can also be used to weld thin rigid sheets or straps up to 5 mm to thicker sheets. Precautions are, however, needed in this process to avoid sticking of work material to the hot wedge. Best of all this process can be used for welding films by using pressure rollers arranged above and below the films being joind together as shown in Fig. 4.39 (c).

Apart from hot wedge method films can also be welded by hot plate, hot strip and thermal impulse methods.

In *hot plate welding*, resistance heated plate is moved over the films to be lap welded. When the desired welding temperature is reached, pressure is applied to accomplish the weld. The films to be welded are laid out on a work plate as shown in Fig. 4.39 (d).

In *hot strip welding*, the strip heater, heated by an electric element, is advanced by rollers and is simultaneously forced by pressure P against the films to be lap welded which are laid out on a workplate as shown in Fig. 4.39 (e). The films can be advanced under the pressure rollers by moving either the welding head or the workplate.

In *thermal-impulse process* the material (films) is raised to the welding temperature almost instantaneously as a strong current pulse is passed through an electric heater. The heater may be of point, strip, or even an odd shaped type. Because the heat can be accurately metered, overheating at the joint is avoided.

In *press welding* heat is transferred to the area to be welded by the hot platen of the welding press. The plastic pieces with their edges scarfed are clamped between the resistance heated press platens as shown in Fig. 4.39 (f). After the workpieces have been raised to the welding temperature, they are kept under the required pressure as the platens are cooled by the water ciculated through the ducts.

Presses usually make butt welds. A typical plastics welding press for butt joints develops fairly high pressures, heats the work locally, and compresses the softened zone from all sides. That is why this technique is also referred to as *static-jig welding*. This technique can butt weld sheets, bars, strips and plates.

Stresses may be developed in welding of plastics especially if the sheets to be welded are large in thickness. To relieve these stresses it is a good practice to anneal the welded articles from a temperature 25 to 30°C below the softening point of the material.

Hot tool welding produces strong welds at a high production rate. This process is applicable to plastics which cannot be joined by high frequency induction welding, for example PTFE (polytetrafluoroethylene), polyethylene, and polystyrene. Butt, fillet and T-joints can be made by this process, Acrylics joined by hot tool welding retain transparency and clarity at and around the joint. It can also be used for welding films for seams of considerable length. When large quantities of welds are required, the hot tool welding method can be easily mechanised.

4.3.4 High Frequency Induction Welding

In H.F. induction welding the workpiece is placed in a high frequency field set up between two metal electrodes as shown for roller seam welding in Fig. 4.39 (c). Only those plastics which are imperfect dielectric can be welded by this process. The few free elctrons existing in such plastics give rise to conduction current when the material is placed in the H.F. field. The work done to displace the charged particles is converted into heat. Some heat is also generated when the field alternates. To increase the amount of heat generated use is made of very high frequency current in the range of 30 to 40 MHz or even higher. Generally no filler material is used. As all the heat is generated directly in the body of the workpiece being welded, welding speed is high and the electrodes are not overheated.

H.F. induction process is used to make spot, static-jig, and seam welds; however butt, fillet, and tee joints are difficult to make. The welds produced are tight and strong. The process can be easily automated to weld films, sheets and tubes. Lap welds by seam welding machines can be carried out at speeds as high as 27 to 65 m/hr.

Among the merits of high frequency welding are high production rate, economy, and satisfactory joints. It can weld material up to 5 mm thick. However, materials with low dielectric dissipation factor like PTFE, polyethylene, and polystyrene are not possible to weld by H.F. induction welding. But polyehylene can be welded by this process by placing a strip of PVC in the joint. PVC being an imperfect dielectric gets heated up under the action of H.F. current and transfers the heat to polyethylene to accomplish the weld.

4.3.5 Friction Welding

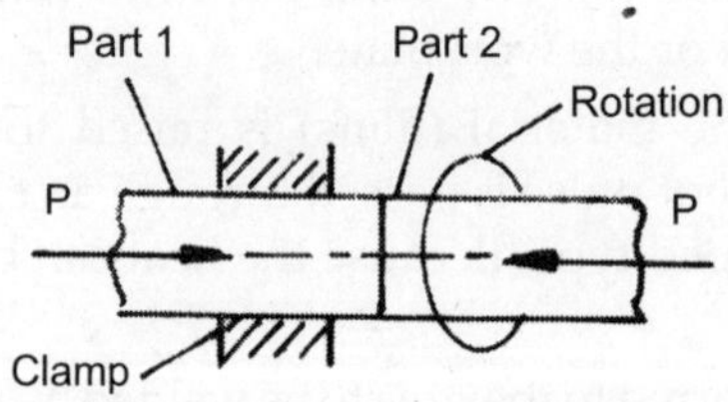

Fig 4.40: Friction welding of plastic rods.

Plastics are friction welded the same way as metals, though normal set-up consists of rotating one piece and keeping the other stationary, as shown in Fig. 4.40, but large pieces can be welded by keeping them stationary and rotating a short insert between them. The quality of the weld depends upon the speed of rotation, the axial force applied, and the amount of plastic deformation involved. This is process is also called *Spin Welding* in polymer welding industry.

Because the heat is generated at the interface the properties of the adjoining material are not affected and the joint has good mechanical properties. Due to heat produced directly on the surfaces being joined this process has the advantage of high welding rate, adaptability to automatic control and usability under field conditions. However the process can be used only if one of the components is cylindrical so that it can be rotated. Also flash formed at the joint means not only the wastage of material but also the additional cost in machining to remove it.

Friction welding of PVC tubes and pipes is well developed. Prior to welding the ends of the tubes are sized by heating the tube ends in oil to 100°C for 3 to 4 minutes and then clamping the tubes in gauge for 3 minutes followed by water cooling to room temperature. The welding is accomplished by rotating one of the tubes in a chuck. The speed of rotation depends on tube diameter, for example, 50 mm diameter tube is rotated at 800 rpm while 80 mm diameter tube is rotated at 600 rpm and the spinning time is 1± 0.5 minutes. After the desired viscous fluidic temperature of 140-160°C is reached the rotation is stopped and a pressure of 20 to 40 N/cm^2 is applied till the weld is cooled to room temperature in about 7 to 10 minutes.

Friction welds in PVC compare in quality with the parent material. Typical joint strength on like materials is about 90% that of the parent material.

In an alternative method of friction welding of plastics, the two parts to be joined are rubbed together at a lower frequency (typically 100–300Hz) and higher amplitible (typically 1–2mm) than ultrasonic welding. The friction caused by the relative motion, combined with the clamping pressure, between the two parts creates the heat which melts the contact areas of the two parts. At this point, the polymeric materials begin to form layers that interwine with one another resulting in a strong weld. At the completion of the vibratory motion the parts remain held together under pressure until the weld joint cools and the melted plastic resolidifies. The frictional movement can be linear or orbital, and the joint design of the two parts should allow this movement.

4.3.6 Ultrasonic Welding

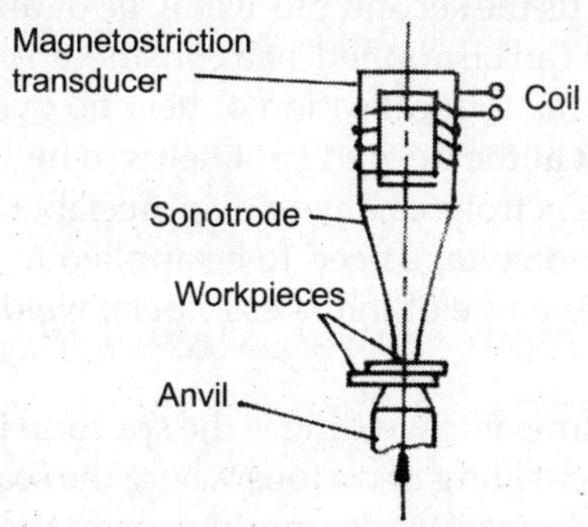

Fig 4.41: Ultrasonic spot welding of plastics.

For ultrasonic welding of plastics the welding machine has the same features as the one for metals. The main element of the welding machine is a transducer, which converts the H.F. energy supplied by ultrasonic oscillator into vibrations. The vibrations are applied to the work though a sonotrode which is set up on an anvil as shown in Fig. 4.41.

The mechanical vibrations applied to the work cause the generation of heat in the plastic material. Pressure is applied to the softened material to complete the joint. Welding takes place the same instant as the H.F. voltage is applied to the transducer coil. The frequency used is up to about 20 KHz.

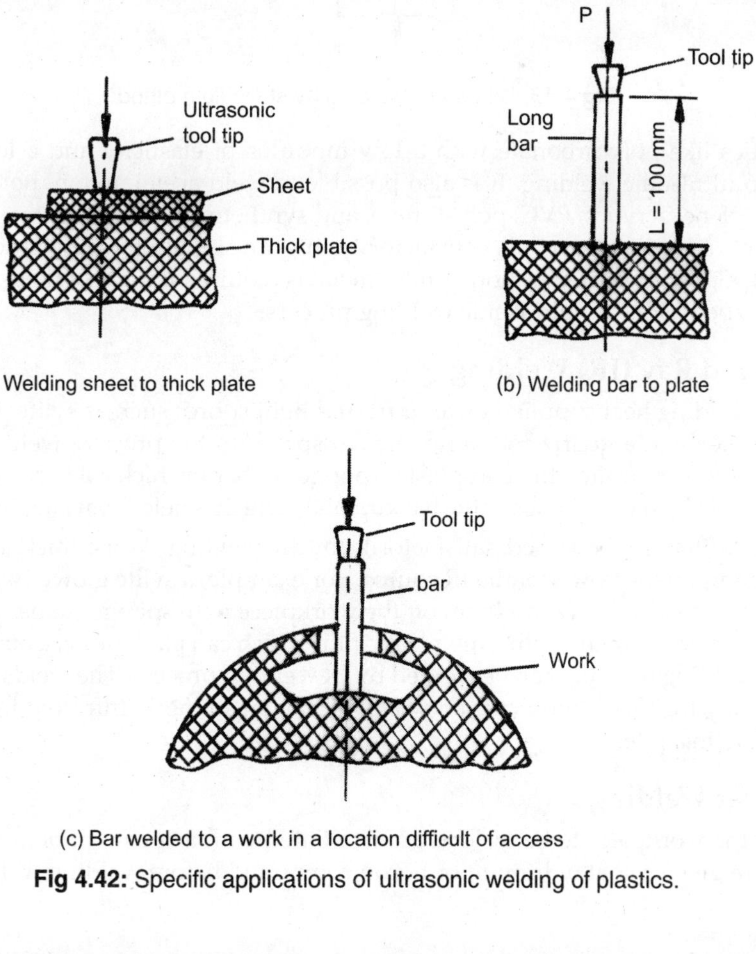

(a) Welding sheet to thick plate

(b) Welding bar to plate

(c) Bar welded to a work in a location difficult of access

Fig 4.42: Specific applications of ultrasonic welding of plastics.

The distinct features of ultrasonic welding include (i) its capability to weld from one side, that is, no second electrode is required, thus the second piece may be of unlimited thickness as shown in Fig. 4.42 (a), (ii) the ultrasonic energy can be applied at a considerable distance from the weld being made as shown in Fig. 4.42 (b), (iii) due to localisation of heat no overheating of the bulk material takes place, (iv) concentration of heat at the desired spot helps in higher welding speed, (v) surface contaminants like grease, graphite, electrolytes have no appreciable effect on weld quality, (vi) no interference in radio reception, (vii) no voltage need to be applied to the welding tool, (viii) a large number of thermoplastics and a wide range of thickness can be welded by ultrasonic welding, (ix) ease of application and automation.

Specific field of application for ultrasonic welding is the spot and jig welding of films and sheets of medium and large thickness, in the sealing of cartons where the seam may be contaminated with grease, and in the packing of various preserves containing current conducting liquids.

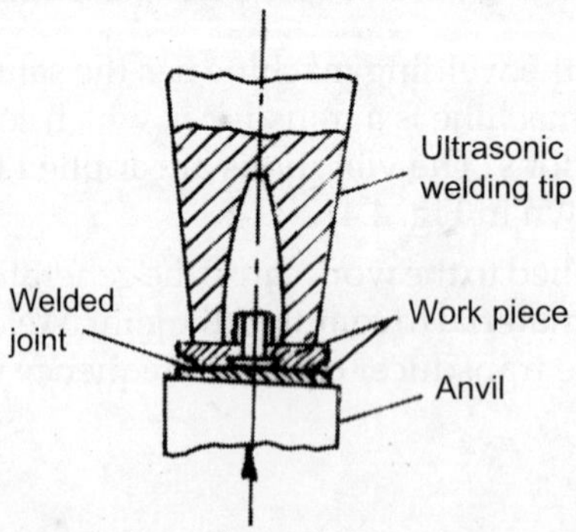

Fig 4.43: Ultrasonic welding by static-jig method.

Rigid plastics like polycarbonate with a low modulus of elasticity and a low melting point respond best to ultrasonic welding. It is also possible to weld acetal, nylon, polypropylene, high density polyethylene, acrylics, PVC, polystyrene, and synthetic textiles by ultrasonic welding. Lap and tee spot joints are made best of all. Satisfactory joints can also be made by static-jig technique, as shown in Fig. 4.43. No edge preparation or filler metal is required for any of these joints. Dissimilar plastics can also be welded by ultrasonic welding process.

4.3.7 Infra-red Ray (IR) Welding

In this process welding heat supplied by an infra-red light source such as sylite glower, a chrome-steel resistance element, a quartz rod lamp, etc. To speed up the process, welding is carried out on a black backing plate from a foamed plastic,sponge rubber or thick rubberized fabric. Welding pressure is supplied by the resilience of the backup plate which is held firmly against the workpiece.

Polyethylene film can be joined satisfactorily by IR welding. Work thickness which can be welded depends upon the power of the IR source. For example, a sylite glower with a temperature of 1200°C kept at a distance of 12 to 14 mm off the workpiece with sponge rubber backing can weld a maximum thickness of up to 2 mm. Any plastic film which can pass into viscous fluidic state and requires a low welding pressure can be welded by IR welding process. The welds produced by this process are usually free from undercuts and have high joint strength. Infra-red light can also weld sheets stacked up in a pile.

4.3.8 Nuclear Welding

In this process the workpiece to be welded are irradiated with a stream of neutrons. The surfaces to be welded are given a coat of lithium or boron compound before welding. When such a coated surface is bombared by neutrons, nuclear reaction takes place resulting in the generation of heat. The heat so produced raises the surfaces to the viscous fluidic state and therefore they can be welded. This process can be used to weld PTFE to polyethylene, polystyrene, quartz, and aluminium.

Nuclear welding has a limitation in that it cannot be applied to materials which become strongly radio-active when irradiated with neutrons.

4.3.9 Laser Welding

This process requires one part to be transmissive to a laser beam and the other part be either absorptive or a coating on that be absorptive to the beam. The two parts are put under pressure while the laser moves along the path to be welded. The beam passes through the first part and is absorbed by the other one or by the coating on that and generate enough heat to soften the interface creating a permanent weld.

Semiconductor diode lasers are typically used for plastic welding. Wavelengths in the range of 808 nm to 980 nm can be used to join various plastic material combinations. Power levels from less than 1 watt 11 to 100 watt are needed depending on the materials, thickness and the required welding speed.

Diode laser systems have the following advantages in joining plastics.

1. Cleaner than adhesive bonding
2. No micro–nozzles to get clogged
3. No liquid or fumes to affect surface finish
4. No consumables required
5. Higher output
6. Can weld at difficult to access points
7. High level of process control.

Requirements for achieving high strength in welded joints include the following:

1. Adequate transmission through upper layer
2. Good absorption by lower part or the coating on it
3. Good wettability between two parts
4. Good joint design as also adequate clamping pressure and contact area
5. Lower power density.

The plastics that can be joined by laser welding include polypropylene, polycarbonate, acrylic, Nylons, ABS, etc.

Specific applications include sealing/welding and joining of medical containers, automobile remote control keys, heart pacemaker casings, syringe tamper evident joints, headlight or tail light assemblies, pump housings, cellular phone parts and catheter bags (bags used for collecting urine in medical treatment).

Transparent Laser Plastic Welding

New *fibre laser* technology allows for the output of longer wavelengths, with the best results typically around 2000 nm, significantly longer than the average 808-1064 nm diode lasers used for traditional laser welding of plastics. Because these longer wavelengths are more readily absorbed by thermoplastics than the infra-red radiation of traditional plastic welding. Common applications are expected to fall mostly in the medical industry for devices like catheters and microfludic devices. The heavy use of transparent plastics, especially flexible polymers like TPU (Thermoplastic polyurethane), TPE (Thermoplastic Elastomers) and PVC, in the medical devices industry makes transparent laser welding a natural choice. Also, the process requires no laser absorbing additives or colorants making testing and meeting biocompatibility requirements significantly easier.

4.3.10 Solvent Welding

In solvent welding, a solvent is applied to the parts to be welded, that can temporarily dissolve the polymer at room temperature. When this occurs, the polymer chains are free to move in the liquid solvent and can mingle with other similarly dissolved chains in the other component. Given sufficient time, the solvent will permeate through the polymer and out into the environment, so that the chains lose their mobility. This leaves a solid mass of entangled polymer chains which constitute a solvent weld.

This technique is commonly used for connecting PVC and ABS pipes, as in household plumbing. The *gluing* together of plastic (polycarbonate, polystyrene or ABS) models is also a solvent welding process.

Dichloromethane (methylene chloride), which is obtainable in paint stripper, can solvent weld PC (polycarbonate) and PMMA (polymethylmethacrylate). Dichloromethane chemically welds certain plastics; for example, it used to seal the casings of electric meters. It is also a component-along with *tetrahydrofuran*— of the solvent used to weld plumbing. *ABS plastic* is typically welded with *Acetone* based solvents which are often sold as paint thinners or in small bottles as nail polish removers.

Solvent welding is a common method in plastics fabrication and used by manufacture of in store displays, brochure holders, presentation cases and dust covers. Another popular use of solvents in the hobby segment is *model building* from injection molded kits for *scale models* of aircrafts, ships and cars which predominantly use *polystyrene* for this purpose.

4.4 Machining of Plastics

Although parts made of plastics are usually produced by moulding and forming processes many high-precision items must often be machined from standard shapes by turning, drilling, milling, sawing, grinding, polishing, etc. This is done to meet tolerances closer than those which can be made by moulding and forming techniques, i.e., below 0.2 to 0.4%. However, the break-even point has been reported to be as high as 15,000 to 60,000 pieces with automated machining. Some parts can be made most economically by slicing them off from extruded shapes, Fig. 4.44. Hard-to-mould forms, such as threads can often be obtained most economically by machining. Machining is also used to remove excess material, which may be in the form of flash, runners, sprues, flanges, etc., that is encountered on almost all moulded and formed plastic parts.

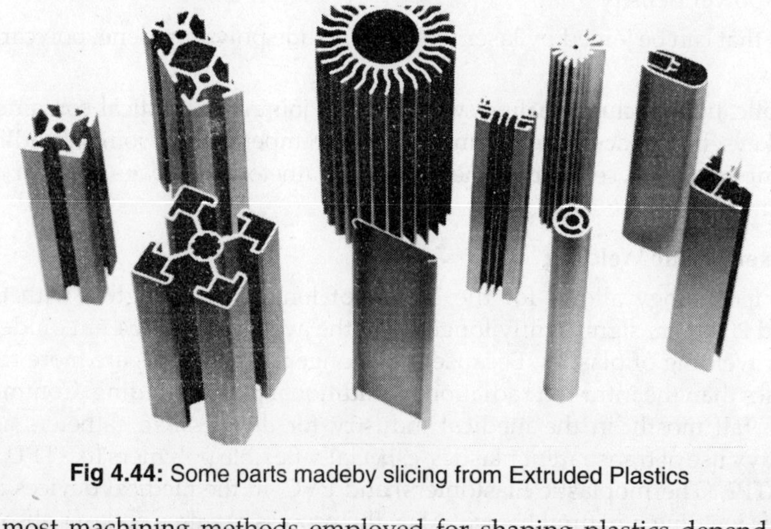

Fig 4.44: Some parts madeby slicing from Extruded Plastics

In general, most machining methods employed for shaping plastics depend on the use of existing machines and tools developed for the fabrication of wood and metals, and little has been done to develop machining equipment or methods especially suited to plastics. Further, there are some principles of machining which apply to plastics alone because properties of plastics are different from those of metals. For example, (1) they do not readily conduct but are easily affected by heat, (2) some plastics contain abrasive fillers, (3) plastics are mostly soft and yielding, and (4) some plastics are quite brittle though soft.

Thermoplastics soften, lose shape, become gumming and clog cutters; thermosetting plastics deteriorate and char at high temperatures. The heat generated is not conducted away rapidly by plastics when they are cut so good results depend upon practices conducive to cool cutting. This calls for tools with sharp cutting edges and smooth polished faces. For sawing, a band saw is preferable to a circular saw because the teeth on a long band saw blade have more time to cool. Air, water, and oil-coolants are commonly used for cutting plastics.

Hard and wear-resistant non-ferrous alloys, cemented carbides, or sintered oxide tools are necessary to cut plastics with abrasive fillers.

Cutting tools stand high speeds well for most plastics because the materials are soft. Machine tools can work to as close tolerances in cutting plastics as with any other material, but in most cases a close tolerance is of no use because the plastic will not retain a size. Many plastics yield and change appreciably in size under cut while some absorb moisture and swell. In some cases, plastic pieces are chilled to make them rigid for cutting.

Some plastics are brittle and chip readily when cut. For these, tools must be sharp and cuts light. Shearing is commonly done with sharp dinking dies. In some cases, plastics are heated before being cut.

When the cut surface of a plastic is examined in detail, it is often rough and cracked or covered with undesirable tool marks. In other cases, it will be found that excessive frictional heat developed under machining pressures has caused burning in case of thermosets, and/or gumming in the case of thermoplastics of the machined surfaces so that further finishing is required.

To understand machining operations for plastics, it is necessary to understand the cutting action which occurs in solid polymeric materials. This involves analysis of chip formation, mechanisms of the forces and deformation developed in the material during cutting, rise in temperature that occurs in machining, etc.

4.4.1 Chip Formation

The analysis of cutting by a single edge tool is a very important first step in the understanding of the machining of plastics. Different types of chips formed during machining of plastics include the following:

1. Continuous flow type
2. Continuous shear type
3. Discontinuous-simple-shear type
4. Discontinuous crack type
5. Discontinuous complex type chip
6. Discontinuous complex type chip (shear with crack).

Brief description of all these types of chips follows.

1. Continuous Flow Type Chips

In this type, the chip deformation is essentially elastic and the chip thickness is nearly equal to depth of cut. This type of chip is called continuous flow type because it is produced by high elastic deformation and not by the plastic fracture. This type of chip is not formed in machining of metals.

Fluctuation in cutting force and surface roughness are small, and the dimensional accuracy of the machined part is fairly good when this type of chip is produced. However, care should be taken to avoid undercut or overcut, which tends to occur when this type of chip is formed because of high elastic formation. Such types of chips are usually observed in slow-speed cutting of materials which show high rubber-like elasticity having large elongation at fracture, such as PTFE or PE. Cutting conditions should be chosen to produce this type of chip.

2. Continuous Shear Type Chip

The cutting action for the formation of this type of chip results in a shear plane which is generated upward from the point of the cutting tool, and the chip is produced by shearing action along this plane. The chip is continuous because of the small shear intervals. The shear plane is produced in the direction of minimum work required to form a chip, and this type of chip is produced when the shear strain on the shear plane is smaller than the limiting rupture strain. Also, characteristic of this type of plastic cutting action is the existence of shear marks along the chip surface and a chip thickness that is generally larger than the depth of cut.

This type of chip is very similar to that produced in machining of metals. Cut surface quality is good when this type of chip is produced. Thus a plastic which produces continuous-shear type chips is considered to have good machinability. This type of chip is frequently observed in the cutting of plastics like polystyrene.

3. Discontinuous Simple Shear Type Chips

This type of chip is produced by shearing action along a shear plane; when the shear intervals are rather large. Such chips occurs when the shear strain at the shear plane becomes larger than the limiting rupture strain. The machined surface is excessively rough, and accuracy of the machined surface is usually inadequate when such chips are produced. Thus, conditions producing this type of chips should obviously be avoided. Such chips are formed in machining of polymethyl methacrylate (PMMA) with a negative rake tool.

4. Discontinuous-Crack Type Chips

This type of chip formation will be observed in cutting brittle materials, such as thermosets (phenolics, polyesters, or epoxy resins), and some of the thermoplastics, such as polymethyl methacrylate (PMMA) or polystyrene (PS), when a tool with an excessively large rake angle and a large depth of cut is involved. In this case a crack occurs at an oblique downward angle, and the discontinuous, blocky chip is produced by the bending moment, which acts on the chip after the crack grows to some length. Thus, this type of chip formation is a type of brittle fracture. This type of chip is never observed when metals are cut, except for some kinds of cast irons.

Many large cracks are observed on the cut surface when this type of chip is produced. The roughness of cut surface is rather excessive, and cut accuracy is poor. Therefore, material which produces this type of chip is rather difficult to machine. However, the improper selection of cutting conditions can also produce this type of chip even in materials ordinarily having good machinability.

5. Discontinuous Complex Type Chip

These chips are produced by the complex stress resulting from the action of a large compressive stress operating in combination with a shear stress when the cutting is done with a large negative rake tool. Dimensional accuracy of the machined part is poor and surface roughness is also high. Such chips may be observed in machining polystyrene with large (30°) negative rake cutting tool.

6. Discontinuous Complex Type Chip (Shear with Crack)

For the formation of this type of chips, a crack occurs at a downward angle from the cutting point - in addition to the separation of the material along the usual shear plane. When the crack grows to some length, a new shear plane forms from the tip of the crack upward to the free surface, and the plastic deformation zone is enlarged. Thus, the shear angle becomes larger than it does in the case of the simple shear, and a discontinuous chip forms under the combined effects of shear with crack. Periodic cracks are left on the cut surface, resulting in roughness and dimensional inaccuracy. Such chips may be observed in machining polymethyl methacrylate (PMMA) with cutting tool having 0° rake angle.

In conclusion it may be said that it is important to select such variables as the geometry of the cutting tool, the cutting speed, and the depth of cut so that the continuous-type chips are produced to obtain good machining results.

Thus, the type of chips formed during machining of plastics provide very important information on how cutting is proceeding. This is true in all kinds of machining, such as turning, drilling, milling, sawing etc.

4.4.2 Machianability of Plastics

Machinability is defined as the ease with which a material can be cut. The ease is mainly determined on the basis of three criteria, viz., power consumed in machining, tool life, and the cut surface finish. Thus, good machinability means lower power consumption, small amount of wear on the tool, and smooth cut-surface. It is defined quantitatively as follows.

$$\eta^1_c = \frac{V_m}{V_w.hp.S_f} \quad ...(4.1)$$

where,

η^1_c = Machinability factor or machining efficiency,

V_m = Volume of the material cut per unit time, mm³/min,

V_w = amount of tool wear per unit time mm³/min,

hp = net power consumed during cutting, kw or kg-m/min,

S_f = roughness of cut surface, mm.

The machinability factor, becomes large and the machinability improves as power consumption, tool wear, and surface roughness decrease. If S_f is not a factor as in rough machining, the definition of machinabilitty changes to,

$$\eta^2_c = \frac{V_m}{V_w.hp} \quad ...(4.2)$$

When tool wear is also negligible, as with high quality tool material, the definition of machinability becomes,

$$\eta^3_c = \frac{V_m}{hp} \quad ...(4.3)$$

Also, since hp is equal to the main cutting force, F_c (kgs) and cutting speed V (m/min) then,

$$hp(KW) = \frac{F_c.V}{6120} \quad ...(4.4)$$

($\because$ 1KW = 102 Kg. m/sec)

and since V_m in the case of turning is equal to 2π.R.d.f,

where,

R = radius of workpiece, mm,

d = depth of cut, mm

f = feed velocity, mm/min.

Putting these values in = n (4.3), we get

$$\eta_t = \frac{2\pi.R.d.f \times 6120}{F_c.V}$$

$$= k\frac{d.f}{F_c.n} \quad ...(4.5)$$

where,

k = 6,120

n = r.p.m. of workpiece ($\because$ V = 2πR.n)

In the case of orthogonal cutting, f is negligible, and = n (4.5) becomes,

$$\eta_s = k.\frac{d}{F_c} \quad ...(4.6)$$

Also, in *orthogonal cutting** the total energy per unit of material removed, called the specific cutting force is $\frac{Fc}{b.d}$ (Kg/mm²), where b is the width of the cut. Thus, the machinability improves as the cutting force decreases. Therefore, to determine machinability of a plastic, one have to use testing apparatus which can measure cutting forces, the amount of tool wear, and surface roughness. In the absence of these equipment, a rough qualitative estimate of mechinability can be made from the type of chips produced in machining a specific plastic material. Thus, machinability becomes better as the percentage of continuous chips increase.

In the light of knowledge gained in preceding sections on chip formation and machinability rating detailed information about turning, drilling, milling, broaching and grinding, etc. on some of the more widely used plastics follows.

4.4.3 Single Edge Cutting of Plastics

This section is divided into two parts, viz., single edge cutting of thermoplastics and single edge cutting of thermosets, and includes the processes of turning, shaping, planing, etc.

Thermoplastics discussed hereunder include the following.

1. Polyethylene
2. Polyvinylchloride (Rigid)
3. Polymethylmethacrylate (acrylic)
4. Polystyrene
5. Styrene Copolymers
6. Polytetrafluoroethylene (PTFE)
7. Polyamide (nylon)
8. Polycarbonate
9. Acetal resin
10. Polypropylene

---

* **Orthogonal cutting:** The essential features of orthogonal cutting are:

1. The cutting edge of the tool is perpendicular to the direction of travel.
2. The cutting edge clears the width of the workpiece on both ends.
3. The tool prepares a surface which is parallel to the work surface.
4. The chip flows over the tool face and the direction of chip flow-velocity is normal to the cutting edge.
5. Only two components of the cutting forces are acting on the tool, which are mutually perpendicular and can be represented in a plane, and
6. The maximum chip thickness occurs at its middle. Such a cutting system is also known as *Two-Dimensional Cutting*.

and the thermosets include:

1. Thermoset laminates
2. Cast thermoset resins.

Brief descriptions of single edge cutting of these thermoplastic and thermoset plastics follows.

4.4.3.1 *Polyethylene (PE)*

This is one of the four most widely used thermoplastics in the world; the other three being polypropylene (PP), polystyrene (PS), and polyvinylchloride (PVC). It has good machinability and can be machined on conventional machine tools.

Continuous chips are always produced when polyethylene is turned. Continuous shear type chips are produced at higher cutting speeds, and continuous flow type chips are formed at lower cutting speeds. The magnitudes of the cutting forces are rather small compared with those produced in cutting other plastics. There is always a critical rake angle for optimum cutting results. It is particularly important to use the proper/critical rake angle for machining PE because it has high elasticity. Because of the elastic nature of PE, the use of improper rake angle will cause the cut surface to springback, thus causing dimensional inaccuracy in the finished product. Critical rake angles for some of the often used cutting conditions for PE are listed in Table 4.2.

Table 4.2: Critical Rake Angles for Cutting Polyethylene

Cutting speed (m/min)	Critical rake angle for depth of cut of		
	0.05 mm	0.10 mm	0.15 mm
0.8	41°	30°	24°
50	35°	25°	20°
100	26°	16°	13°
200	18°	9°	8°
400	10°	4°	3°

In conclusion it can be said that due to the relatively low forces required to cut PE, and because continuous chips can be obtained over a wide range of cutting conditions, PE is said to have excellent machining qualities.

For best results tools with large rake and relief angles having a generous nose radius should be used with a high cutting speed and a fine rate of speed. Under such conditions, machined parts of PE with smooth and accurate cut surfaces can be produced.

4.4.3.2 *Rigid Polyvinylchloride (R-PVC)*

Rigid PVC is relatively easy to machine except when large depth of cut, high cutting speed alongwith a large positive rake angle of the tool is used which result in the formation of discontinuous-shear type chips. The continuous chips are produced under almost any other set of cutting conditions.

In general the chips turn from continuous into discontinuous type as the depth of cut becomes larger than 0.1 mm.

Both cutting force components (F_c & F_t) decrease with rake angle as the latter turns from negative to positive. There is also a critical rake angle for a given set of cutting conditions for rigid PVC. Table 4.3 lists some values for critical rake angle with the change in depth of cut.

Table 4.3: Critical Rake Angles for Different Depth of Cuts for Rigid PVC

Depth of Cut (mm)	Critical Rake Angle (degrees)
0.05	13
0.10	11
0.15	10

Specific cutting forces in Rigid PVC machining are relatively small compared with those in other plastics. They decrease as the depth of cut increases, and also decrease as the rake angle increases from negative to positive. Thus, for R- PVC it is better to choose as large a depth of cut as feasible, and to use a positive rake angle tool to reduce the specific cutting force provided the desired smoothness of surface is achieved and chips are of the continuous type. An accurate and smooth surface can be obtained easily in R-PVC when cutting conditions are properly selected.

In conclusion it can be said that the machinability of R-PVC is rated good-to-excellent because of low specific cutting force, smoothness of the mechined surface, and also the relatively wide range of conditions which produce continuous chips.

4.4.3.3 *Polymethyl methacrylate (PMMA)/Acrylic*

This plastic is widely used as an engineering material and is often fabricated by machining. Although acrylic is easy to machine, improper cutting conditions can cause many problems.

Different types of chips are produced under different machining conditions. However, in general continuous chips are produced as cutting speed and depth of cut become small, and as the work temperature rises. The chips change from continuous to discontinuous type as the rake angle of the cutting tool becomes too positive and the acrylic is cut at relatively high speed.

Chips will also change from the continuous to the discontinuous crack type as the depth of cut or feed rate in turning increases. Very deep cuts should be avoided to prevent rough surfaces, even though cutting time and cost are reduced.

The forces generated in turning acrylic vary with cutting speed, depth of cut, and the rake angle of the tool. It is reported that the cutting forces increase in direct proportion to the depth of cut, and decrease slightly as the cutting speed increases, and as the rake angle increases from negative to positive. In fact there exists a critical angle for a given set of cutting conditions. Table 4.4 shows the critical rake angles for turning PMMA at a cutting speed of 0.8 m/min.

Table 4.4: Critical Rake Angles for Acrylic (cutting speed = 0.8 m/min)

Depth of Cut (mm)	Critical rake angle ((deg.)
0.05	40
0.10	27
0.15	25

The specific cutting forces of PMMA vary with depth of cut, rake angle, and work temperature. It is preferable to use large depths of cut, positive rake angle, and higher work temperature to reduce the specific cutting forces.

In conclusion, the machinability of PMMA ranges from good to excellent.

4.4.3.4 *Polystyrene (PS)*

Polystyrene is relatively difficult to machine compared with PMMA (acrylic) or other thermoplastics; however it can be successfully machined if cutting conditions are carefully selected.

Chips similar to those obtained in machining acrylic are produced when turning polystyrene; chips become tightly curled as the depth of cut decreases, and they get elongated, stick like, as the rake angle becomes negative.

Continuous chips are formed with a rake angle between 0 and 20°, but they turn into discontinuous-crack type when the depth of cut or the positive rake angle of the tool is too large. Further, they change into discontinuous complex type formation when too large a negative rake angle is used. Critical rake angles decrease as depth of cut increases, as is shown in Table 4.5.

Table 4.5: Critical Rake Angles for Turning Polystyrene

Depth of Cut (mm)	Critical rake angle ((deg.)
0.05	25
0.10	18
0.15	15

In cutting polystyrene it is very important to select conditions which produce the continuous-shear type chips.

Both components of cutting force (F_c & F_t) decrease with the increase in rake angle from negative to positive.

Optimum cutting conditions for polystyrene consist of using cutting speeds under 1 m/min., a tool rake angle in the range of –10 to 30°, and a depth of cut under 0.3 mm. Discontinuous crack type chips tend to be produced at higher cutting speeds. Thus, in turning polystyrene the depth of cut and cutting speed should be kept as low as possible.

The region in which continuous chips are produced is not very wide in turning polystyrene. Therefore the mechinability of PS is rated as only fair, it is in fact the worst among thermoplastics, in spite of the low values of specific cutting force.

Other problems faced in turning polystyrene are gumming, cracking, and chipping. Because gumming is caused by the heat generated during machining , auxiliary cooling must be provided. When improper coolant is used, PS may crack after machining. Thus, the selection of the coolant is also important. On the other hand, cracks due to machining may also occur when depth of cut becomes too large. Moreover, chipping caused by chattering will occur when unsuitable fixtures are used for holding the work material; thus suitable collet or rubber mounting methods are helpful.

4.4.3.5 *Polystyrene Copolymers (SAN and ABS)*

Modified polystyrenes, having improved strength and heat resistance, are widely used as engineering materials. They are usually copolymers of styrene with acrylonitrile and butadiene. SAN (styrene acrylonitrile) resin is copolymer of acrylonitrile and styrene, and ABS (Acrylonitrile butadiene styrene). ABS is usually available in three grades viz., ABS-L, ABS-M, and ABS-H to denote low, medium, and high impact strength respectively.

Five types of chips are produced in cutting SAN resin, with cutting tool rake angle varying between –20 and 40°, at cutting speeds in the range of 0.4 to 400 m/min, and with depth of cut varying from 0.013 to 0.105 mm. These are the continuous-shear, discontinuous simple-shear, discontinuous-complex, discontinuous-crack, and discontinuous-gummed types of chips. The last type is produced by the heat generated during cutting, although it is essentially discontinuous.

The same types of chips are generated in cutting ABS-L. However, for ABS-M and ABS-H only the continuous-shear and gummed type chips are produced. Gummed type chips are formed only at a high cutting speed of 400 m/min, with a tool having a negative rake angle of 20°.

The variation in cutting forces with cutting conditions for SAN and ABS resins is similar to that in other plastics. The relationships between the cutting forces and tool rake angles in cutting SAN and ABS resins are also very similar to those for other thermoplastics, when continuous chips are produced. However, if discontinuous chips are formed, the relationships show irregularities similar to those observed in cutting acetal resin. As with other plastics, a critical rake angle exists for each set of cutting conditions in turning SAN and ABS resins.

In conclusion, it can be said that the machinability of ABS-H and ABS-M is excellent, that of ABS-L is good to excellent, and that of SAN is good.

4.4.3.6 *Polytetra fluoroethylene (PTFE)*

This plastic is widely used as a material for electronic, mechanical and chemical parts because of its superior high temperature and chemical resistance, as also its low frictional behaviour. Unlike other thermoplastics, PTFE does not form a free-flowing melt. This, plus the fact that it melts at rather high temperature, makes it difficult to mould it to close tolerances. For this reason, PTFE parts must be machined from rod stock when close tolerances must be met. Also, when product shape is very complex, or when insulating tape is to be produced, machining is the only logical means of fabrication.

Three types of chips are produced in turning PTFE. All the chips produced at low cutting speeds (under 1 m/min) are the continuous-flow type. However, at high cutting speeds the chips vary with the depth of cut and the rake angle of the tool. The chips are continuous at large rake angles, but become discontinuous as the rake angle decreases to low values or becomes negative.

Both the force components (F_c and F_t) are affected by the rake angle of the cutting tool. It is thus important to use critical rake angle when machining PTFE because of its high elasticity, so that the work material is not deformed by tensile or compressive stresses during machining. The critical rake angles for different depths of cut for turning PTFE are listed in Table 4.6.

Table 4.6: Critical Rake Angles for Machining PTFE

Depth of Cut (mm)	Critical rake angle ((deg.)
0.05	20
0.10	12
0.15	8

A special machining problem arises in the production of PTFE tapes which are produced mainly by peeling i.e., by cutting a continuous strip from a rod. Cutting conditions must produce continuous chips because chip is used as the tape.

In general, the machinability of PTFE is excellent because of low values of specific cutting forces and the wide range over which continuous chips are obtained. However, it is imperative to maintain the room temperature as constant as possible below or above transition temperature. PTFE has a phase transition temperature between 18 and 25°C, hence, there is a change in specific volume that accompanies the transition. Thus, dimensions machined below the transition temperature will not hold at service temperature above the transition point, and vice-versa.

4.4.3.7 *Polyamide (Nylons)*

Nylons are used for many engineering components like bearings, rollers, cams, bushings, seals, etc. Many precision machining operations are often required to make the parts from standard nylon shapes, such as rods, cylinders, and flat disks, which are available commercially.

Three kinds of polyamide viz., nylon 6, 66, 610 are widely used and are discussed in this section. Four types of chips, i.e., continuous shear, discontinuous simple shear, discontinuous complex, and discontinuous crack type are produced in machining these nylons. In general, nylon is easy to machine successfully because continuous chips are produced under almost any set of cutting conditions. However, discontinuous chips may be produced when the rake angle, depth of cut, or cutting speed is excessive, especially when nylon 66 is machined.

Both the cutting force components (F_c and F_t) increase with the increase in depth of cut. Cutting forces decrease as cutting speed increases. The rate of decrease becomes large when a tool with positive rake angle is used, as some discontinuous type of chips are produced at high cutting speeds when such a tool is used. As with other plastics, a critical angle exists for each set of machining parameters in turning nylons. Table 4.7 summarizes the critical rake angles for turning the three types of polyamides.

Table 4.7: Critical Rake Angles for Polyamides

Depth of Cut (mm)	Resin	Critical rake angle at cutting speed of			
		50 (m/min)	100 (m/min)	200 (m/min)	400 (m/min)
0.05	Nylon 6	31°	28°	24°	21°
	Nylon 66	30°	27°	22°	–
	Nylon 610	29°	27°	23°	20°
0.10	Nylon 6	25°	22°	17°	13°
	Nylon 66	25°	21°	15°	–
	Nylon 610	25°	22°	16°	15°

The specific cutting forces encountered in turning nylons are relatively small; they decrease with increase in depth of cut and also with the rake angle. Thus, a large depth of cut and a tool with a positive rake angle should be used to reduce the specific cutting forces in cutting nylons. However, special care must be taken to avoid producing discontinuous chips.

Nylon is apt to absorb moisture, which will change its properties and affect its dimensional stability. Suitable annealing methods to relieve residual machining stresses are also important to use after turning.

Also, nylon has a tendency to flex, so care must be taken to prevent the stock from being pushed away from the cutting tool. So, the use of a cutting tool made with a critical rake angle is very important. Coolants, preferably of the soluble oil type should be used for high speed turning, and the coolant itself must be kept as cool as possible.

In conclusion, it can be said that nylons have excellent machinability because of the wide range of conditions which produce continuous type chips, good surface finish, and the small amount ot tool wear. Nylon 66 has the poorest machinability amongst polyamide plastics.

4.4.3.8 Polycarbonates (PC)

These are amongst the more durable plastics and have excellent impact strength, creep and thermal resistance.

Long (10 to 20 m length) and tough chips are usually produced in machining polycarbonate. However, the type of chips produced will depend upon depth of cut, rake angle or cutting speed, as is the case with other plastics.

Almost all chips formed in machining PC at a low cutting speed of 0.5 m/min are rigid, curled, continuous chips, except those formed when a tool with a large negative rake angle is used. The curling becomes tighter and smaller as the depth of cut becomes smaller.

Most polycarbonate chips are long and smooth. But when too large a depth of cut and too large a positive rake angle are used, the chips become discontinuous.

When long chips are produced in machining polycarbonate they are tough and strong. Therefore, if not handled properly, the long chips from the workpiece are apt to get entangled around the lathe chuck, the work material or the tool. The entwined chips are strong and difficult to get rid off. Thus, the handling of the chips is rather an important consideration in the machining of polycarbonates.

Cutting forces vary with rake angle, depth of cut, and cutting speed, similar to those in machining of other plastics. Also, there exists critical rake angle for a given set of machining parameters. Table 4.8 summarizes the critical rake angles for turning polycarbonates.

Table 4.8: Critical Rake Angles for Turning Polycarbonates

Cutting Speed (m/min)	Critical rake angles for depth of cut of		
	0.05 mm	0.10 mm	0.15 mm
50	30°	25°	21°
100	25°	19°	15°
200	22°	15°	11°
400	20°	11°	8°

The specific cutting forces vary with the cutting conditions. A large depth of cut, a high cutting speed, and a positive rake angle should be chosen to reduce the specific cutting forces. Values of the specific cutting force for PC are rather large compared with those for other plastics. However, a wide range of cutting conditions will produce continuous chips, thus the machinability of polycarbonates can be rated as excellent.

Unlike other plastics, gumming does not occur in turning polycarbonate, as the material has high melting point. However, as mentioned earlier, the chips can be a nuisance because of their extreme toughness and tendency to curl, fouling the work material, cutting tool, or the machine tool.

4.4.3.9 *Polymethylene Oxide (PMO)/Acetal Resin*

Acetal has unique properties that bridge the gap between metals and plastics. Apart from high strength and rigidity, excellent dimensional stability, resilience, and toughness. It also exhibits superior resistance to temperature, humidity, and chemical corrosion. The chips are continuous and smooth at small depths of cut but change into thin, discontinuous, needle like chips as the depth of cut increases. The tools having a rake angle near zero produce continuous chips even at the larger depth of cut. In general, chips become discontinuous as cutting speed and depth of cut increase.

Cutting forces vary with rake angle, depth of cut, and cutting speed, as for other plastics the critical rake angle for acetal resin depends on the depth of cut and cutting speed. *Table 4.9* lists critical rake angles for acetal resins.

Table 4.9: Critical Rake Angles for Turning Acetal

Cutting Speed (m/min)	Critical rake angles for depth of cut of		
	0.05 mm	0.10 mm	0.15 mm
50	19°	15°	13°
100	17°	14°	11°
200	15°	11°	8°
400	13°	7°	4°

The range of conditions which produce continuous chips, in machining acetal resin, is rather large. In general, the cutting tool for machining acetal resin should have a rake angle in the range of 0°-20° for optimal cutting conditions.

The specific cutting force varies with cutting conditions. It is reported that a large depth of cut, high cutting speed, and a positive rake angle will reduce the specific cutting force.

Because continuous chips are produced over a relatively wide range of conditions, the machinability of acetal resin is graded as excellent or good to excellent. Acetal can be easily machined on standard equipment. It is seldom necessary to use cutting oils, because of the small amount of heat generated by the low coefficient of friction between chips and tool. A large relief angle and a suitable chip-breaker can also help to obtain good results.

4.4.3.10 *Polypropylene (PP)*

Polypropylene is one of the four commodity plastics (the other three being polyethylene, polystyrene and PVC), so is very widely used. It is reasonably rigid with good creep resistance, the material has adequate load bearing capacity for many applications that may involve machining of components.

Long and transparent chips are usually produced from polypropylene under a rather wide range of cutting conditions. The chip change to discontinuous-crack type as the cutting speed increases, when a large depth of cut and a tool having a large positive rake angle are used. In contrast, the chips made with a negatively raked tool at a small depth of cut and a high cutting speed begin to melt, even though polypropylene has a relatively high melting point of about 170°C.

The cutting forces are affected by the rake angle of the tool; so a critical rake angle exists in machining of polypropylene. *Table 4.10* lists the critical rake angle for turning polypropylene under certain sets of cutting conditions. It indicates that the rake angle decreases as cutting speed and depth of cut increase; as is the case with other plastics.

Table 4.10: Critical Rake Angles for Turning Polypropylene

Cutting Speed (m/min)	**Critical rake angles for depth of cut of**		
	0.05 mm	**0.10 mm**	**0.15 mm**
50	21°	18°	17°
100	20°	17°	16°
200	19°	14°	13°
400	18°	13°	12°

The range of conditions producing continuous and transparent chips is rather wide in machining of polypropylene. However, do not choose too large a depth of cut with too large a positive rake angle, or too small a depth of cut with a negatively raked tool at high cutting speeds; this produces discontinuous chips.

Specific cutting forces decrease as depth of cut increases, and also decrease as the rake angle is changed from negative to positive. Thus, choose a depth of cut as large as possible, a cutting speed as high as possible, and a positive rake angle to minimize the specific cutting force.

In general, the machinability of polypropylene is excellent or excellent-to-good. Special attention should be paid to the selection of cutting conditions to avoid discontinuous-crack type or gummed chips.

4.4.4 *Single Edge Cutting of Thermosets*

For the present discussion, thermosets are divided into two groups, viz., *laminates* and *cast thermoset resins*. This selection is, therefore, further divided into two sub-sections for detailed discussion about these two groups of thermosets.

4.4.4.1 *Thermoset Laminates*

Thermoset laminates consist of resins and a base material or filler. The former can be a phenolic, urea, melamine, epoxy, silicone, or polyester resin; and the latter can be paper, fabric, asbestos, glass or nylon. There are more than 300 different kinds of laminates.

Unlike thermoplastics, these laminated industrial plastics are unglamourous, and having little visual appeal, they are generally hidden from view in the cabinets of television sets or in the interior of automobile bodies. They also find increased use in aerospace, missile, and satellite components. By varying the type of filler and resin binders, it is possible to obtain a wide variety of laminated plastics.

The machining methods depend upon the type of laminate involved. Phenolic-paper, phenolic-glass, phenolic-cloth, and melamine-cloth laminates are representative of those used as industrial laminates; and are discussed in this section. Special attention must be paid to the selection of cutting conditions and tool geometry involved in cutting thermoset laminates, as these materials have complicated physical characteristics related to their resiliency, anisotropy, thermal properties etc.

Several kinds of chips are produced by cutting perpendicular to the direction of lamination. Under various conditions of depth of cut and rake angle at a low cutting speed, some are continuous, with or without curl, while others are discontinuous. Continuous chips occur at small depths of cut and large rake angles. The chips become discontinuous as the depth of cut increases and the rake angle is made negative.

At higher cutting speeds, the chips turn from continuous to discontinuous even at small depths of cut. They also tend to become needle-like or powdery as the depth of cut and cutting speed increase. The chips produced from a cloth-base laminate at higher cutting speed, have almost a discontinuous powderlike form.

In general, the type of chips formed in cutting paper and cloth-base phenolic laminates are classified as continuous-shear, discontinuous-simple-shear, discontinuous-complex, and discontinuous-crack types.

Cutting forces vary with depth of cut, cutting speed, kind of material, and the rake angle. As with other plastics, a critical rake angle exists in cutting phenolic laminates that varies with depth of cut and cutting speed. *Table 4.11* lists values of critical angle for machining phenolic laminates under some sets of cutting conditions.

Table 4.11: Critical Rake Angles for Cutting Phenolic Laminates

Depth of Cut	Laminate Base		Critical rake angles for cutting speed of			
(mm)			50m/min	100m/min	200 m/min	400 m/min
0.05	Paper:	A	36°	32°	28°	26°
		B	31°	27°	24°	22°
		C	24°	21°	18°	16°
		D	34°	32°	29°	27°
	Cloth		26°	22°	20°	18°
0.10	Paper:	A	25°	20°	18°	16°
		B	21°	17°	16°	14°
		C	17°	15°	13°	11°
		D	23°	21°	18°	15°
	Cloth		18°	14°	11°	10°

It is evident from the listed values that the critical rake angle varies not only with the kind of laminate but also with the cutting conditions, such as depth of cut, and cutting speed.

Specific cutting forces for phenolic laminates are rather large compared with those for other plastics, and they vary with cutting conditions. They also decrease as the rake angle changes from negative to positive. It is better to choose a depth of cut and cutting speed as long as possible, and to use a positively raked tool for cutting phenolic laminates, in order to minimize the specific cutting forces. However, under these conditions the chips become discontinuous, resulting in rough machined surfaces.

It has been reported that the surface finish of the cut surface is not affected by cutting speed or rake angle but it improves as the feed rate is reduced. (Note: *Feed rate in turning is equivalent to depth of cut in orthogonal cutting*). It is also reported that the surface finish improves with the increase in nose radius of the cutting tool.

Phenolic laminates are apt to wear the tool material. High speed steel (HSS) is very easy to wear out when phenolic laminates are cut. The tool wear in HSS is much larger than that of the carbide tool. A diamond tool is most economical for exceptionally long runs on laminates in spite of its high initial cost.

Continuous chips of paper and cloth-base phenolic laminates are produced under a rather narrow range of conditions. Based on the large specific cutting force required, rather a poor surface finish, the large amount of tool wear during cutting, and the small percentage of continuous chips produced under various cutting conditions, the machinability of phenolic laminates can be rated only as fair. However, phenolic laminates can be easily machined with good tolerances and good surface finish when optimum cutting conditions are employed.

With regard to other laminates, almost all the chips produced from cloth-base melamine laminates are discontinuous and powderlike. A critical angle seldom exists in cutting melamine laminates. Values of specific cutting force are nearly equal to those for paper-base phenolic laminates. Tool wear in cutting melamine laminates is rather large. Therefore, *machinability of melamine laminates is rated as poor.*

Tool wear in cutting glass-base laminates is very large compared with other laminates, even when carbide or ceramic tools are used. The surface finish of the machined surface is very poor in glass-base laminates. Thus, the *machinability of glass-base laminates is the worst amongst the thermoset laminates, and diamond tools must be used for cutting glass-based laminates in production operations.*

It is more important to reduce cutting speed for glass-base laminates than it is for other laminates. They are usually machined dry. The jet of compressed air is effective, however the resulting glass dust from dry machining may be irritating to the operator skin and eyes. The skin irritation may be avoided by using appropriate protective creams and lotions, and goggles to protect the eyes. However, it is best that some means of dust removal are provided. The best method for that is the provision of a suction system with an air jet directed towards the cutting edge to serve as a coolant and to direct the chip into the suction nozzle.

In conclusion it can be said that the important factors in cutting laminates are the cooling methods, the tool relief angles, and consideration to avoid delamination effects. In cutting laminates, air cooling by a jet directed at the point of cutting is preferable to a liquid coolant. Also, laminated plastics undergo rather large elastic deformation and recoveries, so tool relief angles should be as large as possible. To avoid possible delamination during machining clamp a backup piece to prevent separation by the cutting force. Likewise while turning a laminated tube, it should be mounted so that the direction of rotation is kept the same as that used when the tube was rolled.

4.4.4.2 *Cast Thermoset Resins*

Some of the thermosets such as phenolics, polyester and epoxy resins are weidely used in cast form. Machining methods for cast resins are described herein for polyester and epoxy resins. These plastics are prone to cracking during machining as they are brittle like glass or ceramics. However, when optimum cutting conditions are used, they can be machined easily to close tolerances and without cracks.

Peculiar type of chips are observed when cast polyester is machined. One type is discontinuous and is opaque with rough surface; it is produced with a negatively raked tool used at moderate cutting speed. Another similarly discontinuous type chip is also produced but which is transparent with smooth fractured surface. That is obtained under a rather wide range of cutting conditions, with a large depth of cut and a tool having a large positive rake angle. The latter type is called the discontinuous crack type and is produced by elastic fracture without plastic deformation. Still an other type of chip is the continuous shear type. It is obtained at a low cutting speed and small depth of cut and appears opaque; some of these have cracks along the surface.

In machining cast polyester, continuous chips are produced only at a small depth of cut, however, the chips become discontinuous as both the cutting speed and depth of cut increase. The chip formation is improved with the rise in temperature.

The occurrence of cracking is a peculiar phenomenon in machining cast polyester resin. Continuous chips are formed at low cutting speeds, but the chip becomes discontinuous at high speeds, leading to sudden decrease in cutting force.

The critical rake angles do not appear to exist for machining of polyester resin. This is because the chips are always of the discontinuous-crack type due to the brittle behaviour of the material when large positive rake angle is used.

Values of specific cutting forces for cast polyester resin are rather small. They also decrease as the depth of cut and the cutting speed increase, and as the rake angle changes from negative to positive. However, the cut surfaces will be rough.

In conclusion, the machinability of cast polyester resin is relatively poor because of the narrow range of conditions which produce continuous chips and the roughness of the machined surface even at small values of specific cutting force. It is important to choose a low cutting speed and a small depth of cut, with a tool having a 0° or slightly negative rake angle, for satisfactorily machining of cast polyester resins.

Very similar phenomena to those for machining cast polyester are observed in machining cast epoxy resins. Continuous chips are produced only at a small depth of cut and a slow cutting speed, with a tool having a zero or slightly negative rake angle.

The depth of cut must be smaller than 0.02 mm, the cutting speed below 50 m/min, and the rake angle between 0° and –10°.

The machinability of cast epoxy resin is also rated as poor because of the narrow range of conditions which produce continuous chips, and the roughness of the machined surface even at small values of specific cutting force.

4.4.5 Drilling of Plastics

Drilling is often performed on plastic mouldings and laminated parts. However, plastics are rather difficult to drill without some damage. Because their properties differ from those of metals, many difficulties not encountered in drilling metals, like gumming, burning in the drilled hole, cracks around hole edges, or the growth of cracks after drilling can occur in the drilling of plastics.

In drilling operation friction occurs between the outer surface of the drill and the walls of the drilled hole. This close clearance makes the chips hard to eject from the hole as the hole becomes deep. Because plastics have low thermal conductivity, high coefficient of thermal expansion, low softening temperatures, and large elastic recovery characteristics, the drilled surface tends to expand inward into the hole, which further increases the friction between the drill and the hole. Chips softened by the generated heat weld easily to the drill. These phenomena affect the appearance of the hole, cutting forces of thrust and torque, and the types of chips generated. The appearance of the hole is also affected by the shape of the drill. Some of the types of drills used in drilling plastics are shown in Fig. 4.45.

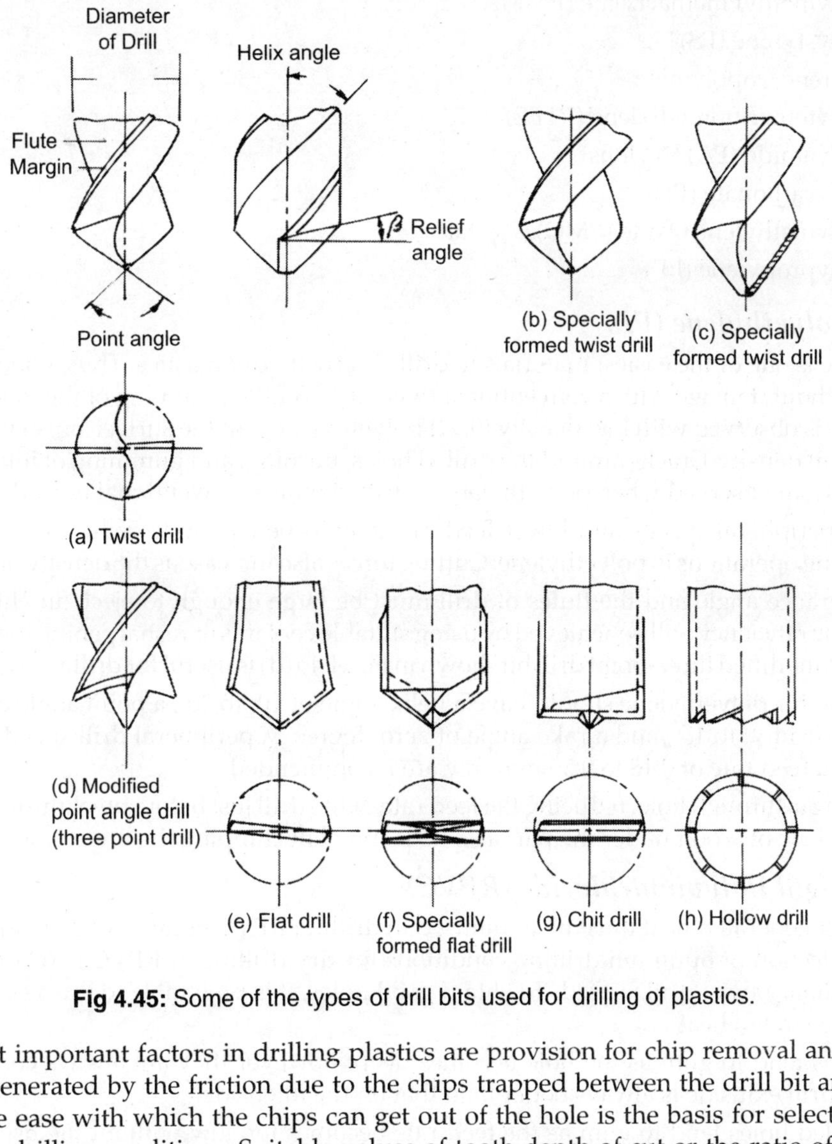

Fig 4.45: Some of the types of drill bits used for drilling of plastics.

The most important factors in drilling plastics are provision for chip removal and avoidance of the heat generated by the friction due to the chips trapped between the drill bit and the work material. The ease with which the chips can get out of the hole is the basis for selecting the drill bit design or drilling conditions. Suitable values of tooth depth of cut or the ratio of feed rate to peripheral speed of drill must be chosen. Many cracks will occur around the hole edges if the tooth depth of cut is too large; gumming or burning can occur if the tooth depth of cut is too small. In selecting a drill design, allow adequate clearance on the cutting edge and use drills with large helix angle as well as wide and polished flutes. Also, the point angle of the drill bit will have a definite effect upon the value of tooth depth of cut.

In the light of above stated facts the drilling of thermoplastics and thermosetting resins for some widely used plastics are briefly described in the following sections.

4.4.6 Drilling of Thermoplastics

Drilling characteristics of the following thermoplastics are included in this section.

1. Polyethylene (PE)
2. Rigid polyvinyl chloride (RPVC)
3. Polymethyl methacrylate (PMMA)
4. Polystyrene (PS)
5. Styrene Copolymers
6. Polytetra fluoroethylene (PTFE)
7. Polyamide (PA)/Nylons
8. Polycarbonate (PC)
9. Polymethylene Oxide (PMO)/Acetal
10. Polypropylene (PP)

4.4.6.1 Polyethylene (PE)

Polyethylene is one of the easiest materials to drill among thermoplastics. Thus, good holes can be obtained without damage with a conventional twist drill. A little gumming of the inner surfaces of drilled holes is observed with low density (0.92) polyethylene. But the surface finish improves with the increase in density. Cracks around the drilled holes, burning, and gumming of inner surfaces of holes are seldom observed when polyethylene is drilled with a conventional twist drill.

Higher peripheral speeds and lower feed rate should be chosen to reduce the cutting forces during drilling operations in polyethylene. Cutting forces also increase as the density of PE increases.

The clearance angle and the flutes of drill must be large enough to eject cut chips readily. A higher cutting efficiency will be achieved by using suitable coolants or resharpening the cutting edge frequently. A modified three-point drill bit shown in Fig. 4.45 (d) is useful for drilling in polyethylene.

A drill bit for polyethylene should have a helix angle of 10° to 20°, a point angle of 70° to 90°, a lip relief angle of 9° to 15°, and a rake angle of zero degree. A peripheral drill speed of 600 to 900 m/min and a feed rate of 0.18 to 0.25 mm/rev are recommended.

In drilling a through hole, reducing the feed rate of the drill just before breakthrough minimises burning. A block of wood under the part also helps prevent chipping the underside.

4.4.6.2 Rigid Polyvinylchloride (RPVC)

The effects of generated heat must be minimised in drilling RPVC because of its tendency to gum. Thus, the selection of optimum drilling conditions for dry drilling of RPVC is relatively difficult when an ordinary twist drill is used. To obtain good holes, it is preferable to use a suitable coolant to remove generated heat.

The chips tend to gum as the hole becomes deep; however they are always continuous. The hole on the drill exit side is always better than that on the inlet side.

The drilled holes tend to gum as the feed rate becomes too low without the use of a coolant. Gumming is minimised in wet drilling, even when small feed rate is used.

Drilling conditions for obtaining good holes without swelling or melting of the chips around the hole edge requires that instead of reducing the rotational speed of the drill it is better to increase the feed rate.

Finally it can be said that for achieving good results in drilling RPVC, the heat generated during the operation must be minimized. For holes having a depth greater than twice their diameter, drills with extra-large flutes should be used. Drills should be ground to reduce end resistance and the area of cutting surface should be reduced to approximately one-third that of the normal amount to lessen heat buildup.

4.4.6.3 *Polymethylmethacrylate (PMMA)/Acrylic*

Acrylic is rather easy to drill because of its good machinability. However, gummed, opaque holes will result if drilling conditions are not suitable. Since acrylics are widely used for their transparency, drilled holes must usually be as transparent as possible.

Chips from the same drilling conditions tend to gum at the end of the drilling process, although they are throughout of the continuous type. The range of optimum drilling conditions which will produce a good hole, exhibiting no cracks or gumming, becomes narrower as the depth of the hole being drilled increases. In actual operation it is a good practice to pull the drill up frequently when deep holes are being drilled. It is also a good practice to place some material like wood beneath the acrylic plate to avoid cracks around the exit side of the hole. The occurence of cracks around the exit side of the drilled acrylic plate is frequent with constant-load feed conditions, but not under constant feed rate conditions.

Cutting forces such as torque and thrust are relatively large when acrylic is drilled. Holes drilled by the specially designed drill, shown in Fig. 4.46, decrease in roughness compared with holes made by the conventional drills. Because of the lack of friction between the specially designed drill and the hole surfaces, little heat is generated , thus good, smooth holes are produced.

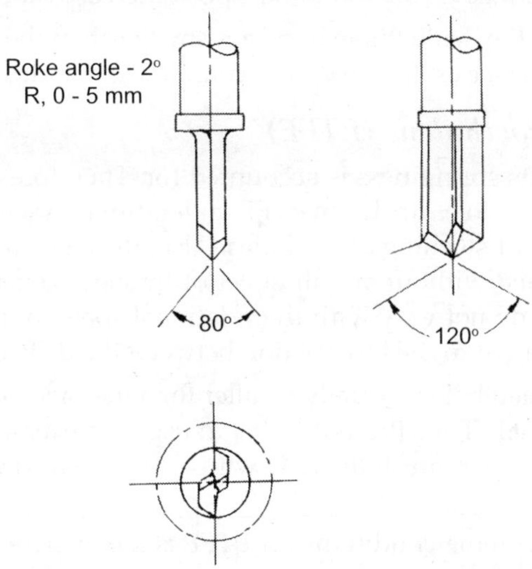

Fig 4.46: Specially designed drill bits for drilling acrylic/PMMA.

In conclusion it may be said that slow, spiral-twist drills with polished flutes as wide as possible (to help in removing continuous ribbon-like chips) are useful for drilling acrylics. The recommended point angle of the drill varies between 55° and 140°, depending on the size of the the flute and on the ratio of the hole diameter to hole depth. The wider point angle results in a shorter cutting edge and narrow chips. The lip clearance angle should be ground to 12° to 20°, depending on the hole depth.

Removal of chip is no problem in drilling shallow holes, and no coolant is needed, although a jet of compressed air directed into the hole as it is being drilled is helpful. Water or a soluble oil-water coolant can also be used as the hole becomes deeper.

4.4.6.4 *Polystyrene (PS)*

Polystyrene is one of the most difficult materials to drill. The chips produced and the drilled holes tend to gum severely under dry drilling conditions; thus it is hard to drill polystyrene without any coolant. When suitable coolant is used, chips are continuous without gumming. The selection of the type of coolant to use is very important in drilling polystyrene because when petroleum type coolant is used, cracking often occurs after drilling. It is very important to efficiently remove the chips and the heat generated by the drill.

The range of optimum drilling conditions for polystyrene is relatively narrow-much narrower than that for acrylics.

With regard to the shape of the drill, the point angle must be in the range of 60° to 90°, the rake angle should be 0° or slightly negative, the clearance angle should be relatively large–between 12° and 15°, the helix angle must be in the range of 40° to 50°, and the flutes must be as wide as possible and highly polished.

4.4.6.5 *Styrene Copolymers*

Two of the well known styrene copolymers are SAN (Styrene Acrylonitrile) and ABS (Acrylonitrile-Butadiene Styrene). ABS is further available designated as ABS-L, ABS-M and ABS-H according to their impact strengths, which are low, medium and high respectively.

Chips formed from SAN and ABS-L gum at low feed rate, however there is no gumming of ABS-M and ABS-H chips. Gumming in SAN resin can be avoided when a suitable coolant is used. Gummed chips tend to clog the hole edges on the inlet and exit sides. The inner surface of the hole will also be affected by the heat generated as the rotational speed increases and the feed rate decreases.

In SAN the diameter of the drilled hole becomes smaller than the diameter of the drill used as the peripheral drill speed increases; however it becomes larger in drilling ABS resins.

4.4.6.6 *Polytetrafluoroethylene (PTFE)*

PTFE is easy to drill if its springiness is accounted for. Therefore, a sharp edged drill must be used to minimize springback. Holes drilled in PTFE under different sets of drilling conditions do not show any cracks, gumming or swelling of the holes. Also, the chips formed from PTFE are always continuous, long, and formed without any ill effect of the heat generated. Cutting forces during drilling are very small and do not vary with the rotational speed of the drill. The small values of the drilling forces are attributed to the low friction between the drill and the work material.

The diameter of the drilled hole is usually smaller than the size of the drill used because of the springy nature of the material. Thus it is advisable to use *successive drilling technique* i.e, drill first with an undersized drill and then finish the hole with a correct sized drill to eliminate the effect of springback.

The range of optimum drilling conditions for PTFE is relatively wide.

With regard to the shape of the drill bit, a point angle of 70° to 90°, a rake angle of 0° or slightly negative, a relief angle of 9° to 15°, a helix angle of 10° to 20°, and wide, deep, highly polished flutes are recommended for PTFE containing fillers such as mica or gaphite which wear away the drill bit rapidly.

4.4.6.7 *Polyamides (PA)/Nylons*

Polyamides, i.e., nylons can be drilled satisfactorily with a conventional twist drill, provided its thermal characteristics, high toughness, and water absorption characteristics are taken care of. The selection of the drill design and drilling conditions is based upon the usual principles of minimizing heat generation and ensuring chip removal. Gumming of the inner surface of the drilled hole, cracking around the hole edges, and swelling are rarely observed in drilling nylon 6, 66, and 610.

Chips formed from nylon are always long and continuous and the inner surfaces of the drilled holes are good. The only unusual phenomenon observed in drilling nylon is the relatively large thrust force created during drilling. The range of optimum drilling conditions for nylon is relatively wide; thus good holes can always be obtained with a conventional drill bit under usual drilling conditions. However, the diameter of the drilled holes is usually smaller than the diameter of the drill bit used. That may be taken care of by adopting the *successive drilling technique.*

With regard to drill design, the specially designed drill shown in Fig. 4.45(f) improves the smoothness of drilled surface in nylons. In such a drill bit the point angle should be between 70° and 90°, and a lip clearance angle of 9° to 15°. The drill should have polished flutes at an angle of about 17° and the lands should be one-fourth the width of the heel. For long runs, carbide tipped drills are recommended.

Since low feed rates result in overheating, feed should be as high as possible, consistent with the desired surface finish. For drilling deep holes, the drill need to be pulled back frequently, as in the case of acrylics, to clear chips and prevent excessive heat buildup, and a suitable coolant fed directly to the proper point to achieve optimum results.

4.4.6.8 Polycarbonates (PC)

Good holes can be drilled easily in polycarbonates with conventional twist drill. However, gumming of the surface of the drilled hole can occur, in spite of the relatively high melting point of the resin, when drill geometry or drilling conditions are not suitable. The difference between good and gummed holes can be checked by dipping them into carbon tetrachloride (CCl_4). Gummed holes will produce fine cracks along the axis of the hole when dipped into CCl_4 for just about 15 seconds. However no cracks occur in good holes. Polycarbonate chips are always smooth and continuous; also very tough.

Gumming also tend to occur as the point angle decreases because the tooth depth of cut decreases as point angle decreases. The appearance of the drilled hole becomes less acceptable as the rotational speed increases and the feed rate decreases. Since cracking around the hole edges is not a problem in drilling polycarbonates one can use as small a peripheral speed as desired and as large a feed rate as possible.

Gumming of drilled holes in polycarbonate can also be avoided by the use of proper coolant. If gumming occurs along the whole length of the drilled hole in dry drilling, it will decrease by the use of compressed air and petroleum as coolant; and it almost disappears with the use of rapeseed oil.

4.4.6.9 Polymethyl Oxide (PMO)/Acetal Resins

Good holes can be drilled in acetal with a dry, conventional twist drill. Drilling acetal is very similar to drilling nylons.

The chips formed in drilling acetal are always long and continuous. They do not gum even under high rotational speed or low feed rate of the drill bit. No cracks occur around either the inlet or exit sides of the holes. No gumming of the inner surfaces of the hole is observed when a constant feed rate is used. However, the holes tend to have a smaller diameter than the diameter of the drill used, at relatively high feed rate.

With regard to the drill design, it is recommended that the drills have a point angle of 60° to 90°, a clearance angle of 10° to 15° and have highly polished wide flutes.

4.4.6.10 Polypropylene (PP)

Polypropylene is easy to drill with a conventional twist drill and equipment similar to that used for other thermoplastics, especially that used for polyethylene or polyamide. Swelling around the hole edges at the inlet and exit sides will often occur because of the high elastic deformation of material. Use of fixtures for drilling polypropylene is very important for avoiding elastic deformation; the accuracy of the drilled hole will depend upon the fixturing pressure.

Swelling is reduced as the feed rate increases, but is not reduced by varying rotaional speed of the drill. Swelling can also be reduced when a constant load feeding system is used. Optimum drilling conditions for polypropylene require high drill feed rate. The inlet sides of the drilled holes are always better than the exit sides. Cutting forces in drilling polypropylene are small compared with those in drilling other plastics.

4.4.7 *Drilling of Thermosets*

Two types of thermosets discussed here are:

1. Thermoset laminates, and
2. Cast thermosets.

4.4.7.1 *Thermoset Laminates*

In comparison with drilling of thermoplastics, thermoset resins are relatively more difficult to drill without the occurrence of swelling, splitting, or cracking. Therefore, the selection of optimum drilling conditions for thermosets is more critical. Laminates are drilled differently than other plastics because of their non-homogeneous structure. Drilling at right angles to the lamina tends to cause swelling or cracking around the edges of the hole, while drilling parallel to the lamina tends to cause splitting of the workpiece. Splitting can be overcome by increasing the point angle of the drill and and by clamping the work material in a jig or vise in such a way that it is supported by the side pressure. Also because of the nature of laminates, they have a tendency to close in when the drill is removed, therefore diameters of the drilled holes are usually smaller than the diameter of the drill used. To drill the required hole size in a laminated plastic slightly oversized drill should be used.

The cutting forces during drilling perpendicular to and parallel to the lamination show a distinct difference between them. Thus, the chips produced by parallel and perpendicular drilling are very different. Almost all chips formed by perpendicular drilling are continuous but become discontinuous when the drilling is done in a parallel direction.

Since specially formed drills can eliminate the cracking around the edge of the hole, thus drill design is very important for obtaining good drilled holes in thermoset laminates. Fig. 4.47 shows typical drills used for drilling laminates.

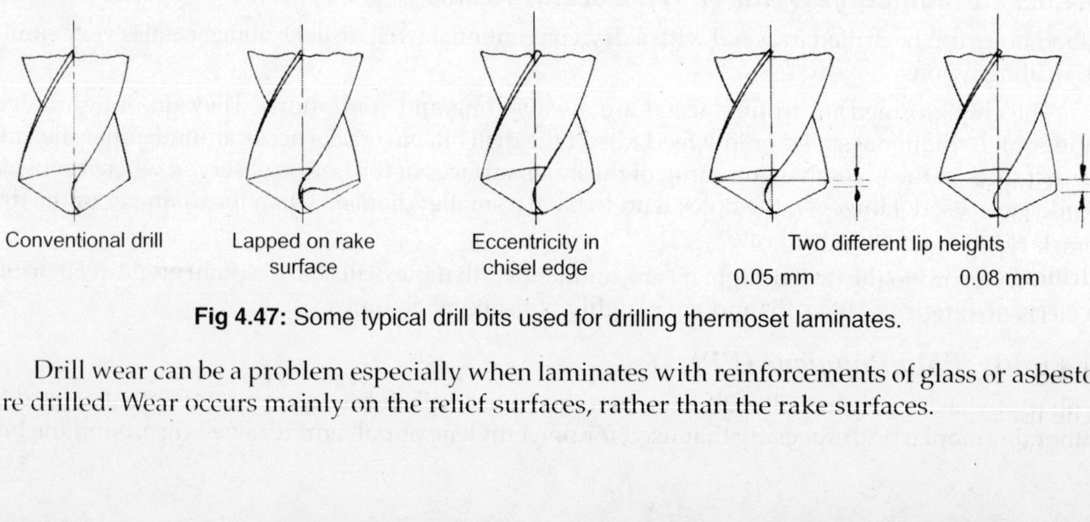

Fig 4.47: Some typical drill bits used for drilling thermoset laminates.

Drill wear can be a problem especially when laminates with reinforcements of glass or asbestos are drilled. Wear occurs mainly on the relief surfaces, rather than the rake surfaces.

The temperature rise occurring on the inner surface of a drilled hole in phenolic paper-base and glass-base laminates varies according to the depth of the hole and the feed rate. Temperature increases as the feed rate decreases.

The use of coolants also has a great effect on the temperature rise. Moreover, the temperature rise that takes place when phenolic laminates are drilled is usually larger than in the case of PVC or other thermoplastics. This further shortens drill life.

Optimum drilling conditions for laminates are based on the swelling and cracking observed around the hole edges. Swelling indicates partial delamination immediately adjacent to the drilled hole and the lifting of layers in the laminates.

Chips formed from cloth-base phenolic laminates are always discontinuous and powder like. However, swelling or cracking around the hole edges are rarely observed. Thus, the optimum drilling conditions for a cloth-base laminate are relatively wide compared with those for a paper-base laminate. Also, paper-base melamine iaminates behave like cloth-base phenolic laminates with regard to chips and the appearance of the drilled holes.

4.4.7.2 *Cast Thermoset Resins*

Thermosetting resins are more difficult to drill than thermoplastics in that cracking around the inlet or exit edges of the hole is common in the drilling of cast polyester resin as well as urea moulded parts, cast phenolic, cast epoxy, or other brittle resins. The size of the cracks is affected not only by the drilling conditions, but also by the shape of the drill. As regards the drilling conditions, feed rate is the major variable. In drill design, the point and rake angles of the drill have the largest effect on the size of the cracks.

Cracks become large as the point angle becomes large at high feed rate. However, the cracks increase slightly as the helix angle and the peripheral speed increase.

The range of optimum drilling conditions will be wider when the specially formed drills, shown in Fig. 4.48 are used, or when the work material is preheated to a suitable temperature to semi-plasticize the thermoset; that is also done for hot punching.

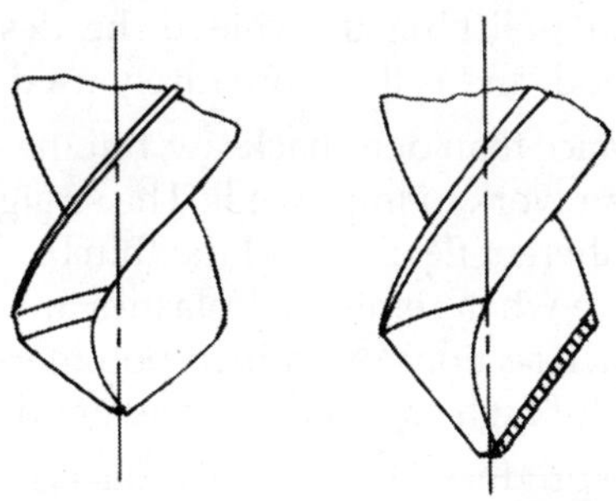

Fig 4.48: Special drill bits for drilling cast thermosetting resins.

Chips become continuous at work temperature in the range of 50° to 75°, even with large feed rates. However, they become discontinuous when the temperature becomes too high. Discontinuous crack type chips are produced at high feed rates. These phenomena differ from those occuring in drilling of thermoplastics.

The drilling characteristics of cast epoxy resin is very similar to those of cast polyester resins.

4.4.8 Sawing and Other Parting-Off Operations

The use of plastic sheets as a raw material has increased considerably over the past few decades. Because most processes for producing plastic sheet (extrusion, casting, hot-press lamination) yield an initial product which has relatively rough edges, a parting-off operation is almost imperative to finish the edges of the sheet or to cut the larger sheets into smaller ones of the desired edges.

Straight line parting-off is also used for cutting blanks needed for thermoforming; curved cuts are required when an irregularly shaped piece of sheet has to be cut from the raw sheet material for some specific product. Parting-off may also be used to provide a finished straight edge on one side of a fabricated item.

The parting-off techniques normally used for plastics were mainly adapted from the methods used in the wood working industries. However, as plastics are different in their cutting characteristics, such wood working techniques have not proved very satisfactory for plastics, both from quality and economic point of view. Most of these techniques are based on the use of methods employing circular saw, band saw, jig saw or some other types of saw. However, sawing is not the best method for cutting plastics as that usually results in excessive wear, cut-surface roughness, frictional heat generation, and saw-cut inaccuracy. Often another finishing operation, such as filing or sanding is required to achieve a surface finish of the desired quality. That leads to additional cost in manufacturing.

4.4.9 Parting-off Methods for Plastics

The methods commonly used for parting-off plastic sheet include the following:

1. Sawing
2. Abrasive cut-off wheel
3. Shearing
4. Thermal cut-off Methods

Brief descriptions of these methods follows.

4.4.9.1 Sawing

Out of the sawing methods, *manual and power sawing* are the least desirable. The reciprocating action of the saw blade produces a very rough cut with a considerable amount of cracking. This necessitates extensive amount of polishing to achieve the desired surface finish. It is also an inefficient and slow cutting method, and is therefore rarely used industrially.

Band saw produces better surface than does hacksaw, but the surface finish is still not good and requires some additional finishing work to improve it. The straightness of the cut and its accuracy are also inferior, because of the inherent flexibility of the thin blade of the band saw. Also, the wear of the band saw blade is excessive when reinforced plastics are cut. Band saw, however, is useful for cutting plastics when very thick sections of unfilled polymers are to be cut and when irregular and curved cuts are to be made. But high cut accuracy and good surface finish cannot be obtained.

The quality of the cut surface produced by a *circular saw* is of better average quality than other saws described above. The cut surface finish can be further improved by using the circular saw with larger number of teeth per unit length, for example 2 teeth rather than 1 tooth per lineal centimeter. Using such a circular saw it is possible to produce cut surface that does not require further finishing. The circular saw is most suitable for making straight cuts in plastic sheet, however rapid wear of the saw teeth necessitates frequent resharpening.

In general, sawing produces a relatively rough cut surface; the tool wear is excessive, especially when plastic containing abrasive filler is cut; the frictional heat generated and power consumed are both usually high. The three sawing methods are shown in Fig. 4.49.

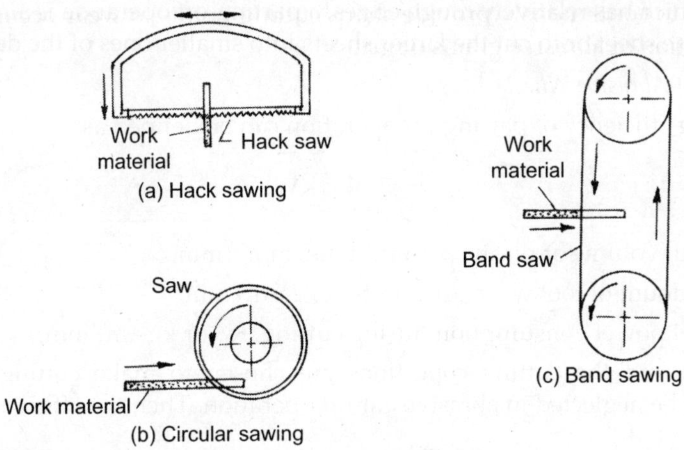

Fig 4.49: Three methods of sawing plastics, (a) Hacksawing, (b) Circular sawing, and (c) Band sawing.

The cutting force in circular sawing increases with cut-off rate and the thickness of the sheet being cut. Cutting force also increases as cut height increases. Thus, the cutting force decreases when sawing is done as near the centre of saw spindle as possible.

The total work done in cutting is minimized as the cutting force decreases, resulting in little heat generation during cutting, thus maximizing tool life. Therefore a low cut-off rate, a high peripheral saw speed, and a small cut-off height must be chosen to minimize the cutting force. With regard to the selection of a saw blade, it is preferable to make the pitch of the teeth as large as possible. The teeth should have positive rake angle and a suitable set to further reduce the cutting force.

In general, a circular saw produces a relatively rough cut edge with chipping or cracking. When sawing is done under conditions to avoid the rough cut, a large amount of heat will be generated, resulting in degradation or burning in thermosets and gumming in thermoplastics. The result will depend upon the depth of cut per tooth.

The tooth sharpening method used for saw blade is also very important to obtain good results. The saw should be sharpened without side deflection or eccentricity, individual cutting edges being kept uniform. Proper *tooth-set* will also reduce the side friction between the saw and the work material.

The saw blades used for cutting plastics are usually made of carbon steel or quenched high strength steel (HSS). Carbide-tipped saws are widely used for cutting plastics containing abrasive fillers and have longer life than do steel saws. Diamond-tipped saws are used successfully for cutting glass or asbestos-reinforced plastics in large scale production operations.

Circular sawing is still widely used for cutting unfilled thermoplastics because it has a high cut-off rate without significant tool wear and is more convenient to use.

4.4.9.2 *Abrasive Cut-off Wheels*

Abrasive wheel cutting of plastics involves the use of an abrasive wheel which is composed of a large number os small and very hard mineral abrasives held together by a bonding material, similar to the grinding wheel used for machining metals but having very small thickness. The cut-off or abrasive wheel is composed of three components viz., the abrasive, the bond, and the voids between the abrasive grains. The abrasive acts as the cutting tool, the bond holds the abrasive grains, and the voids provide the chip clearance.

The majority of grinding wheels use synthetic abrasives, such as aluminium oxide and silicon carbide. The ideal abrasive should possess a high resistance to point wear but should be friable enough to fracture when seriously dulled. A properly designed abrasive wheel must decompose at a rate sufficient to continuously provide sharp cutting points to wear occurs. This is called the self-sharpening action of abrasive grains or grit.

Cutting Efficiency of Abrasive Wheel

The machining efficiency of parting-off operation can be defined as,

$$\eta_1 = \frac{V_m}{V_w.(KW)} \quad ...(4.7)$$

Where,

V_m = Cut volume of plastic per unit time, mm³/min.

V_w = amount of tool wear per unit time, mm³/min.

K_W = net power consumption during cutting, K_W or kg– m/min.

The cut-off wheel and the cutting conditions are chosen to make cutting efficiency as high as possible. V_w can be neglected in abrasive cut-off operation. Then the efficiency η_1 becomes

$$\eta_2 = \frac{V_m}{(KW)} \quad ...(4.8)$$

Where,

V_m = $vbB \times 10^3$, mm³/min

v = cut-off rate, m/min

b = thickness of the sheet, mm

B = thickness of cut-off wheel, mm

K_W = F_t. U/6120 (∵ 1 KW = 102 kg–m/sec)

F_t = the tangential cutting-off force, kg

U = peripheral speedd of wheel, m/min

$$\therefore \eta_2 = \frac{\nu bB \times 6120}{Ft.U} \quad ...(4.9)$$

It is reported that abradability increases slightly with increasing cut-off rate for thermosets. Thus, one may choose as large a cut-off rate as possible without damage to the cut surface. On the other hand, the abradability of thermoplastics decreases as the cut-off rate increases. Thus, for thermoplastics a high-speed cut-off rate is not advisable. The proper conditions for each application need to be determined by trial and error. However, the available data indicates that usually a cut-off rate ranging from 0.5 to 3.0 m/min will be satisfactory but will depend largely on the plastic to be cut, and its thickness. An abrasive wheel peripheral speed of between 2500 and 3500 m/min is usually satisfactory for most plastics. The actual value will depend upon the diameter of the cut off wheel, the contact area between the wheel and the work material, cut-off rate, and the grain depth of cut.

Cut-off Wheel Selection

Resin and rubber bonded wheels are usually used for cut-off operation for plastics. The appearance of the cut surface made by rubber and resin bonded wheels is not much different. However, the heat generated by the rubber wheel is higher than that of the resin-bonded wheel, resulting in some degradation, gumming, or burning. Hence, the power required, and the wear of the wheel are smaller, and the surface quality is better, in the case of the resin-bonded wheel. Therefore, resin-bonded wheels are recommended for plastics, especially when operated dry.

The abrasive wheel design that has proved most efficient in cutting plastics is shown in Fig. 4.50. Basically it consists of an abrasive and conventional bonding-agent wheel formulation sandwiched between two metallic films. Another special feature of the wheel is the gridlike waffle pattern, which is embossed into the metal side walls of the wheel and is simultaneously moulded into the abrasive wheel composition itself.

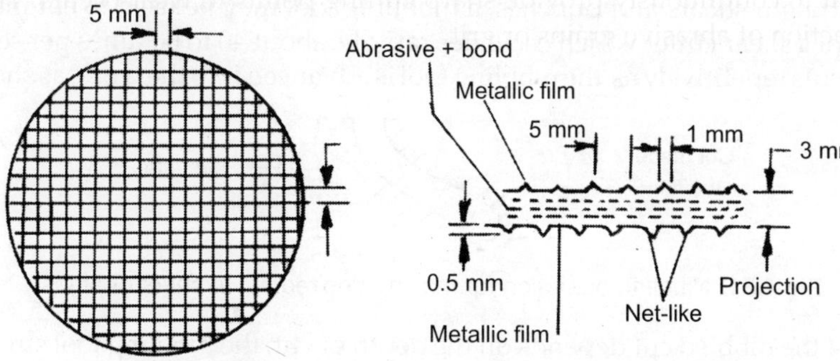

Fig 4.50: Abrasive cutoff wheel for efficient.

The gridlike pattern on the sides of the wheel is effective in reducing the friction between the side walls of the wheel and the cut surface. In addition, the metallic film on the wheel surface is found to be effective in conducting the heat away from the cut and dissipating the same to the atmosphere by means of conduction and convection. Because of the lower operating temperatures possible with this type of wheel, gumming and burning of the cut surface are almost totally eliminated leading to improved cut surface quality. The wheel is efficient in cutting not only plastics but also other non-metallic materials like glass, ceramics, asbestos, cement, carbon, etc.

4.4.9.3 *Comparison of Abrasive Wheel and Circular Saw for Cutting Plastics*

The relative merits and limitations of these two widely used methods for cutting plastics may be assessed from the following features.

1. The heat generated by sawing is usually two to three times that of abrasive cutting.
2. The life of an abrasive wheel is considerably longer than that of a circular saw. This is because the saw teeth wear faster due to decreased hardness of the tool caused by the heat generated during cutting. However, the abrasive wheel wears slowly, because of the extreme hardness of the abrasive.
3. Sharpening a circular saw is much more complex and expensive than dressing an abrasive wheel. Even if the wheel gets loaded, it is relatively easy to redress, compared with resharpening the circular saw.
4. By proper selection of the wheel it is possible to avoid gumming and loading of the wheel to bring the maintenance of the abrasive wheel to a bare minimum.
5. Abrasive wheels are generally less expensive than circular saws. Similarly, carbide-tipped circular saws are considerably more expensive than ordinary steel-toothed saws; however, the former have a significantly longer life than the latter.
6. The cutting force of an abrasive wheel at the high cut-off rates is smaller than that of a circular saw for thermosets. However, reverse is true for thermoplastics.
7. The abrasive wheel becomes more effective with use, as the sharp cutting edges of the abrasive get exposed with wear and that increases the cutting ability of the wheel.

Overall abrasive cutoff method is superior to circular sawing, especially for thermoset laminates, from the stand point of surface finish, cutting accuracy, tool life and economy.

4.4.9.4 *Shearing Methods*

Punching and straight line shearing operations are widely used in plastics fabrication industry. However, unless proper precautions are taken, the cut surface is prone to significant chipping and cracking. It is possible by heating to get materials which shear clearly without chipping. Unless that is done, additional finishing of a sheared surface is often necessary to get acceptable quality finish.

Apart from straight shears and punches, a nibblling tool may be used to cut plastics. Nibbling involves the use of a shear cutter which oscillates rapidly, about 40 to 60 times per second to make a series of shallow cuts repetitively as the nibbling tool is advanced into the work, as shown in Fig. 4.51.

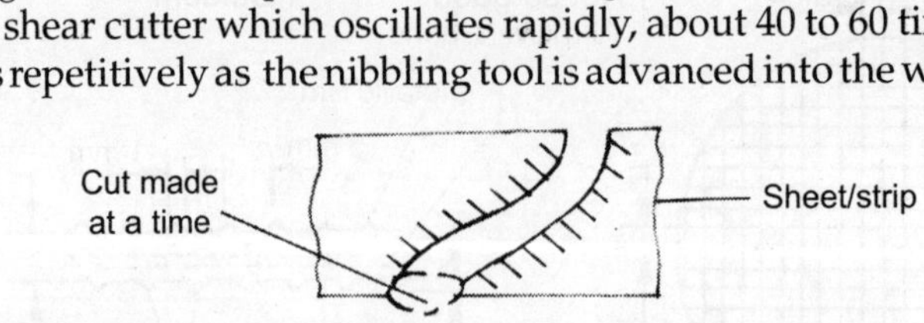

Fig 4.51: Nibbling operation for making desired cut in sheet or strips.

The quality of the nibbled cut depends on the depth of cut, the nibbling tool strokes per second, and the feed rate of the material into the cutter.

Nibbling is an exceptionally useful cutting method for thin plastic sheets. It has a high cutting rate, involves little wastage, can produce very smooth cuts requiring no further finishing, and can be used to cut along irregular or curved lines in plastics. However, it is not suitable for cutting very thick sheets of plastics.

4.4.9.5 Thermal Cut-off Method

Thermal cut off methods generally used for cutting thermoplastics involve the use of a *hot wire* or a *hot air or gas flame torch*, as shown in Fig. 4.52, to melt the plastic and thus part it. Because such methods involve a good possibility of burning the plastic material, and because internal stresses can be set up in the material near the cut edge due to sequential heating and cooling, these methods are not widely used for cutting plastics.

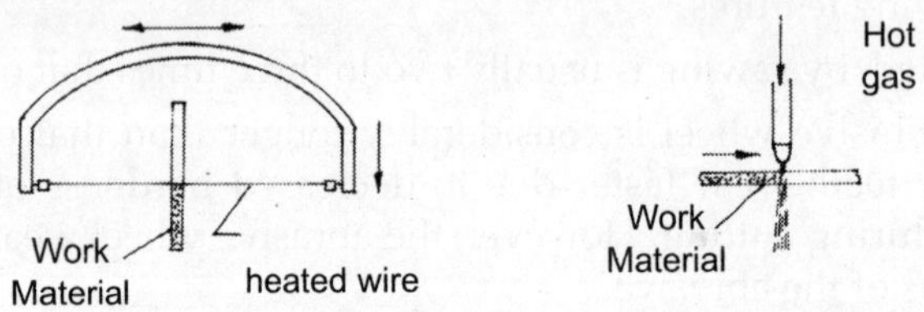

(a) Heated wire cutting (b) Hot gas cutting

Fig 4.52: Two thermal cut-off methods for plastics, (a) Heated wire cutting, and (b) Hot gas cutting.

The only major use of thermal cutting has been in cutting polymethyl methacrylate (PMMA) or foamed polystyrene, wherein a hot wire is used in a way similar to using a jigsaw.

Table 4.12 summarises the merits of different cut off methods for plastics.

Table 4.12: Comparison of Cutting-off Methods Used for Plastics

Cutting action	Cutting-off methods	Cutting-off rate	Roughness of cutoff surface	Degeneration of cutoff surface	Wear of tool	Analysis*
With tool..	Hack sawing	Slow	Poor	Seldom	Large	X
	Band sawing	Fairly quick	Not so good	Seldom	Large	B
	Circular sawing	Fairly quick	Good under optimum condition	Large in some cases	Large	A
With abrasive grains	Abrasive cutting off	Quick	Very good	No	Seldom	A
	Ultrasonic cutting off	Very slow	Not so good	Severe	....	X
	Abrasive blasting	Very slow	Fairly good	Scarcely	....	X

Shearing	Simple shearing	Quick	Not so good	Seldom	Fairly large	A
	Nibbling	Quick	Good under optimum conditions	Seldom	Seldom	B
Heating...	Frictional heating	Fairly quick	Not so good	Severe	....	(A, B)
	Electrical heating	Slow	Good for thermoplastics	Severe	....	(B)
	Gas heating	Slow	Poor	Severe	....	X

*Guide to symbols:

A: Cutting-off method for straight cut only.

B: Method for irregular forms as well as straight cut.

X: Not used on industrial scale.

( ): May be used in some cases.

4.4.10 Milling of Plastic Laminates

The milling of plastic laminates is very similar to that of milling brass. Because of the laminated structure of the plastic laminates, down cut milling, as shown in Fig 4.53 *(a)*, is always used to prevent any tendency towards delamination. The cutter should have a nagative rake of about 10°. The width of a ridge between two milled slots should be greater than the depths of the slots. This will keep the ridge safe from breaking.

The slot ends should be designed to be round instead of square, as shown in Fig. *4.53 (b)*. This makes a stronger part and will permit the use of an end mill to make the rounded end, instead of hand filing. Corners on machined or punched laminated parts, should be rounded to prevent delamination and fractures, Fig. 4.53 (c).

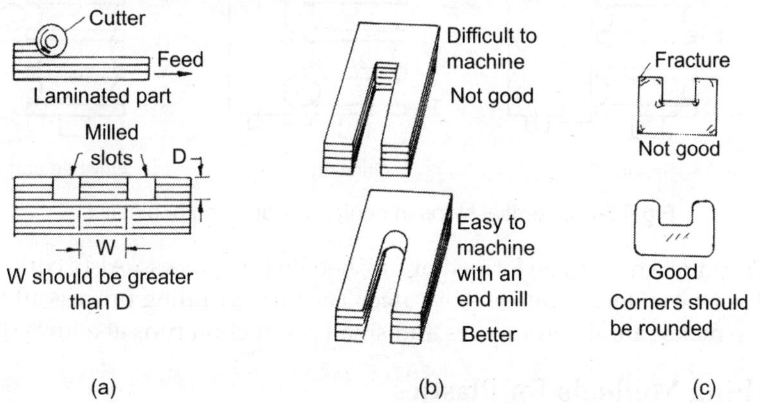

Fig 4.53: Methods for successful milling of plastic laminates.

Laminated plastics can be machined on standard wood or metal working equipment. On glass-based laminates, the cutting tools get dulled rapidly. Diamond or tungsten carbide tools give more satisfactory life.

4.4.11 Grinding of Plastics

Grinding machines used for metal working are frequently used for finishing plastics when the tolerances of parts is relatively tight. Operations such as surface and centreless grinding are used mainly for plastics. The grinding of thermoplastics is rather difficult because of their low melting point, and results in clogging of the grinding wheel. The grinding of thermosets, especially those including abrasive filters such as glass, asbestos, or mica, is strongly recommended in practice, because of the difficulties involved in machining such materials with conventional tools.

The grinding wheel with open-grain spacing and low grade is preferred for grinding plastics. It should be used with sufficient volume of coolant especially when grinding thermoplastics so as to prevent overheating and wheel loading. The grain size of the grinding wheel is determined by the finish desired.

4.4.11.1 *Centreless or Form Grinding of Plastics*

Form grinding of plastics is a unique application of centreless grinding, designed to produce cylindrical and spherical shapes from a variety of materials, including reinforced plastics.

In centreless grinding, the workpiece in the form of a rod is passed between two wheels, supported on an angular work rest blade, as shown in Fig. 4.54. The large formed work-wheel removes the stock, while the smaller regulator wheel serves to rotate the work piece at the correct speed. The workpiece is ground down to size, after which the wheels are opened and the finished piece is ejected. A liquid coolant such as water is used to keep the workpiece and wheels cool. A superb soft and silky finish is obtained by this process.

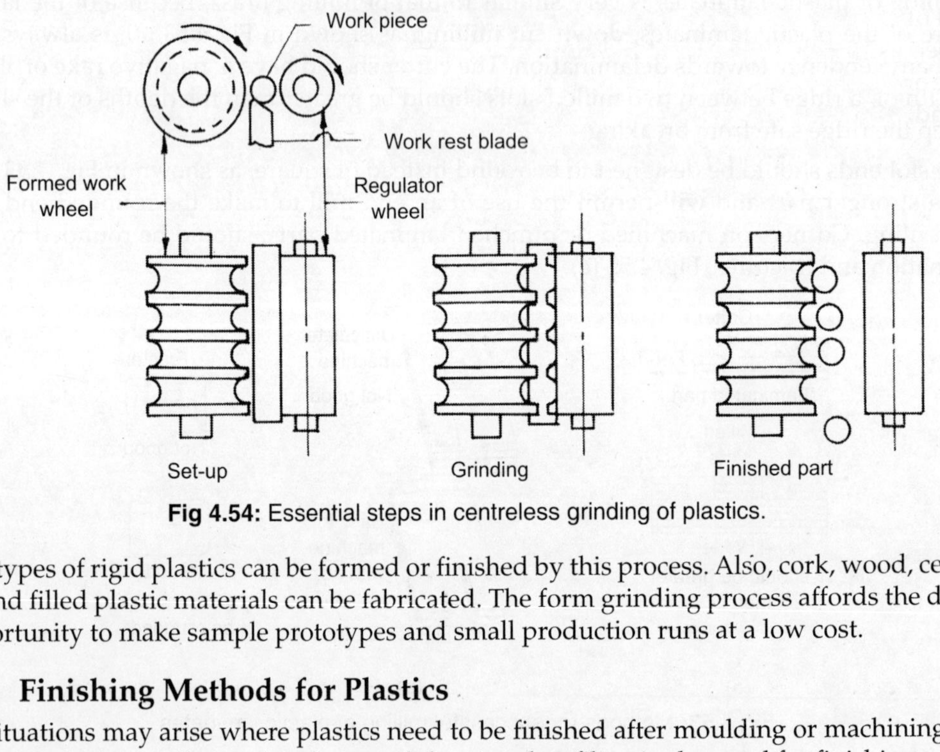

Fig 4.54: Essential steps in centreless grinding of plastics.

All types of rigid plastics can be formed or finished by this process. Also, cork, wood, ceramics, glass, and filled plastic materials can be fabricated. The form grinding process affords the designer an opportunity to make sample prototypes and small production runs at a low cost.

4.4.12 Finishing Methods for Plastics

Many situations may arise where plastics need to be finished after moulding or machining. These finishing methods usually involve the use of abrasives but filing is also used for finishing moulded articles, for smoothening, for deburring, and for finishing the edges and corners of sheets of plastics. Apart from grinding, that has been described earlier, the other industrially important methods include the following.

1. Filing
2. Abrasive belt grinding (Sanding)
3. Barrel finishing

4. Blast finishing
5. Buffing.

Brief description of these methods follows.

4.4.12.1 *Filing*

Both thermoplastics and thermosets can be filed successfully. However, selection of the proper file is of great importance, since the shape, cut size, and pattern of the file will determine the ease and speed of stock removal, as well as the appearance of the filed surface.

Thermoplastics can be filed easily and can be prepared for final polishing by filing alone if the job is done correctly. The use of ordinary metal finishing files on thermoplastics is unsatisfactory because of excessive clogging. A mill file with deep, single-cut, coarse, curved teeth (commonly known as *vixen file*), as used for aluminium, magnesium, and other soft metals, is very effective for thermoplastics. This type of file produces a shaving action with its sharp teeth that will remove thermoplastics smoothly and cleanly. With regard to the shape of the files, half-round files are useful for notches or grooves. The file should be cleaned frequently with a wire or fibre brush in order to prevent clogging. Long sweeping strokes and light pressure should be used.

Parts moulded from thermosetting plastics by compression, transfer, and plunger moulding always require some finishing operations for removal of flash from the parting line and also for removal of gates. Stock removal by filing is influenced by the pressure acting on the file and by the filing speed. Fig. 4.55 shows the correlation between stock removal per unit time, filing speed, and pressure on the file for filing a paper-base phenolic laminate. The stock removal increases as the filing speed and pressure increase. However, files tend to wear rapidly when filing thermosetting plastics.

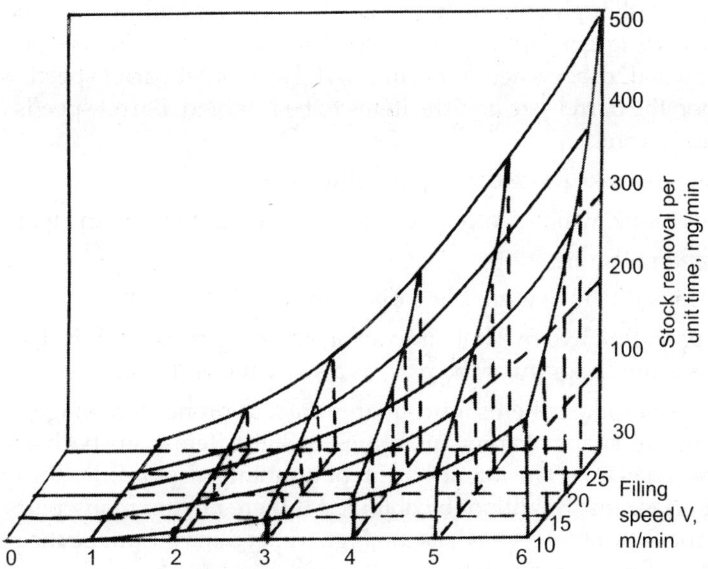

Fig 4.55: Effect of process parameters on stock removal in filing of paper-base phenolic laminates.

4.4.12.2 *Abrasive Belt Grinding (Sanding)*

In some cases, the marks left by gate cutters, or flash at a parting line, must be removed from a moulded component. Sanding is commonly the last step in finishing plastics before buffing. Where sanding is planned, a coarse, wet or dry abrasive belt is used first. This is followed by belts of successively finer grits until a satin finish is obtained. It is often advisable to start with medium fine grits and use progressively finer grits. Care must be taken to ensure that all scratches left by the previous coarser grits are removed by the succeeding belts of finer grit. Finer abrasive belts are usually used more often for thermoplastics rather than for thermosets.

Machine sanders for finish work should be run with a coolant and with extremely light feed to reduce frictional heat and to prevent clogging the abrasive belt, particularly when sanding thermoplastics. Care should be taken in filtering the coolant to prevent abrasive grit contamination of the surface of the moulded part.

4.4.12.3 *Barrel Finishing*

Barrel finishing involves tumbling the pieces in a barrel with some other abrading material(s). It is used to remove sharp corners, flash, and stumps of gates of moulded parts, and to polish surfaces. It is the cheapest method of achieving these aims. The equipment is not expensive, and the only labour involved is in the loading and unloading of the parts to be finished. Parts are placed in a rotating barrel with tumbling media and compound for a specific length of time. Maple wood pegs, felt blocks, ivory nut chips are usually used as media; the tumbling compound is generally composed of sawdust, oil, and abrasive particles. Tumbling has been applied successfully since long to thermosets, however it was much latter that know-how was made available to achieve successful barrel finishing of thermoplastics like polymethymethacrylate (PMMA), PTFE, polystyrene, and polyamide resins.

Tumbling is carried out in two ways, viz., *wet tumbling* and *dry tumbling.* The former is used almost exclusively for the removal of excess stock, while the latter is used for all operations including polishing. Horizontal and oblique barrels are used in wet tumbling. The horizontal barrel is generally octagonal or hexagonal in shape, while the oblique barrel is tapered and has eight to twelve sides. Although oblique barrel is easier to load and unload, the horizontal barrel is preferred because of its larger capacity and better tumbling action. Barrels are made of steel and often have lining of wood, rubber, neoprene, or vinyl. Horizontal barrel speeds vary from 20 to 38 rpm, depending upon the barrel size and the items to be tumbled. Barrel speeds in dry tumbling is kept between 28 and 32 rpm.

Three types of actions take place during tumbling, viz.,

(i) *Chipping or cutting*, which removes burrs by breaking them off the workpiece,

(ii) Grinding and polishing,

(iii) Lustering - usually by dry tumbling operation.

The number of parts that will fit safely into a barrel is determined by the barrel size, the size of the part, weight, shape and fragility of the part, as also by the end result desired.

Parts of simple shapes like spheres and squares are less prone to damage, and better finished than those with intricate shape. Larger quantities of light items can be barrel finished in one operation than the heavier ones.The intended aim of tumbling also decides the quantity of parts to be tumbled. Good surface finish will not be obtained if there are too many parts in the barrel. It is imperative to use proper tumbling speed, as too high a speed may damage the parts and too low a speed will not produce the desired finish.

The vibratory and high speed planetary motion of barrel can produce better surface in shorter time. However, it is essential to find out the optimum tumbling parameters by trial and error before making the production runs, as adequate data may not be available for all types of plastics.

4.4.12.4 *Blast Finishing*

This process involves blasting or bombarding moulded plastic parts with tiny nonabrasive pellets to remove the flash by impact. There are two ways to do that. One method involves the use of compressed air to propel the pellets, while the other uses a rapidly rotating wheel to hurl the pellets on to the part. Normal flash with a thickness of 0.15 to 0.20 mm is best removed with pellets of plastics and agricultural products like crushed nut shells, instead of abrasive grit. They have enough mass to be propelled with sufficient energy to break off the flash or burr but are not large or sharp enough to break, nick or scratch the moulded part.

Blasting speeds vary from 3000 to 4200 m/min, according to the kind and size of pellets used; the faster the speed, the rougher the surface and more the wear of the pellets. It is easy to deflash thermosetting plastics. However, blast finishing is more difficult for softer materials. Therefore, to blast finish thermoplastic parts it is preferable to harden them by cooling to below freezing point by using some refrigerant like dry ice.

Blast finishing of plastics is usually more economical than barrel finishing when blasting is done under suitable conditions.

4.4.12.5 Buffing

Buffing is the process of removing surface defects and gate marks to achieve a shining lustrous surface on plastic parts. The buffing operation consists of three steps, viz., ashing, polishing, and wiping.

Ashing is done with a ventilated wheel of open construction, made up of alternate layers of say 450- and 200 mm diameter muslin disks. In this way, an ashing wheel about 100 to 125 mm wide may be built up. The ashing wheel is kept dressed with a slurry of pumice (powdered solidified lava) and water during the buffing operation. The plastic part is held lightly against the wheel and kept in constant motion to prevent burning or uneven ashing. It is essential that the wheel be operated slowly enough to retain charge of the slurry.

The *polishing operation* is performed in a similar manner and on a similarly constructed buffing wheels as used for ashing, the difference being that the wheel is operated dry and a polishing compound is applied to one-half the surface of the wheel, the other half remains untreated. The plastic part is first held against the treated half of the wheel for polishing and then moved to the untreated side so that the polishing compound will be wiped off.

For *production buffing* of parts of simple contors, automatic buffing machines are available which are custom made. All dry buffing and polishing operations require an efficient exhaust system. Suitable sheet metal hoods should enclose as much of the wheel as is practicable and should be connected to the exhaust piping. The exhaust should pass through a dust collector rather than released to the open air.

In the ashing and polishing of thermoplastics overheating must be avoided, since it may soften and distort the surface into ripples. Hence it is necessary to avoid excessively hard buffing wheels, excessive wheel speed, or excessive pressure of the work against the wheel. Although data is available for buffing of specific thermoplastics like polyethylene, acrylics, polystyrene, nylons, acetals, polypropylene, etc. but it is beyond the scope to include all those in this text.

For polishing thermosets, greaseless compositions on a muslin buffing wheel are recommended for the removal of surface defects, light or residual flash, marks from machining left by the belt sander. Although conditions and materials differ somewhat with the individual job, it is a common practice to use a No. 220 greaseless compound on a full-disk loose muslin buffing wheel at a speed of 1,500 m/min. Two types of buffing procedure are followed in buffing thermosets. These are :

1. Cutdown buffing
2. Luster buffing.

1. *Cutdown buffing* is a procedure preliminary to final luster buffing that converts a dull sanded surface into one with a smooth semigloss. In cases where a very high luster is not needed, cutdown buffing becomes the final operation.

   For cutdown buffing of thermosets, compositions of waxes are used; the compound is essentially a fast-cutting buffing powder in a grease in a grease binder. A dry fast-cutting bar with no free grease is well adapted to this operation and produces a minimum of buffing dirt. Sewed, pocketed, or full-disk loose buffs are used, depending on the imperfections to be removed. Speeds of buffing range from 1200 to 1800 m/min.

2. *Luster Buffing*compositions or waxes are composed essentially of the finest abrasive buffing powder, such as levigated (paste of very fine powder) aluminium in a grease binder. The powders are finer and contain less grease than those used for cutdown buffing. Loose muslin buffing wheels at speeds from 1200 to 1500 m/min are generally used. If compositions with excess grease have been used, the soft residual film of grease can be removed by being wiped with a clean, dry, soft buffing wheel to expose the lustrous surface.

Like for thermoplastics, data are available for achieving desired results in buffing of thermosetting plastics like phenolic and urea formeldehydes, etc. but such detailed description is beyond the scope of this text.

4.4.13 Miscellaneous Manufacturing Processes for Plastics

There is a large number of miscellaneous manufacturing processes employed for producing the desired product from a given plastic. Some of the more widely used processes include the following.

1. Punching
2. Staking
3. Broaching
4. Riveting
5. Snap-in and Snap-on
6. Silk screening
7. Pultrusion
8. Thread Cutting
9. Filament Winding
10. Spray up

Brief description of these processes follows in this section.

4.4.13.1 Punching

The term punching is often used to describe the production methods of making laminated parts by blanking, piercing, and shaving, or combination of these operations. Holes of nearly every geometrical shape have been punched in laminated sheets, however odd-shaped holes should be avoided unless the laminate stock is thin enough to allow the holes to be punched.

Punch-grade laminates can be divided into hot and cold punching groups. The heating range of laminates can be from 38° to 138°C (100° – 280°F). The heating may be accomplished by plates heated by steam or electricity, ovens, hot liquid baths, infrared ray lamps, or dielectric heating equipment. The laminated plastics should be heated as rapidly and uniformly as possible and should not be kept hot longer than necessary.

Punchable laminates can be cold punched in maximum thickness ranging from 0.8 to 3.2 mm and hot punched in maximum thicknesses ranging from 2.4 to 3.2 mm.

Punching is the forcing of a hole in a sheet, and *blanking* is the punching out of a piece from a sheet. Fig. 4.56 shows a pierce or punch, and a blank die and a typical part produced by this die. Where both punching and blanking operations are involved, the part is first punched and then blanked. Because of the resilience (elastic yield) of laminates, punched holes tend to be smaller than punch diameter and blanks tend to be larger than the diameter of the blanking die. With hot punching grades of plastics, thermal contraction results in holes and blanks smaller than the corresponding diameter of punch or die.

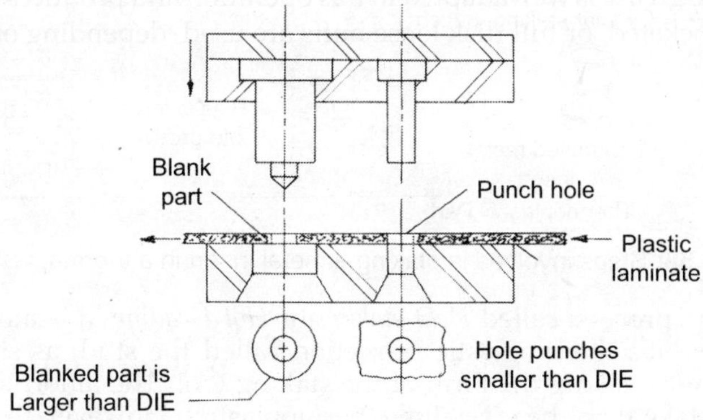

Fig 4.56: Die set-up for punching and blanking of plastic laminates.

The diameter or a diagonal of a hole to be punched should be no less than the thickness of the stock, as shown in Fig. 4.57 (a). Holes in laminated plastics should be no closer to each other or to the edge of the part than the thickness of the material, Fig. 4.57 (b). Square or rectangular holes should have their corners even farther apart than the thickness of the stock in order to prevent cracking between the holes, Fig. 4.57 (c). If it is necessary to place holes closer together and/or closer to the edge than the limits specified, the holes may be drilled.

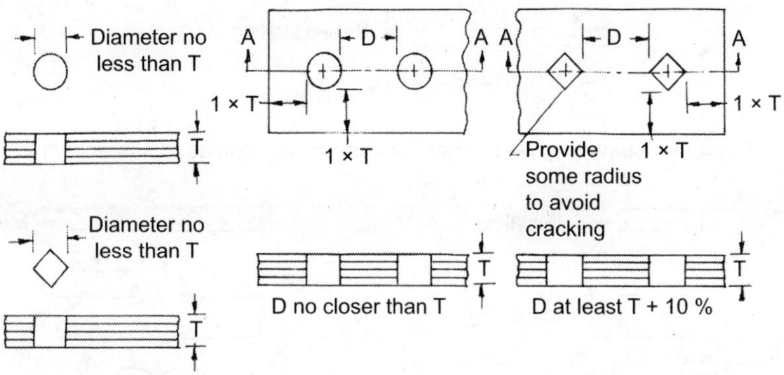

Fig 4.57: Limiting conditions for Punched Holes in laminated plastics.

Dies for Punching: Dies used for punching laminates are of the same general types as those used for punching metals, except that the clearance between the punch and the die should be less than those normally employed in punching metals. Progressive dies are usually preferred, because they are more economical to make and permit higher production rates. For standard tolerances, the die hole should be no more than 0.100 mm larger than the punch, giving a clearance of 0.050 mm all around. For clean punching of laminated plastics it is recommended that closer tolerances between punch and the die be held.

4.4.13.2 *Staking*

In this process, a metal insert is held in place by cold flowing a thermoplastic material into an external undercut on the insert, as shown in Fig. 4.58. The metal insert should be knurled and slightly oversize to stop it from twisting loose when in use. An interference fit of about 0.0250 mm between the outer diameter of knurl and the moulded hole is generally used. The metal insert is pressed into the moulded hole and a circular staking punch comes in contact with the plastic and forces the thermoplastic material into the groove.

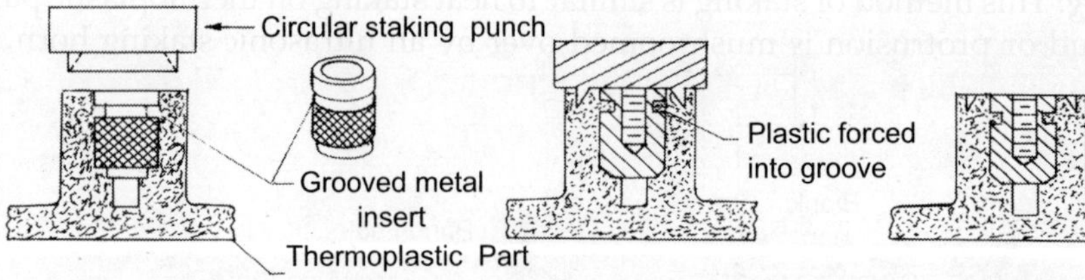

Fig 4.58: Steps involved in staking a metal insert in a thermoplastic part.

In a variant of the process, called *Heat staking* or *Hot Heading*, a heated metal staking tool is pressed down and over a thermoplastic projection called the stud, as shown in Fig. 4.59. The thermoplastic stud melts, taking the form of the staking tool. The amorphous thermoplastics are much easier to heat stake than the crystalline thermoplastics. However, this type of staking is not as fast as ultrasonic staking.

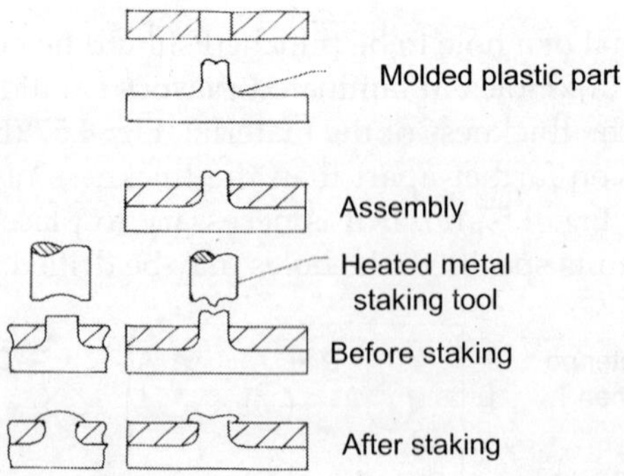

Fig 4.59: Steps involved in heat staking a thermoplastic assembly part.

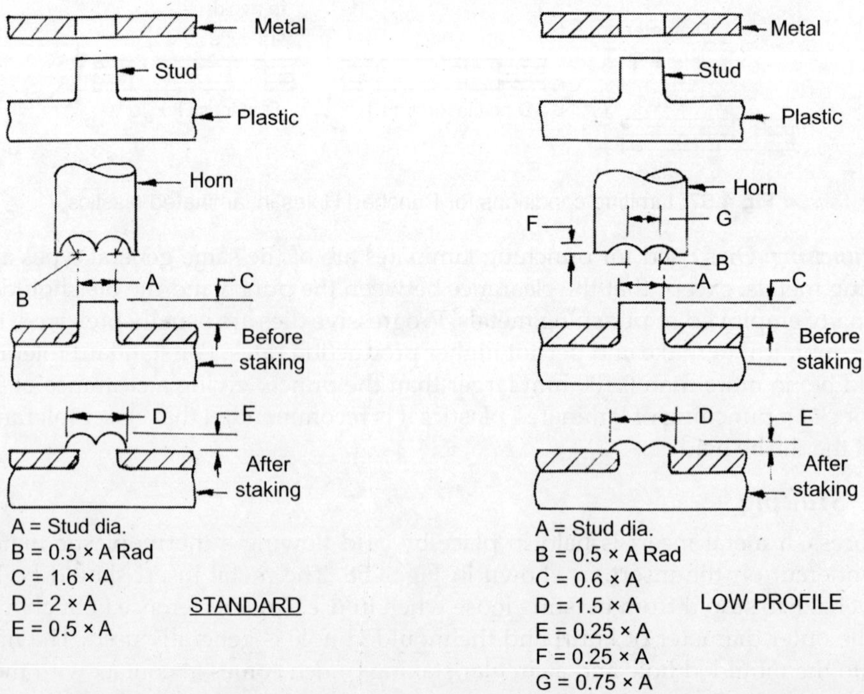

Fig 4.60: Standard and low profile Ultrasonic Staking of thermoplastics.

Ultrasonic Staking: This method of staking is similar to heat staking on thermoplastic parts, except that the plastic stud or protrusion is mushroomed over by an ultrasonic staking horn, as shown in Fig. 4.60. Ultrasonic staking is much faster than heat staking. Also, the head of the staking tool remains cool and eliminates the sticking and cleaning problem that is always present in heat staking. The assemblies are tighter than those formed by heat staking, due to the fact that there is no recovery in the staked thermoplastic material. Also, there is less chance of material degradation.

4.4.13.3 Broaching

In this process, a metal stamping with saw teeth on both sides of the insert broaches through the plastic as the part is assembled, as shown in Fig. 4.61. It is necessary to develop the hole size to determine the exact amount of undercutting that will be necessary for the flow of plastic material into the teeth to provide the desired pull out strength. An interference fit of about 0.380 to 0.500 mm per side, between the insert teeth and diameter of the hole in the thermoplastic material is found to be adequate for most applications.

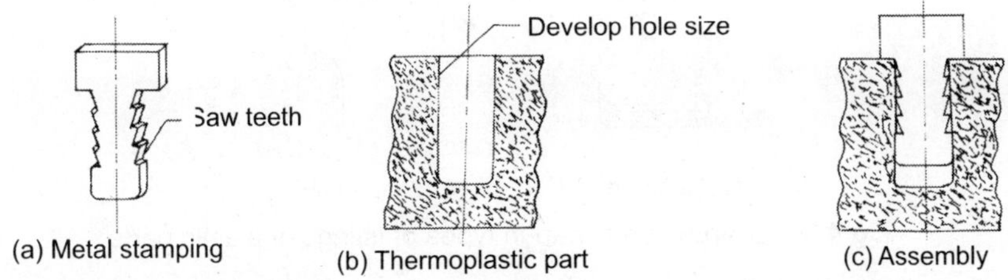

Fig 4.61: Metal and thermoplastic polymer assembled by broaching.

4.4.13.4 Riveting

Riveting is perhaps the oldest method of fastening plastic parts together. Very few fasteners can match the advantages of the tubular and split rivets. Rivetting can be done manually or with the help of bench rivetting equipment, but they do not carry the precision tolerances of metal screws and inserts. However, rivets are not considered to be the best type of fastener to withstand tension. Their great advantage is that they are inexpensive and easy to install. Fig. 4.62 illustrates a semi tubular rivet. The minimum distance of the rivet from the edge should be three times the shank diameter of rivet, and the clinch allowance should be 0.6 times the shank diameter.

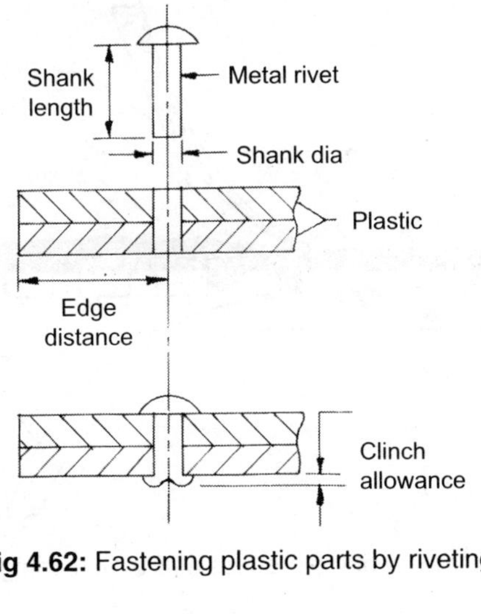

Fig 4.62: Fastening plastic parts by riveting.

4.4.13.5 *Snap-in and Snap-on Fittings*

Fig. 4.63 illustrates a snap-in and a snap-on fittings. Snap-in fittings require a tough, stiff plastic engineering material like nylon and acetal. The softer plastics such as polyethylene, polypropylene, fluorocarbons, flexible vinyls, etc. are not recommended for this type of fittings. In the snap-in fit, the part design should provide sufficient flexibility for the prongs to flex and snap back when they are inserted into the joining part. The strength of the part is limited by the amount of undercut that can be produced by injection moulding. Snap-on fits are used with more flexible materials in food containers, etc.

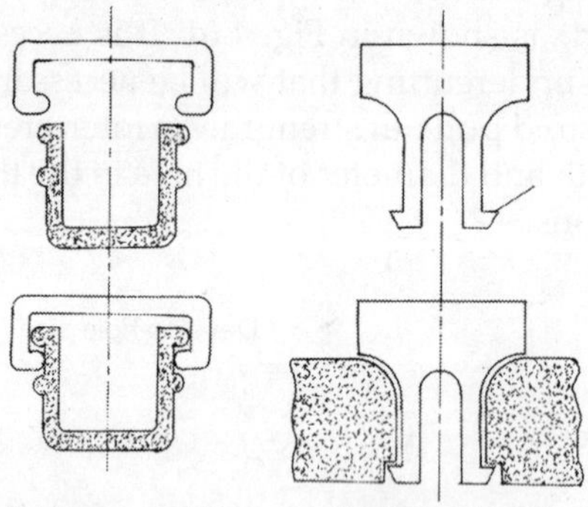

Fig 4.63: Snap-on and snap-in types of fittings of plastic parts.

4.4.13.6 *Silk Screening*

Silk screening is a decorating process used to force paint or ink through a stencil fabric, commonly called a *silk screen*, onto the plastic part that is to be decorated, as shown in Fig. 4.64. The screen consists of a taut woven fabric securely attached to a rectangular frame and carefully marked with a stencil in a manner that allows the paint to be pressed through the screen only at areas where the stencil is open. For most applications a screen with 230 perforations per square inch (35 perforations/sq.cm) is adequate. The stencil fabric is generally nylon or stainless steel. The screen is placed above the plastic part to be decorated, and a flexible rubber squeezee forces the ink or paint through the openings in the screen onto the surface of the plastic part. The rubber squeegee ranges in hardness from 50 to 80 durometer, Shore type A.

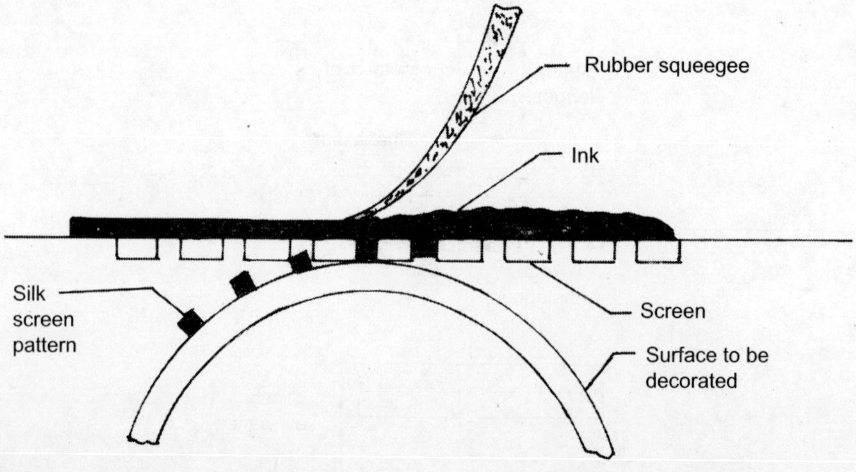

Fig 4.64: Essential elements of the process of Silk Screening for decorating plastics.

Multicolour decorating and intricate designs can be used in this process by using a series of different screens. Silk screen decoration is done on a flat or curved surface but does not give the three dimensional depth effect. Although the inks used for this process are expensive, the overall costs are generally economical.

Fig 4.65: Silk Screen painted lamp Shades.

Fig. 4.65 shows a lamp shade that has been silk-screen decorated. The injection moulded shade is made from high density polyethylene, and the surface is flame-treated before being silk-screened.

4.4.13.7 Pultrusion

Pultrusion is a special polymer processing technique that combines continuous reinforcing fibres (roving yarn) and a resin matrix in a suitable die, to produce uniaxially reinforced profiles with outstanding longitudinal resistance. Modifications to the basic process now allow the incorporation of some transverse reinforcements, either continuous by winding or discontinuous as mat.

The essential elements of the pultrusion as shown in Fig. 4.66, include a filament-dispensing station, a resin bath for fibre impregnation, a guiding device to position the numerous strands, and a heated forming and curing die where the fibre/resin system acquires the desired shape and cures to a solid-state under pressure. The process requires the application of a rather strong pulling force which is applied to the solid protruded profile by a gripper cum puller device. The device features two or more pairs of contoured clamps mounted on endless tracks and taking turns in pulling a set distance releasing and returning to the starting point.

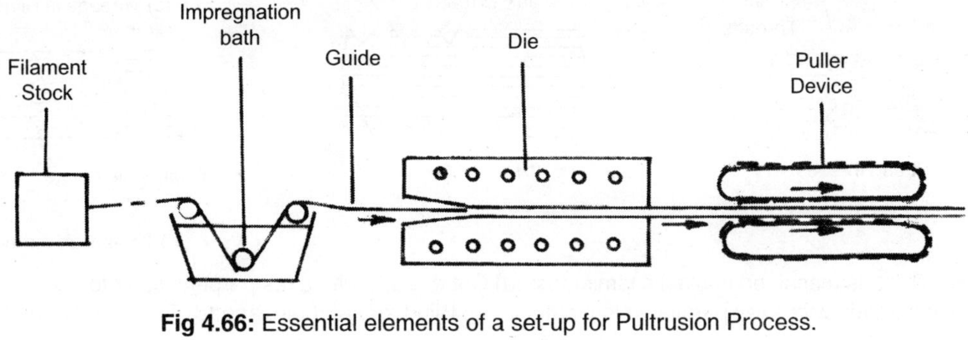

Fig 4.66: Essential elements of a set-up for Pultrusion Process.

The heart of the process is the die which is typically 75 cm long, and involves relatively high pressures as well as frictional forces. It is made of hardened steel and has a tapered entrance that facilitates the compaction of the composite. Pultrusion speeds are primarily limited by curing within the die and are thus slow, about 1-2 m/min. Precuring in the die followed by final curing in a suitable tubular oven, can speed up the process particularly for thick cross-sections. Large pultrusion machines can involve pull load capabilities as high as 10 tons.

Pultrusion products are basically straight, with a uniform cross-section. However, modifications can be made to produce slightly curved or twisted products.

Typical applications of pultrusion include cylindrical products, such as fishing rods, golf clubs and arrow shafts, antennas, and oil well sucker rods. Basically rectangular products like sail battens and snow mobile track stiffeners, channel profiles such as guide rails, cable trays, and ladder rails. Hollow profiles such as ladder rungs, pole-line or cherry picker booms, and windmill blades are also made by pultrusion.

A variant of the process, called *Pulforming* forms a heated, but not fully cured *pultrudate* in suitable moulds mounted on endless tracks or on a carousel. The cross-sectional area of the product must be nearly uniform along its length, but the cross-sectional shape can be varied significantly. This is of particular value in the case of leaf springs for example, where the moment of inertia can be widely varied to match the bending moment.

Roll-trusion is another variant of Pultrusion that forms profiled continuous fibre-reinforced products in the gap between contoured counter-rotating heated rollers. The pull is considerably less than in the stationary die pultrusion process, but the forming zone is short and most of the curing must take place in the absence of compacting pressure.

4.4.13.8 *Thread Machining of Laminates*

Threads are generally machined and not moulded in laminated plastics. Unlike moulded threads, machined threads do not start abruptly, nor do they come to an abrupt end. Machined threads start and stop with a feather edge. The feather edge in laminated plastics is weak. If the axis of a thread is perpendicular to the laminations, Fig. 4.67 (a),the threads will have maximum strength. If the threads are cut on the outside diameter of a tube or a rod, Fig. 4.67 (b), the mating part will exert a stress in direction parallel to the axis of the tube or rod, the threads of the laminated plastic will tend to shear off or delaminate.

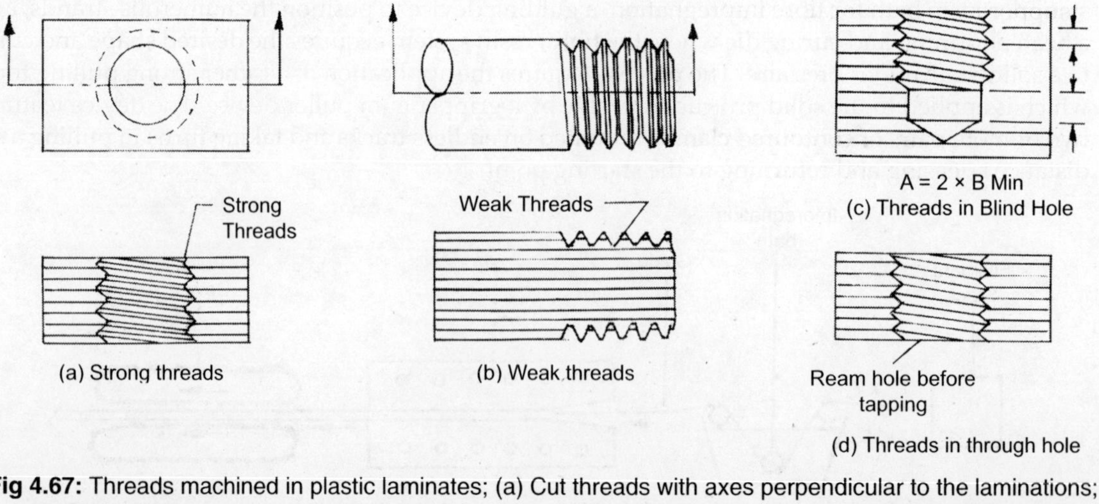

Fig 4.67: Threads machined in plastic laminates; (a) Cut threads with axes perpendicular to the laminations; (b) Cut threads with axes parallel to laminations, (c) Blind tapped hole, and (d) Through tapped hole.

When drilling or tapping parallel to the laminations, always clamp the workpiece between two supports to prevent splitting. This is not necessary when working perpendicular to the laminations, although a backup plate to prevent chipping results in a cleaner hole. A blind tapped hole should have sufficient clearance at the bottom, Fig. 4.67 (c), in order to prevent delamination/stripping of the threads. Through tapped holes Fig. 4.67 (d) should be reamed before tapping, as that will lead to the edge of the cut thread to be much smoother.

4.4.13.9 *Filament Winding*

Filament winding is an FRP processing techniques that involves the mechanical winding of resin-impregnated continuous filament over a mandred or core, as shown in Fig. 4.68.

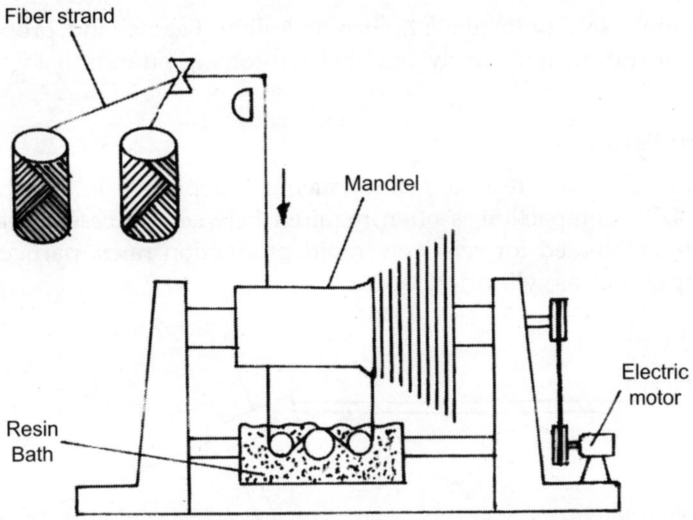

Fig 4.68: Set-up for Filament winding for producing high-strength products from fibre-reinforced plastics.

The basic elements of a filament winding machine normally include a filament-dispensing unit, say in the form of a spool(s), a resin bath to impregnate the continuous fibre strands, a carriage mounted guide whose axial motion is controlled through a screw mechanism, and finally the supports and drive spindle for the rotating mandrel/core.

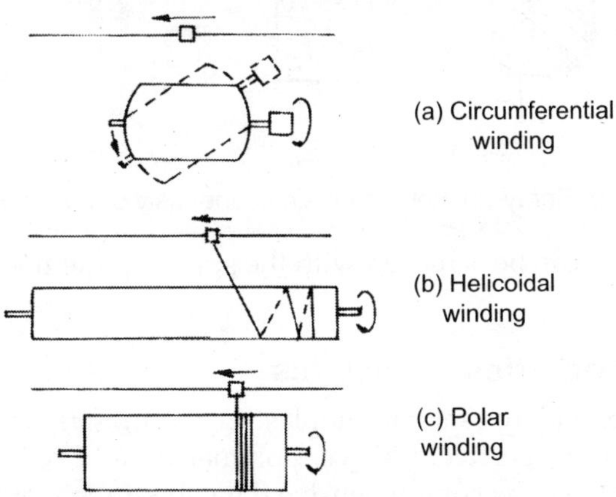

Fig 4.69: Three methods of Filament winding; (a) Circumferential, (b) Helicoidal, and (c) Polar method of winding.

After winding and curing the composite material at room temperature or by infrared ray heating or by the use of an autoclave, the part is mechanically removed from the cylindrical or tapered mandrel. Filament winding operation can be carried out by three methods of increasing complexity, viz., cicumferential or hoop winding, helical winding, and polar winding as shown in Fig. 4.69. Winding rates can range from 5 kg of composite per hour for complex parts to as much as 250 kg per hour for large pressure tanks.

Applications for filament winding include horizontal and vertical storage tanks for chemicals, oils, fuels, etc. and stack liners of structures. Pressure or crush-resistant tubings, pipes, and conduits, as well as pressure bottles, tanks and vessels, rocket motor cases, and shotgun barrels, are also made by this process.

Parts that require axial reinforcement, such as hollow beams, and propellers, can be made by winding tapes featuring transversly oriented reinforcing strands held together by stitched polymeric fibres.

10.23.10.10 Spray-up

In this process, fibre glass and resin are simultaneously deposited in a mould by spray gun(s), as shown in Fig. 4.70. Compaction is often required between successive passes. Spraying and compaction can be automated for relatively rapid production rates, particularly in the case of simple convex shapes such as cylindrical tanks.

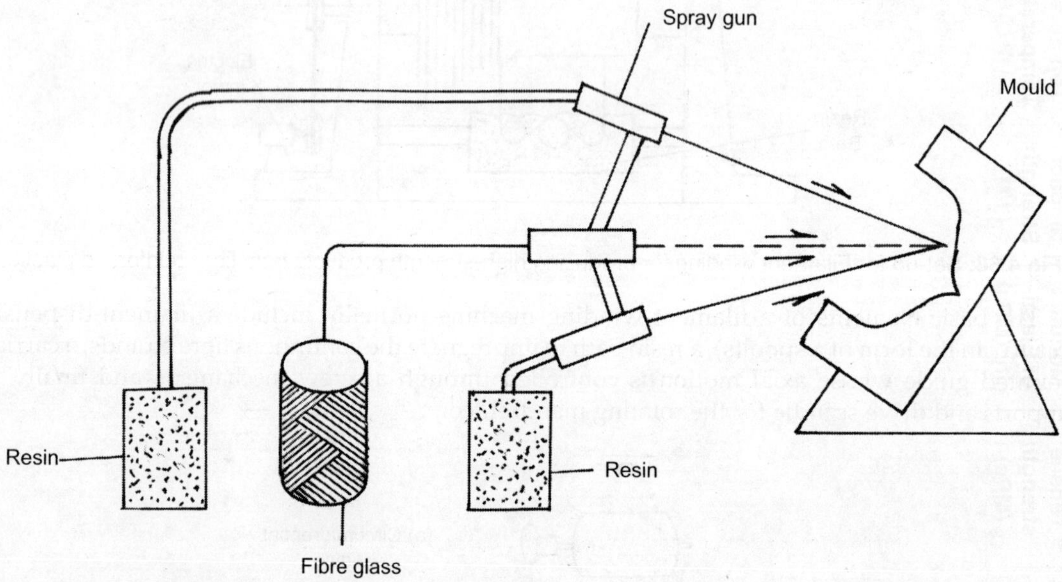

Fig 4.70: Schematic of Spray-up method for simultaneously depositing fibre glass and resins.

Faster production rates can be achieved with the spray-up method for the fabrication of boats and other large objects.

4.5 Summary of Properties of Plastics

The properties of the more widely used thermoplastics, thermosets, and elstomers are summed up in Tables 4.12, 4.13 and 4.14 respectively. Newer polymers that have been marketed more recently are not included in these tables as comprehensive data about their properties are not available in the standardised form.

Table 4.13 : Characterictics of Major Thermoplastics

Class	Trade names	Tensile strength (MPa)	Modulus of Elasticity (MPa)	Impact Strength (J/M)	Top Service Temperature (°C)	Advantages	Limitations	Typical Uses
ABS: Acrylonitrile butadiene styrene	Abson Cadco Cycolac Marbon Seilon	35-55	690-2760	107-535	120	Varieties offer wide choice of physical, chemical, and electrical properties at moderate cost	Burns but self-extinguishing grades available; long weather exposure changes appearance	Auto trim, impellers, cases, piping, helmets, knobs, handles, grills, and appliance housings
Acetals	Celcon Delrin	55-69	2585-2825	75	85-105	High strength, stiffness, and endurance, and solvent and heat resistance; lowest cost bearings	Fair to poor resistance to burning, radiation, weathering, and strong acids and bases	Auto trim, impellers, housings, piping, bearings, gears, cams, levers, fans
Acrylics: Polymethyl methacrylate (PMMA)	Acrylite Lucite Perpex Plexiglass	41-72	1930-3450	10-43	60-95	Moderate strength; good optically from clear to colored; good electrical resistance	Soft with low heat resistance	Lenses, signs, name-plates, decoration, display novelties, dials, glazing, bottles
Cellulosics: Cellulose acetate Also plus butyrate Cellulose propionate Ethyl cellulose	Ethocel Forticel Lumarith Plastocele Tenite	10-60	380-2760	21-590	50-95	Good strength, toughness, transparency, surface gloss, chemical resistance, and moldability from various grades	More quality at higher costs; mostly used for low loads and temperatures	Knobs, handles, appliance housings and trim, glazing, eye shades, packaging

Contd...

Chlorinated Polyether	Penton	41-69	900	1765	120	Good strength to 250°F (121°C) temp., water, and chemical resistance, and moldability	High cost	Piping, lining, pumps, bearings, valves, meters, tanks
Fluorocarbons: Tetrafluoroethylene (TFE) Chlorotrifluoroethylene (CFE) and others	Fluorthene Polyfluoron Teflon	17-45	415-1725	135-215	205-260	Outstanding inertness and temp. and elect. resistance; low friction	Low strength but reinforceable; high cost	Bearings, piping, electrical insulation, enamels, nonstick-high-temp. coating
Phenoxies	Bakelite	65	2895	80-135	77	Good ductility, creep resistance, low-temp. properties, and processability	Low heat resistance	Aircraft ducts, piping and conduits, computer components
Polyamides	Nylon Ultramid Versalon	41-82	1450-2830	32-855	100-150	High strength and rigidity; high impact, temp., electrical, and chemical resistance	Softened by alcohols and glycols and absorbs water	Cloth, bristles, sutures, bearings, tubing, gears, cams, gaskets, insulation
Polycarbonates	Lexan Merlon	62-72	2345	640-855	120	High strength, dutility, rigidity, and electrical resistance from - 355° to 250°F (–215° to 121°C) transparent	Fair chemical resistance, some heat aging	Safety shields, lenses, covers, sight gages, helmets, gears and cams, forms
Polyethylenes	Alathone Ethylux Polythene	10-38	90-1240	75-1070	93	Tough from -70° to200°F (–57° to 93°C) good chemical, moisture, and electrical resistance; low friction and easy processing	Many grades and quality varies with cost	Housings, piping, ducts, containers, insulation, housewares, toys, coatings, film, packaging

Contd...

Polyimides	Kapton Vespel	72	3170	38	315	Strong and rigid with excellent heat, abrasion, creep, and radiation resistance	Low impact strength; available only as coatings, films, or solid forms	Motor, wire,circuit, and instrument insulation; circuit boards
Polypropylenes	Escon Propylux Tenite	8-20	1035-1515	32-800	120	Good rigidity and resistance to heat, chemicals, moisture, and electricity; easy processing	Impact strength at sacrifice of rigidity or processability	Portable TV cabinets, piping, inside auto trim, luggage, hinges, packaging
Polystyrenes	Cerex Loralin Lustron Styron	17-76	1585-3450	13-590	50-100	Good electrical, heat, and stain resistance; hardness, stability, and moldability; many varieties	Tendency to load crack; ultraviolet degradation	Piping, dials, knobs, toys, packaging, high freq-insulation, battery boxes, dental plates, dinnerware, auto & appliance parts
Vinyls: Polyvinyl acetate alcohol chloride etc.	Chemaco Compar Elvanol Resistoflex Saran Vinylite	7-76	4-3725	13-107	138	Very flexible to rigid; good flame, electrical, chemical, oil, abrasion, and weather resistance in various grades; easy to process; colorful	Some heat degradation; some grades permanently indented; stiff at low temperatures	Floor and wall covering, upholstery, rainwear, tubing, toys, insulation, phono. records, safety glass interlayers

Table 4.14: Characteristics of Major Thermosetting Plastics

Classification		Trade names	Tensile strength (MPa)	Impact strength (J/m)	Advantages	Limitations	Typical uses
Aldehydes	Phenol Formaldehyde Phenol Furfural (Phenolics) Most widely used plastic	Bakelite Durez Resine Textolit	24-83	13-1765	Low cost; good heat and electrical resistance, rigidity, strength, hardness, and stability; low moisture absorption; can be made fairly resistant to chemicals; easily molded, cast, or laminated	Colors generally confined to blacks and other dark shades; severely attacked by strong acids and strong alkalis	Handles, pulleys, wheels, fuse blocks, plugs, coil forms, laminates, shellmolds, radio cabinets
	Urea-Formaldehyde (Ureas)	Bakelite Beetle Plaskon	35-70	13-1875	Physical properties comparable to phenolics; easily molded in large or small sizes; high resistance to organic solvents	Not cast; relatively high molding pressure; not highly weather or moisture resistant; poor resistance to strong acids and alkalis	Cabinets, toilet seats, electric devices, buttons, containers

Contd...

Aldehydes	Melamine-formaldehyde (Melamines)	Catlin Cymel Melmac	35-70	13-642	Excellent electrical and heat resistance; good strength; hardest plastic; good colorability and stability; low moisture absorption	Higher cost; not cast; fairly high molding pressures	Switch panels, moldings, buttons, ornaments, kitchen and dinner ware, lighting fixtures and switchgear
Polyesters: Alkyds (Glyptal) Allyl cast resins		Dacron Kriston Laminac Mylar Therma-flow	20-70	16-642	Exceptional dielectric properties; passes high-frequency radio waves; low moisture absorption; resists heat to 400°F (240°C); good colors; suitable for many operations; good weather, chemical, and flame resistant varieties	Higher cost; compromise among some properties necessary	Radar domes, ignition, radio, and T.V. parts, aircraft glaze, boat and car bodies, reinforced, structural shapes
Epoxides: Epoxy resins Epichlorohydrin bisphenol		Araldite	20-83	20-1605	Good chemical and electrical resistance; low shrinkage; adheres well to metal, glass, etc; not affected by most organic solvents	High cost; mostly available in liquid form for casting, foaming, and reinforced molding and laminating	Tools and dies, precision castings, high strength laminates, encasing electrical parts, adhesives, coatings and points

Table 4.15: Characteristics of Major Elastomers

Names	Properties (see foot note)	Features and particular uses
Natural rubber Natural Polyisoprene NR	A:R B:17-24 C: 7.5-8.5	Excellent physical properties; good resistance to cutting, gouging, and abrasion; low heat, ozone, and oil resistance
Isoprene Synthetic polyisoprene IR	A:R B:14-55 C:4.0-6.0	Same as natural rubber but requires less mastication; auto tyres, power belts, hoses, gaskets, seals, rollers
GR-S or Buna S Styrene-butadiene, SBR	A:R B:1.4-2.1 C: 4.0-6.0	Good physical properties when reinforced; excellent abrasion and water resistance; not oil, ozone, or weather resistant
Butyl Isobutylene isoprene, IIR	A:R B: 17-21 C: 7.5-9.0	Excellent weather and heat resistance; low gas permeability; good chemical, ozone, and age resistance; fair strength and resilience; tyre inner tubes, steam hoses, and insulation
Polybutadiene Cis-4 BR	A : R B : 1.4-7.0 C : 4.0-10.0	Overall properties like rubber and SBR but better abrasion and weather resistance and resilience; usually used in blends
Ethylene propylene, EPR (terpolymer = EPT)	A : R B : 7.0 (Max.) C: Poor	Good mechanical properties when reinforced; exceptional sunlight, oxygen, and ozone resistance; good electrical and temperature properties; for insulation, footwear, weather stripping
Neoprene Chloroprene CR	A : S B : 21-28 C : 8.0-9.0	Excellent ozone, heat, weather, and flame resistance and mechanical properties; good oil and chemical resistance; for oil hoses, tank linings, and insulation
Buna N, Nitrile Acrylonitrile-butadiene, NER	A : S B : 3.5-6.2 C : 4.5-7.0	Excellent oil and good chemical resistance; fair mechanical and poor low-temp. properties; carburetor and gas tank and pump parts, gaskets, printing rolls
Thermoplastic elastomers Thermolastic (EVA)	A : S B : 4-21 C : 5.0-11.0	Good elastic and mechanical properties; superior flexibility at low temperature; processing fast and at low cost
Thiokol (PS) Polysulfide	A : S B:>7.0 C: 4.5-6.5	Outstanding oil, solvent, and water resistance; other properties fair to poor; for gaskets, diaphragms, hoses, and life jackets
Hypalon (HYP) Chlorosulfonated-poly-ethylene CSM	A: S B: 28 C: 6.0 (max.)	Excellent ozone, weather, and acid resistance, and color stability; fair oil and low temp. resistance; for chemical and petroleum hoses, connectors, etc., shoes, and flooring
Urethane (PU) Polyurethane diisocyanate	A: S B: >35 C : 5.4-7.5	Exceptional abrasion, cut, and tear resistance; high strength, modulus, and hardness; good oxygen, ozone, and sunlight resistance; especially for vibration dampening and sound deadening; low moist-heat resistance

Contd...

Chloro-butyl Chloro-isobutylene-isoprene IIR	A: T B: 17-21 C: 7.5-9.0	Properties similar to butyl with service temps. to 400°F (204°C) and good oil resistance when blended; for inner tubes and curing bladders
Silicone Polysiloxane SIL	A: T B: 4-9 C: 1.0-5.0	Temps. -120° to 600°F (–84° –316°C); high oxygen, ozone, and radiation resistance; high compression set; low strength, wear, and oil resistance; insulation; seals, gaskets
Viton (FLU) Fluorocarbon elastomers	A: T B: >14	Temps. -40° to 600°F (–40° –316°C); outstanding oil and chemical resistance, especially at high temps.; good mechanical properties; for aircraft and industrial equipment
Acrylic rubbers Polyacrylate ABR	A: T B: 1.4-2.8 C: 4.5.7.5	Excellent temp., ozone, and oil resistance; poor water and low temp. resistance; for seals, gaskets, hose, and O-rings

Code for designation of properties :

A. ASTM type: R for no resistance to oils; S for specific resistance to oils; T for prolonged exposure to abnormal temperatures and compounded oils.
B. Pure gum tensile strength range - MPa
C. Pure gum elongation at rupture - 100%.

Table 4.13 to 4.15 can be made use of for selecting suitable polymer for given service conditions as is borne out by the case study that follows in the next chapter.

5 CASE STUDY: AUTOMOTIVE PRODUCTS

5.0 General

Plastics and rubbers have long been used in the automotive industry, but in the past few decades in particular, their use has grown considerably in volume and sophistication. The areas of application have expanded from such classical ones as electrical insulation or elastomeric components to include appearance and functional elements.

An automobile is made primarily of the following three classes of materials: *Metals*, which comprise steel as formed sheets or machined parts as well as cast iron and aluminium alloys; *Polymeric materials* which include plastics, rubbers, paints, varnishes, and textiles; and finally *glass.*

The weight-based percentage use of polymeric materials in automobiles can be broken down into plastics about 40 wt%, elastomers including tyres about 40 wt%, and paints and varnishes about 15 wt%. Tyres are estimated to account for about 80 wt% of the automobile elastomers.

The different sections of a common automobile (passenger car) that involve the use of polymeric materials include,

1. Major Structural Elements
2. Powertrain and Underhood Elements
3. Interior Elements
4. Exterior Elements

Polymeric materials used as different components in all the automobile body parts are described briefly in the following sections.

5.1 Major Structural Elements

These elements include:

1. Body Frame
2. Bumper System
3. Suspension System
4. Vibration and Noise Control System.

5.1.1 Body Frame

Automobiles were earlier manufactured with a rigid and heavy metal chasis onto which the body was attached. Currently, most automobiles have a self-supporting body frame consisting of pressed and welded sheet metal onto which sheet metal body panels are attached, however polymeric body panels are rapidly being introduced. Fig. 5.1 shows the frame and related structural elements of a common automobile car.

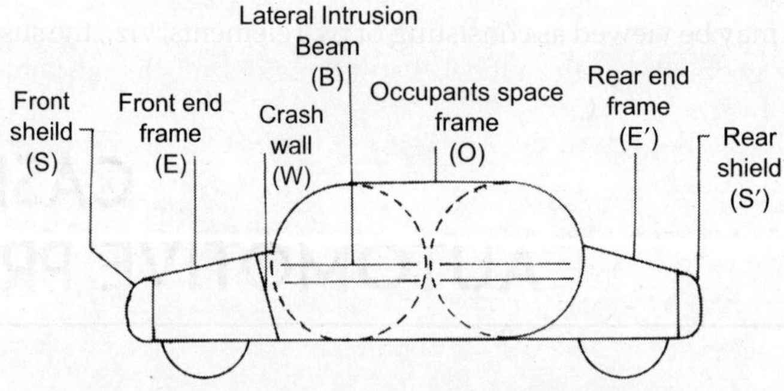

Fig 5.1: Automobile frame and related structural elements.

Increasing concern for occupant's safety is associated with the search for a very resistant and undeformable occupants space frame (O), combined with front (E) and rear (E′) end-frames that support functional elements and can collapse in a controlled manner in the case of a serious accident, with the maximum possible absorption of energy.

One of the solutions involves, the use of filament wound composite occupants' space frame to which moulded composite front and rear end-frames are fastened. Bumpers (S and S′) attached to the front and rear end-frames, a crashwall (W), and composite door intrusion beams (B) complement the occupants protection.

5.1.2 Bumper system

Bumpers are designed to meet two major mechanical constraints. One is that they show little or no sign of damage after a minor impact, secondly they must contribute adequately to a protection of the front and rear-end integrity upon impact at speeds of 4 and 8 km/hr respectively.

All-metal, chrome-plated bumper systems have been practically eliminated, and a variety of distinct designs involving polymeric materials have been introduced.

The present trend seems to be for all-plastic integral designs that are injection-moulded from a variety of high performance plastics capable of withstanding impacts even at low temperatures. These include PAT, PBT, ABS, PC, PA-6, HIPP and A (ABS/PBT). In these designs the energy absorption is often achieved through the honeycomb type design of a back element that is easily moulded. Multicomponent box-beam system can be assembled by friction or ultrasonic welding, as also by chemical bonding.

5.1.3 Suspension System

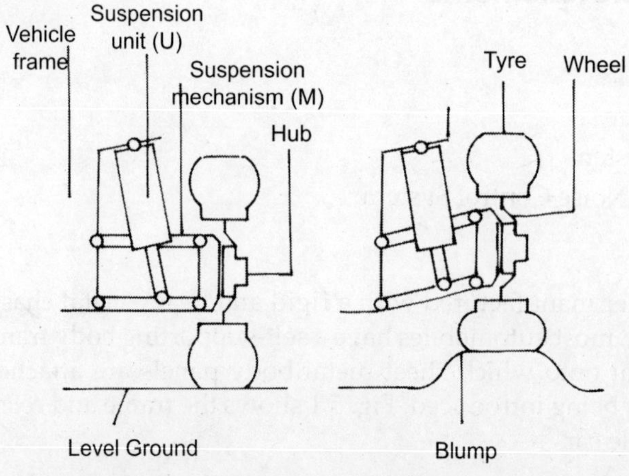

Fig 5.2: Automobile Suspension System

A suspension system may be viewed as consisting of two elements, viz., the suspension mechanism (M) that controls the geometry of the wheel displacement, and the suspension unit (U), which controls the amplitude of the displacement and its damping, as shown in Fig. 5.2. In some cases the two systems may be indistinguishable.

Traditionally, metals have been used almost exclusively for suspension elements, including the all-important springs. However, once popular metal leaf springs have almost totally been replaced by metal coil springs and tension bars. But there is now much development in polymer-based composite, i.e., fibre reinforced plastic (FRP) springs, with variable cross-section leaf springs made by special filament winding or pultrusion process. Corrugated FRP springs, somewhat equivalent to coil springs in terms of their possible location, are also being tried.

Polymeric materials and their associated fabrication techniques are expected to facilitate the implementation of sophisticated, though inexpensive, pneumatic/hydraulic systems which have the best potential for comfort and road holding compromises.

5.1.4 Vibration and Noise Control

Engines are mounted on appropriate elements of the body frame through rubber mounts normally consisting of a rubber block sandwich bonded to two metal mounting plates. The visco-elasticity of the rubber and the geometry of the block are carefully designed to filter engine vibrations and noise. NR and CR (polychloroprene rubber) rubbers are most commonly used for this purpose. Other noise producing or vibrating elements, such as exhaust system, are also insulated from the body frame through rubber mounts.

Other noise producing elements include tyres, cooling fan, vibrating panels, aerodynamic flow etc. Sound-deadening coatings which earlier were based on asphalt, increasingly involve polymers such as PVC, EVA, EPDM, and IIR; which are frequently combined with inorganic fillers.

In the interior, rubber floor mats, floor carpets and their backing material contribute to noise reduction, and the introduction of polymer-based floor panels have a favourable effect on noise reduction.

5.2 Powertrain and Underhood Elements

Powertrain includes all the elements from engine to tyres which serve to move and stop a car. The major items forming the power-train and underhood elements include the following.

1. Engine
2. Transmission
3. Driveline
4. Fuel/Air System
5. Liquid Cooling/Heating System
6. Electrical Systems

Brief description of the use of polymeric products in these elements/systems follows.

5.2.1 Engine

Elements shown in Fig. 5.3, such as the cam cover and the oil underpan, are moulded or stamped from glass-reinforced plastics like PA, PAT, RIM-Epoxy. Light weight plastic (e.g. PEEK) or composite construction is particularly promising for reciprocating parts such as moulded piston skirts, filament wound or pultruded connecting rods, pultruded valve stems or push rods, and possibly valve springs. Timing gears and sprockets are commonly moulded from RTP plastics such as PA, PP, PMO, PAT with cord-reinforced elastomeric belts.

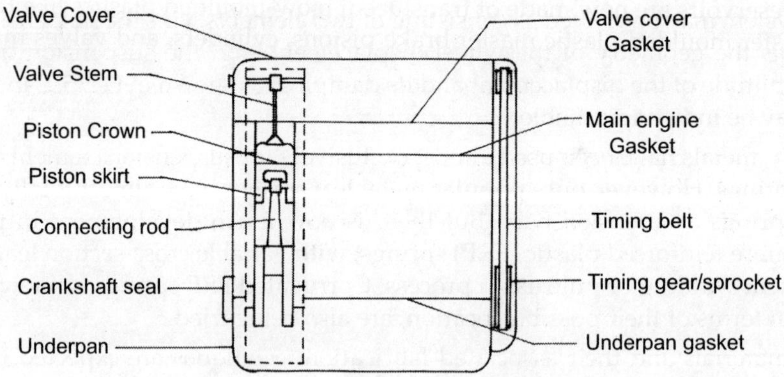

Fig 5.3: Automobile Engine Elements.

Elastomeric low-pressure gaskets or seals have long been used for crankshafts, valves, valve cover, underpan, etc. Resistance to oil must now be achieved at higher temperatures in the presence of different additives for applications such as engine-oil lubricated gearbox, turbocharged engines, etc. High performance or speciality conventional elastomers are used (e.g. CR, FKM), and the newer high-performance thermoplastic elastomers have considerable potentials. The high pressure main engine gasket normally involves a thermosetting binding resin.

5.2.2 Transmission

Plastics even of the reinforced type, are unlikely to compete with metals for compact torque–transmitting gear systems in either manual or automatic transmission. They (e.g. PMO, PA) are used, however, for some moulded internal control parts. Reinforced thermosets are used for thrust converter reactors in automatic transmissions that may involve temperatures as high as 200°C (392°F), as well as for reverse clutch cones in manual transmissions. Transmission covers and underpans are also now made of moulded or stamped reinforced plastics with elastomeric gaskets and seals.

5.2.3 Driveline

The major elements of a driveline are:

(a) Driveshaft,

(b) Braking system, and

(c) Wheels.

As these elements are controlled through Steering Mechanism so they are also discussed in this section.

(a) Driveshaft: A majority of modern cars are provided with front wheel drive (FWD), which is associated with short driveshafts, and the compulsory use of constant velocity universal axle joints. FWD axle joints require bellow-shaped flexible covers (boots), which contain the lubricating greese and are subjected to large cyclic stresses and deformations. For this application, high performance conventional rubers (e.g. CR, silicone) are being replaced by new thermoplastic elastomers that can be produced by faster production methods.

For rear drive long drivershafts, composite driveshafts made of glass and graphite/carbon are successfully replacing long metal shafts both for weight saving as also for noise, vibration and hardness (NVH) reduction.

(b) Braking System: Automobiles use hydraulic braking system; the brake fluids and their additive can be aggresive to certain polymers.

Brake fluid reservoirs are now made of translucent blow-moulded plastics like HDPE, PA, etc. Injection or transfer moulded plastic master brake pistons, cylinders, and valves made up of PA, PMO, PF, etc., have also been introduced.

Rigid metal brake lines are being substituted by plastic tubings of say PA, and flexible brake hoses normally made of rubber/cord braided construction to withstand the fluid pressure and the road hazards.

Thermoset plastics like PF have been used for brake cylinders and pistons, and are also used as binders for brake pads.

For hand brakes, plastic cable coduits help reduce rust problems associated with road salt used in sub-zero temperature regions.

(c) Wheels: Wheels consist of a rim, flange and hub; the rim is fitted with a tyre. Being non-resilient elements they should be made as light as posible.

Rims are most commonly made of pressed steel welded elements or, occasionally, of a cast aluminium alloy.

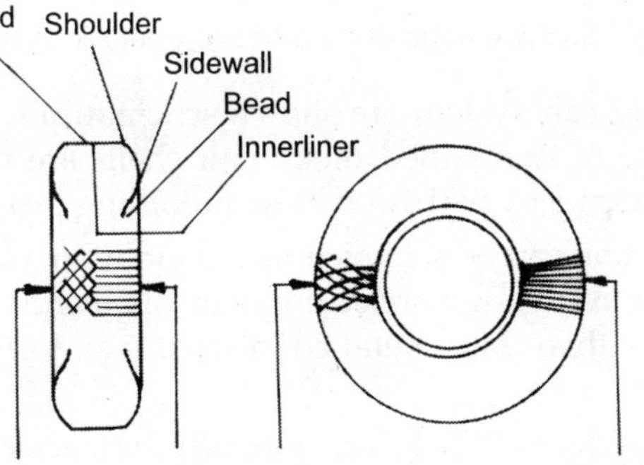

Fig 5.4: Constructional features of an automobile tyre.

A tyre, constructionally shown in Fig. 5.4, with its multiple role of cushioning, steering, traction, etc. is an item of high importance. Conventional tyres are made by complex tyre-building process that involves different rubber compounds for the tread, the shoulders, the sidewalls, the bead area and the innerliner or inner tube, as well as a network of reinforcing cords (belts) arranged radially and circumferentially in modern *radial tyres* or at an angle in conventional *bias tyres**.

Polymeric tyre cords (natural cotton, rayon, nylon - PA, polyester - PET, aramid** kevlar) have been used throughout the long tyre history. The use of nonpolymeric tyre cords (steel or glass) has been limited primarily to belts in radial tyres. Attempts have also been made to produce tyres by casting rather than building from a single material such as polyurethanes (PU).

(d) Steering Mechanism: Engineering polymers are used in steering mechanism with applications involving pleated elastomeric covers (boots) associated with the large displacement rack and pinion steering systems.

5.2.4 Fuel/Air System

Most of the internal combustion engines employ liquid fuel like petrol, diesel, and gasohol. Regular petrol contains led-based additives, while unleaded and premium petrol contain higher amount of aromatic hydrocarbons. Fuel injection, requiring fuel recirculation, introduces hydroperoxides

* *Bias tyres* have diagonal cords across threads of woven fabrics.

* *Aramids:* Synthetic aromatic polyamides which are capable of fibre formation are called aramids. They have extra-ordinary strength and can withstand high temperatures up to 250°C (482°F).

(sour fuel). Gasohol fuels contain a fraction of alcohol for cost reduction and boosting the octane rating. Polymeric material used in contact with such fuels must be resistant to chemical reaction and permeation of volatile fuels through them.

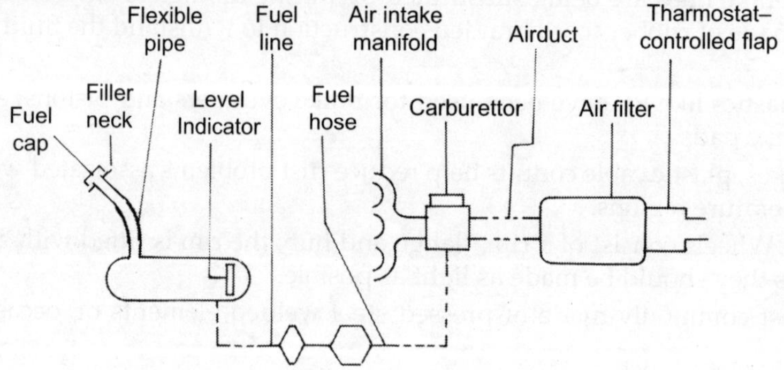

Fig 5.5: Schematic of auromobile Fuel/Air System.

The major elements of fuel/air system are shown schematically in Fig. 5.5. Fuel tanks which have traditionally been made of two joined metal half shells are now often made in one piece HDPE (high density polyethylene) by extrusion blow moulding process.

Reduction of fuel permeation can be achieved by fluoridation, sulfonation, chloro-sulfonation, or irradiation of the PE by introducing barrier polymers as blends or co-extrudates. Plastic fuel tanks are considerably lighter than their metal counterparts and easier to manufacture in space-saving contoured shapes.

For safety reasons, filler necks of PE, etc. are normally connected to the tanks through flexible connecting hoses made of fuel-resistant elastomers. Fuel caps are often made of plastics (e.g., PA-RTP). The level indicator device features plastic elements such as a metal coated tape of say polyimide.

Rigid metal fuel lines are increasingly being replaced by semi-rigid extruded plastic lines (e.g.; PA-11,12). Fuel filters feature a plastic housing (e.g., PA) a polymeric filtering medium, for example polyvinylidene chloride (PVDC).

Fuel pump housings and diaphragms or impellers, depending on the type, may involve engineering plastics like PMO, PBT, etc. Fuel hoses must be flexible and involve higher pressures; they often consist of an inner liner of fuel-resistant rubber (e.g., NBR), a fibre reinforcement (braid) and a rubber cover (e.g., CR, CSPE).

Carburettors with their intricate shapes are well suited for a switch from cast metal to plastics. Non-automotive carburettors have already been made with plastic bodies (e.g., PAT), and floats (e.g., PMO). Automotive carburettor outer linkage mechanisms are often moulded from plastics (e.g., PA, PMO), reducing the number of parts and the assembling time.

Air filter is the major element of the engine air intake system. Both housing and cover of air filters are commonly made of plastics (e.g., PP, PA, PBT), often with a reinforcing filler. The thermostat-controlled flap and its mechanism are also increasingly made of plastics.

Air-intake duct, being flexible, is normally made of an elastomeric polymer and often involves bellows easily formed by blow moulding in the case of a thermoplastic.

For the complex rigid airintake manifold, reinforced plastic materials (e.g., PA-RTP, BMC) formed by injection or transfer moulding with fusible metal cores are being evaluated.

For the exhaust systems, high temperature and corrosion-resistant plastics (e.g., PPS, PI) are finding acceptability for relatively small mechanical parts in exhaust gas recirculation (EGR) or ECD (emission control device) systems.

5.2.5 Liquid Cooling/Heating System

Most automotive engines are now liquid cooled.This method is superior to air cooling in terms of noise and temperature control; and is also convenient for heating the occupants' space. Since sealed liquid cooling systems are the norm, the liquid must have permanent antifreeze properties for cold weather. Ethylene glycol, for example, which is widely used with water, can be aggresive to metals in particular, calling for the addition of corrosion inhibitors, which, in turn, can be aggressive to other materials. Materials are normally chosen to resist temperature as high as 120°-135°C (250°-275°F).

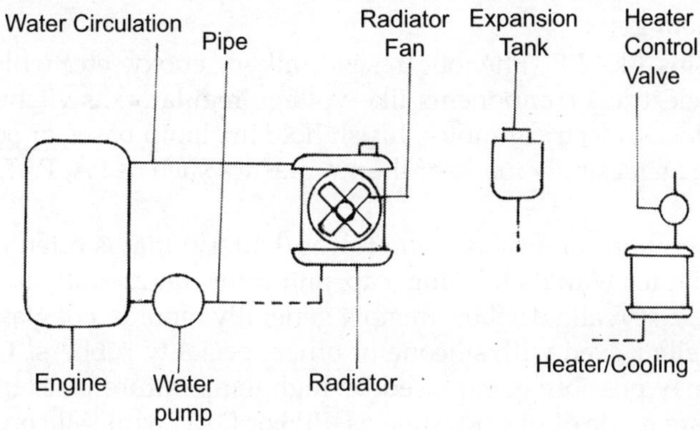

Fig 5.6: Constructional features of liquid cooling/heating system of an automobile.

Fig. 5.6 shows the constructional features of an automobile liquid cooling/heating system. This uses injection-moulded plastic end-caps (headers) (e.g., PA, or PPO-RTP), and blow-moulded transparent or translucent plastic (e.g., PP) for expansion tanks which were earlier made of glass.

Fans, mounted on thermostat-controlled electric motors, are generally injection-moulded one-piece plastic unit (e.g., PP or PA-RTP). Moulded plastic fan shrouds often incorporate motor-supporting brackets of PP or PA-RTP.

Moulded plastics (e.g., PF) are being used for water-pump housing and impeller. Liquid circulation conduits preferably have a large diameter and smooth inner surface to reduce drag and must have adequate flexibility and a specific contoured shape. Thermoplastic elastomers (TPE) are increasingly replacing coventional rubbers. For the suction side, collapse is prevented by the use of built-in corrugations or embeded stiff spirals.

The engine cooling liquid is also used for interior heating; a valve made of injection-moulded plastic (e.g., PA-RTP) controls the flow of hot liquid through the heater, nearly similar in construction to the radiator.

5.2.6 Electrical Systems

Generally, electrical circuits of automobiles operate at 12 volts, sometimes with high current (e.g., starter motor, front lighting); electrical resistance is then of primary concern. The ignition circuit, on the contrary, involves high voltages, thus arc and track resistance become important characteristics.

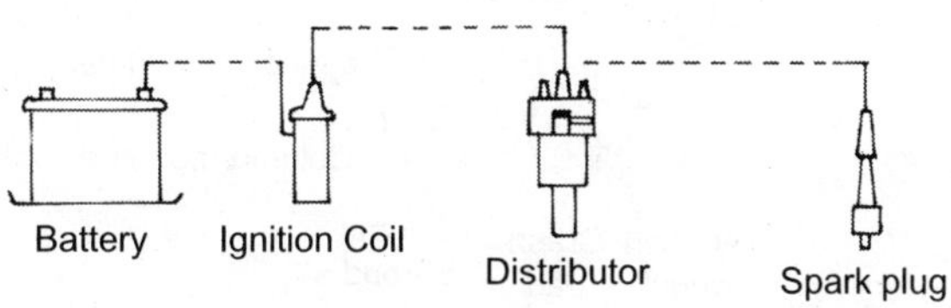

Fig 5.7: Electrical Ignition System of an automobile.

The ignition circuit of an automobile, shown in Fig. 5.7, consists of a storage battery, ignition coil, distributor, and a spark plug. The battery casing, which was earlier made of hard rubber is now almost exclusively made of high impact PP (Polypropylene). Easy moulding PP resins have allowed much thinner walls, as well as built-in separators, making the batteries more compact. The top cover, vent caps and the tray are also made of plastics.

In general, an average automobile features over 100 m of wires and cables. Low voltage insulation/jacketing is normally made of PVC; PA is occasionally used. Individually insulated wires are normally gathered into wiring harnesses. Associated moulded connectors often involve PA resins with or without glass.

Thermosetting resins like PF (Phenolic resins), alkyd, epoxy, etc. which were traditionally used for a variety of electrical components like voltage regulators, switches, fuse boxes, relays, coil forms, terminal blocks, electrical motor, brush holders, lamp bases or sockets, printed circuit boards (PCBs) etc,; are increasingly made of thermoplastics such as PA, PAT, PP, PC, PSU, with or without glass.

The high voltage ignition coil involves the tower (battery) that is often made of PF thermoset and the cap. The distributor with its housing, cap, and rotor traditionally involved cast metal and thermosetting resins (e.g., PF, alkyd), they are now generally moulded of glass-reinforced PAT. The ignition wires are usually jacked with silicone or other speciality rubbers. The ceramic insulating body of spark plugs may possibly be replaced by high temperature-resistant plastics, such as PI. Rigid insulating caps are made of plastics such as PPS or DAP, while silicone rubber is often used for the elastomeric boot.

5.3 Interior Elements

The interior elements are those fitted within the occupants space and include the following items.

1. Dashboard Area
   (a) Dashboard
   (b) Instrument Cluster
   (c) Steering
2. Seating
3. Interior Panels
4. Interior Accessories
5. Flooring
6. Luggage Space/Cargo Area.

Brief description of the plastics used in these elements is included in this section.

5.3.1 Dashboard Area

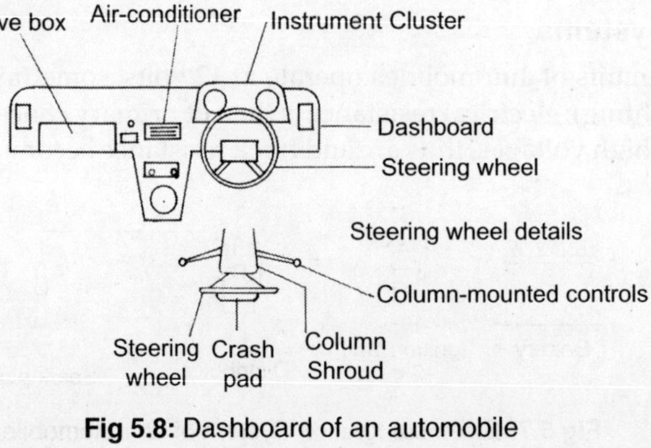

Fig 5.8: Dashboard of an automobile

The dashboard, shown in Fig. 5.8 extends accross the occupants space under the windshield. It hides and shields the functional elements and acts as passive safety element designed to minimise injuries to occupants in case of head-on collision.

(a) **Dashboard:** The dashboard construction consists of a complex injection-moulded frame (e.g., PS, PP, ABS, PC, often RTP form), supporting a semiflexible PU foam padding covered with a thermoformed weather-resistant, plasticized PVC skin. Moderate impacts are absorbed by the skin padding system, while the controlled collapse of the frame can help absorb energy in severe collisions.

Climate control systems (ventilation and heating) make extensive use of plastics that must be dimensionally stable at fairly high temperatures. Hidden fans and ducts often involve filled or reinforced moulded resins (e.g., PP, PA), while vent grille, louvers and surrounds are moulded to attractive finishes (e.g., PA, PPO, ABS). Control panels and mechanisms increasingly use plastics for facings, levers, cable-conduits, etc.

Ash trays and cigarette lighters, that require temperature and incanescence resistance up to 600°C (1100°F), have parts made of suitable plastics (e.g., PF and PI), replacing metals and even ceramics.

(b) **Instrument Cluster:** The instrument cluster is often a separate element that consists of a housing (e.g., PS, ABS, PPO, PMO) and a transparent facing (e.g., acrylic), surrounded by a bezel/rim. Conventional mechanical instruments (speedometer, odometer, tachometer) make considerable use of engineering plastics, ideally suited to such lightly loaded applications and requiring no lubrication (e.g., PMO, PA, often in RTP form).

(c) **Steering:** Conventional steering wheels consists of a metal frame over which rigid (e.g., PP, CAB) or semirigid (e,g,, PVC) plastics are moulded; *friendly* soft steering wheels are increasingly used, which features a self-skinning PU foam cover. Plastic and composite steering wheel frames have also been introduced.

In the steering column, injection-moulded plastic functional elements are now replacing die-cast zinc-alloy parts (e.g., PA-RTP). Shrouds (cover plates) have long been made of moulded plastic (e.g., ABS).

Steering column-mounted controls also feature moulded plastic hollow shafts with a larger cross-section and multi-functional knobs (e.g., PMO).

5.3.2 Seating

Automobiles now almost exclusively feature front bucket seats with adjustable/reclining backs. Rear seats are more often of the fixed type with folding backs. A typical bucket-type seat, shown in Fig. 5.9, features the horizontal seat, the reclining back, and the headrest. Each element of a conventional seat comprises a rigid metal frame, a metal suspension often consists of flat snake-like metal springs, a cellular (foam) elastomeric padding (cushion), and a protective/aesthetic covering (upholstery). A cellulosic liner is often inserted between padding and covering. Coverings currently show a trend away from leatherette (PVC- or PU- coated fabrics) towards wear-resistant woven or knitted fabrics, and valvets (e.g. acrylics).

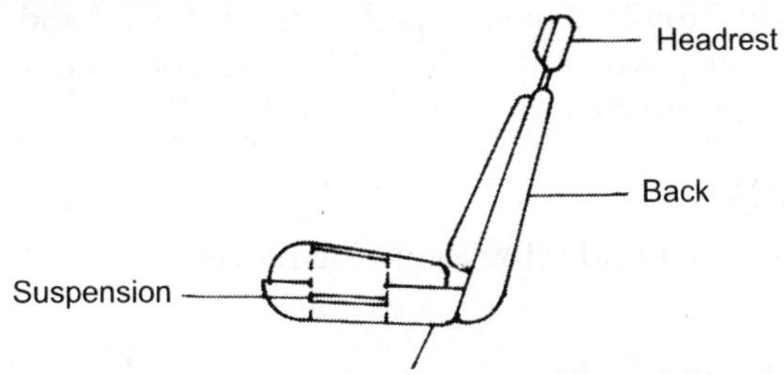

Fig 5.9: Typical Front Seat of an automobile.

5.3.3 Interior Panels

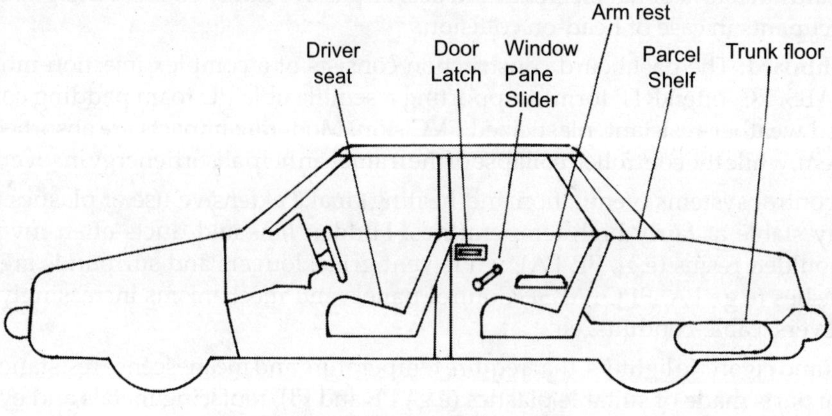

Fig 5.10: Automobile Interior Elements.

Interior panels, Fig. 5.10, comprise primarily door panels, roof liner, roof pillars (posts) covers, and window surrounds. The door panels are generally moulded from impact-resistant plastics (e.g., ABS, PP); this permits the incorporation of features such as latch, ash tray or speaker recesses, arm rests, etc. Roof liners are usually one-piece elements, comprising a semi-rigid foam backing and a formed skin, which are bonded to the roof.

5.3.4 Interior Accessories

Door latching system make use of engineering plastics, sometimes in RTP form or with chrome plating for classical appearance (e.g., PA, PAT, PMO). Bezels, handles, and even keylocks are now made of plastics.

Door window opening devices (regulators, cranks) also make use of plastics. Besides handles, the internal mechanisms now include much lighter plastic brackets and levers, and a novel system has been introduced that involves a preforated plastic tape and sproket wheels.

5.3.5 Flooring

Traditional floor mats made of heavy conventional rubber stock (e.g., SBR) are most conveniently produced in flat form and many become less essential with the introduction of moulded reinforced plastic floor panels that are resistant to corrosion. Floor carpets can now be made with a thermoformable backing to give them the appropriate contour. Fibre and backing (e.g., PP, PA) are resistant to winter salt and slush in cold climates, and are fully washable.

5.3.6 Luggage compartment/Dicky

Luggage compartments or dickies need sturdy load floors, which were traditionally made of sheet metal as part of the car body. Moulded or stamped fibre-reinforced and ribbed plastic load panels are now being used particularly where it is desirable to remove them occasionally (spare wheel well or storage component covers, etc.) (e.g., HDPE, PP, RRIM).

5.4 Exterior Elements

Exterior elements can be grouped under the following heads:

1. Body Panels
2. Exterior Accessories and Trims
3. Weather Stripping

4. Glazing
5. Window Wiper/Wind Screen Washer System
6. Lighting Systems.

5.4.1 Body Panels

Fig. 5.11 and 5.12 show the exterior body panels, and front and rear exterior elements respectively.

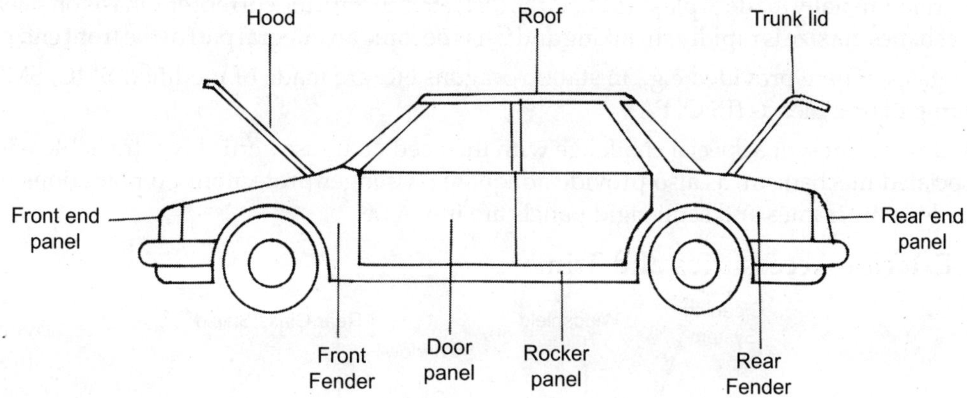

Fig 5.11: Automobile Exterior Body Panels.

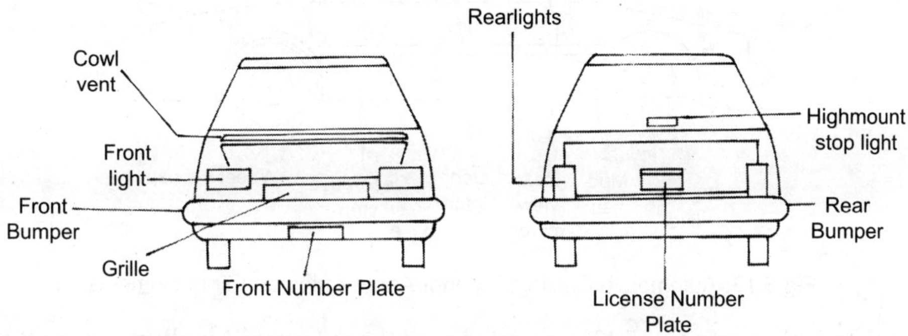

Fig 5.12: Automobile Front and Rear Exterior Elements.

Large body panels classified as horizontal panels are generally considered a more difficult challenge than vertical panels, partly because they are more visible (hood in particular) and because they are more likely to become distorted under their own weight or as a result of temperature changes. Movable parts (hood, trunk, lid) normally have stiffening back elements bonded to the main panel. Examples of plastics used for hoods include SMC, PBT, and PA with suitable fillers. FRP roofs have been in use since 1950's and integrated roof racks (PAT, PA, PVC) are also used in some categories of vehicles. Cowl vent panels (screen, grille) are now routinely made of injection-moulded thermoplastics which are particularly suited to the complex shapes. SMC has been used for rear decks.

Large vertical body panels include front fenders, door panels, and rear quarter panels; they do not require the same rigidity and dimensional stability as horizontal panels. A variety of materials and processes have been used for the purpose such as PRIM-PU, other RIM systems, SMC, and new engineering thermoplastics blends based on PC, PA, and ABS for injection moulding.

Plastic fender liners (e.g., HDPE, PR-RTP) were introduced as far back as 1970's as a protection for metal fenders against road projections.

Front-end and rear-end vertical panels often have similar requirements. Rigid panels normally serve also as structural elements to support front and rear lighting systems, grille, number plate holder, etc. In such cases, they can be made of SMC, as well as PA-RTP or PC-RTP, PRIM, etc.

Valance* and rocker panels, because of their low location, are particularly subjected to impacts. Sometimes they are integrated in the bumper system or part of the relevant body frame and are made of thermoplastic elastomer.

The grille is usually made of plastics like ABS, PBT, etc. but with the current emphasis on sleek aerodynamic shapes, its size is rapidly shrinking and it has become an integral part of the front end panel.

Rear gates, where provided e.g., in station-wagons,etc. are made of modified BMC, SMC, and engineering plastic blends (PPO, PA).

Side doors represent a special challenge with the need to house vertically retractable windows and associated mechanism, as also provide adequate passenger protection. Combinations of rigid reinforced plastic frames and semi-rigid panels are in vogue.

5.4.2 Exterior Accessories and Trim**

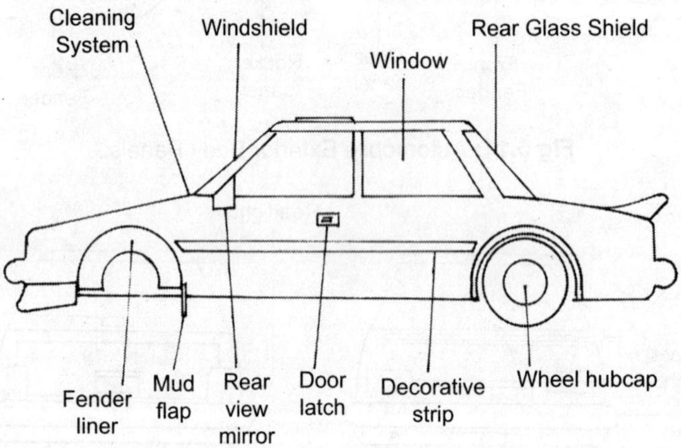

Fig 5.13: Automobile Glazing, Exterior Accessories and Trim elements.

A variety of small elements, Fig. 5.13, associated with the automobile body and its protection have long since been made of polymeric materials. They include fender extensions made of thermoplastic PU, PA, or SMC, and mud flaps made of conventional rubbers or thermoplastic elastomers.

Underbody fairings are made of light, impact-resistant plastics in areas sufficiently distant from the exhaust system.

Door handles and bezels (rims) are commonly made of plastics like PET/PBT, PA, etc.

Rearview mirrors feature a streamlined housing (e.g., ABS, PA-RTP), mounted on the body and sometimes a mechanism for adjustment from the interior.

Wheel covers (hub caps) are now commonly made of plastics (e.g., ABS, PA, PPO) which must resist braking heat, and, preferably, baking oven temperatures of 90-120° (194° – 248°F).

Styling involves the use of decorative trim strips made of aluminium covered with plastic (e.g., CAB) and quite often by plastic alone. They may include weld seam strips, side mouldings, bumper strips, and window and light surrounds. They were initially clipped through holes in body panels but are now increasingly adhesion-bonded.

---

* *Valance* is a short curtain or hanging frill.

** *Trim:* Decorative material.

5.4.3 Weather Stripping

Glazing i.e. transparent panels or windows, have traditionally been made of glass panes. Fixed windows are mounted in moulded rubber sealing surrounds (gaskets). For safety reasons, side and rear windows are made of tempered safety glass, while wind-shields normally consist of laminates featuring PVB (polyvinyl butyral) interlayer, sandwich-bonded between two glass panes. Safety glass, featuring a scratch and abrasion-resistant PU film on the inside surface is being considered which would afford a better protection against lacerations of occupants in case of breakage due to accidents.

A considerable weight advantage could result from the adoption of plastic glazing for, at least, side and rear windows. Polymers such as PC or acrylic with suitable surface treatment or coating (silicones, PU), which are already in use for sun roof windows, are the possible candidates.

The preference for flush-mounted glazing has led to the disappearance of conventional moulded rubber sealing surrounds in favour of bonding techniques. In some cases, a suitable polymeric surround is moulded directly over the glass panel.

5.4.4 Windshield Wiper/Washer system

Windshield or rear window wiper blades have tradionally been made of an elastomeric material, but conventional rubbers (e.g., NR) are being replaced by thermoplastic elastomers. Wiper arms are also now moulded in plastic (e.g., PET/PBT).

Windshield/rear window washer system make extensive use of detergent and antifreeze-resistant plastics for the bottle (e.g., PE), the tubing (e.g., PVC) and the motor/pump device.

5.4.5 Lighting Systems

Considerable changes have affected the design and fabrication of front lighting systems (head lights). Aerodynamic and styling considerations are leading to the replacement of conventional glass-based sealed beam systems by combination of inter-changeable conventional or hologen lamps (bulbs) with socket, reflector, housing, seal, and lens that can all be made of plastics. Examples include PSU (lamp socket); ABS/PC (housing); PAT and PA (reflector); and PC, PES, and PAR (lens). Some challenging problems associated with front lighting systems, involving high temperatures up to 215°C (420°F), include thermal stability, the metallization of the reflector, as also abrasion and impact resistance of the lenses. Rear lighting systems (tail lights) involve lower temperatures, and polymeric materials used include acrylics, PC (lens), PP-RPT, ABS, and PAT for housing.

Sidemarker light now tend to be incorporated into front and rear lighting assemblies, but new high-mount stop lights are being introduced that are often interior-mounted.

The provision of interior courtesy lights is straight forward but major changes are taking place in the area of indicator lights, with the introduction or generalization of LED displays and fibre optics that are reported to increase local temperatures.

5.5 Future Trends

Over the past few decades, the use of polymeric materials (plastics, rubbers, and composites) in automotive applications has grown steadily. The areas of application have expanded from such traditional ones as electrical insulation or elastomeric components to include appearances and functional elements, many of them highly technical, which now abound in the occupants space and under the hood.

The introduction of large polymeric exterior components, initiated on a large scale in 1970's, is rapidly gaining acceptance, and the near future will probably see a real break through with massive conversion to polymeric body panels. The near future is likely to see polymer-based composites widely introduced as structural materials for vehicle frames and other structural components, while high temperature-resistant plastics and composites will play an increasing role in engine construction.

*****

6 RECYCLING OF PLASTICS

6.0 INTRODUCTION

Plastic was invented in 1907 in the form of Bakelite and soon spread to almost every aspect of human life. There is hardly any place without plastic in one form or the other. In fact it is unimaginable to think life without plastics as they find use right from a paper clip to aerospace and satellites.

Plastic bags introduced to supermarket in 1977 are indispensible but are causing havoc in the form of waste scattered all over the world not only on land but also in seas and oceans as well. In this period of little move than a century of their existence nearly 10 billion tons of plastics have been produced and most of these are still in existence with a plastic mass of the size of Calafornia state floating in sea. This has resulted in immense damage to the land and sea environment and in fact are causing serious diseases like Cancer by leeching into foods particularly when they come in contact with them at elevated temperatures like in microwave oven heating. Thus in spite of their positive role in making human life comfortable they causing considerable damage to environment as also human and animal life. To reduce their ill effects it is essential to *reduce, reuse* and *recycle* the plastics. For every ton of plastic that is recycled, an estimated 7 yards (6.4 meters) of landfill space is saved.

6.1 Types of Plastics

All plastics are not of the same type and all of them are not recyclable. So, it is essential to know the types of plastics and which ones of these can be recycled. To achieve that aim, it is essential to identify a given plastic when it is discarded and is dumped as waste. A most comprehensive system to facilitate identification of plastics for recycling them has been developed in USA which is now being used extensively and is being adopted worldwide in its entirity or with some modifications.

The US *Society of the Plastic Industry* (SPI), introduced the *Resin Identification Code* (RIC) system in 1988. RIC system was designed to make it easier for workers in materials recovery and recycling facilities to sort and separate the plastic items according to their resin types. Plastics must be recycled separately, with other like materials, in order to preserve the value of the recycled material and enable its reuse for other products after being recycled.

In its original form, the symbols used as part of the RIC consisted of arrows that cycle clockwise to form a triangle 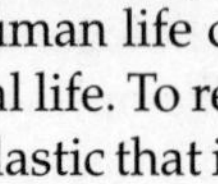

that encloses a number, N. The number broadly refers to the type of plastic used in the product, in *chronological order of when that plastic became recyclable*. So far there are seven numbers which signify as follows.

1. Signifies that the product is made out of *Polyethylene Terephthalate* (PET) used for items like beverage bottles, cups, other packaging, etc.

2 Signifies High Density Polyethylene (HDPE), used for bottles, cups, milk jugs, etc.
3 Signifies Polyvinyl Chloride (PVC), used for pipes, sidings, flooring, etc.
4 Signifies Low Density Polyethylene (LDPE), used for plastic bags, tubing, resistant sacks, etc.
5 Signifies Polypropylene (PP), used for auto parts, industrial fibres, food containers, etc.
6 Signifies Polystyrene (PS), used for plastic utensils, styrofoam, cafeteria trays, etc.
7 Signifies plastics, other than above 6 types, such as acrylic, nylon, polycarbonate, polylactic acid (PLA), etc. used for headlight lenses, safety shields, eye lenses, etc.

6.2 Description of SPI Code

The Seven SPI code symbols with their resin type and typical uses are shown in *Fig. 6.1.*

Table 6.1: SPI Code Symbols, Resins and some Typical Uses

Symbol	1 PETE	2 HDPE	3 PVC	4 LDPE	5 PP	6 PS	7 OTHER
Resin	polyethylene terephthalate	high density polyethylene	polyvinyl chloride	low-density polyethylene	polypropylene	polystyrene	other plastics, including acrylic polycarbonate, polyactic fibers, nylon, fiberglass
Uses	soft drink bottles, mineral water, fruite juice container, cooking oil Packaging	milk jugs, cleaning agents, laundry detergents, bleaching agents, shampoo bottles, washing and shower soaps wraps	trays for sweets, fruit, plastic packing (bubble foil) and food foils to wrap the foodstuff	crushed bottles, shopping bags, highly-resistant sacks and most of the wrappings	furniture, consumers, luggage, toys as well as bumpers, lining and external borders of the cars	toys, hard packing, refrigerator trays, cosmetic bags, costume jewellery, CD cases, vending cups	

Understanding the seven plastic codes will make it easier to choose plastics and to know which plastic to recycle. For example, water bottles that display number 3 or 5 in triangle cannot be recycled in most cases. A 3 indicates that the water bottle has been made from PVC (Polyvinyl Chloride), a 5 means that it has been made from PP (Polypropylene); these two materials are not accepted at most of the recycling centres.

6.2.1 Detailed Description of SPI Code Numbers

The significance of each one of the seven numbers used in SPI Code are detailed as follows.

1 **No. 1 : PET/PETE (Polyethylene Terephthalate)**

PETE

PET or PETE was introduced in 1940 and is one of the most commonly used plastics. It is used for producing crystal-clear beverage bottles such as those for Coco-cola and Pepsi and also for some packaging. It is intended for single use applications; repeated use increases the risk of leeching and bacterial growth. It is difficult to decontaminate it, and proper cleaning requires harmful chemicals. PET may leach carcinogens.

PET plastic is recyclable. It is crushed and then shreded into small flakes which are then reprocessed to make new PET bottles, or spun into polyster fibre. This recycled fibre is used to make textiles such as fleece garments, carpets, stuffing for pillows and life jackets, and similar other products.

The products made of SPI Code No. 1 (PET) plastic should be recycled *but not reused*.

2

No. 2 : HDPE (High Density Polyethylene)

HDPE

HDPE was discovered in 1953. It is a stiff plastic which was first used for pipes in storm sewers, drains, and culverts but is now also used to make milk jugs, detergent and oil bottles, toys, and plastic bags. HDPE *is the most commonly recycled plastics* because it does not break under exposure to extreme heat or cold and is considered one of the safest plastics. It is easy and cost effective to recycle HDPE for secondary uses.

HDPE is very hard-wearing and does not break down under exposure to sunlight or extremes of heating or freezing. So, it finds use in making picnic tables, plastic lumber, dustbins, park benches, bed liners for trucks and many other items requiring durability and weather-resistance.

HDPE products are reusable and recyclable.

3

No. 3 : PVC (Polyvinyl Chloride)

PVC

PVC is one of the oldest synthetic materials in industrial production which was a chance discovery first in 1838 and then in 1872.

PVC is a soft, flexible plastic used to make clear plastic food wrapping, cooking oil bottles, teething rings, toys, and blister packaging for different consumer products. It is also used as the sheathing material for computer cables, to make plastic pipes and parts for plumbing. Because PVC is relatively impervious to sunlight and weather, it is used for making window frames, garden hoses, arbors, raised beds, etc.

PVC is called a *poison plastic* because it contains numerous toxins which it can leech through out its entire life cycle making it harmful to human life and the environment. Almost all products made from PVC need virgin material for their manufacture and less than 1% of PVC is recycled making it perpetually dangerous.

Although PVC plastic is not recyclable, it can be repurposed. PVC products should not be reused for applications with food or for children's use.

4

No. 4 : LDPE (Low Density Polyethylene)

LDPE

LDPE was the first polyethylene to be produced. It has less mass than HDPE; that is why it is considered a separate material for recycling.

LDPE is used for shrink wraps, dry cleaner garment bags, squeezable bottles and plastic bags for packaging bread. Most of the grocery bags are made from LDPE. Some clothing and fruniture also use this material. LDPE is less toxic than other plastics, and relatively safe for use.

LDPE products are reusable but not commonly recycled. When recycled, LDPE is used for plastic lumber, landscaping boards, garbage can liners and floor tiles. However, products made from recycled LDPE are not as hard or rigid as those made from recycled HDPE.

5

No. 5 : PP (Polypropylene)

PP

PP was discovered in 1951. This plastic is tough and light weight, and has excellent heat-resistance properties. It serves as a barrier against moisture, greese and chemicals. That is why it is used as a liner in a cereal box to keep it dry and fresh. It is also used for making disposable diapers, pails, plastic bottle tops, margarine and yogurt packages, potato chips bags, straws, packing tape and ropes.

PP is considered safe for reuse but recycled on a very limited scale. Recycled PP is used for making battery cases, brooms, bins, trays and landscaping border stripping.

6

No. 6 : PS (Polystyrene)

PS

PS which was discovered accidently in 1839 is an inexpensive, light weight and easily formed plastic with a wide variety of uses. It is most widely used for making disposable styrofoam drinking cups, egg cartons, plastic picnic cutlery, plastic CDs and DVDs, foam packaging and widely used *peanut* foam chips used to fill shipping boxes to protect the content during transportation. Polystyrene is also used for making rigid foam insulation and underlay sheeting for laminate flooring for homes and offices.

Since polystyrene is ultra-light weight and structurally weak it breaks up easily making it more harmful to the environment. Beaches all over the world are littered with pieces of polystyrene, end angering the health of marine animals. It accounts for about 35% of landfill in USA.

Polystyrene may leech styrene, a cancer causing chemical into food products especially when heated in a microwave oven. PS is more often reused than recycled. While the technology for recycling PS is available the market for recycling is small. PS *should be avoided as far as possible.*

7

No. 7 : Miscellaneous Plastics

OTHER

Miscellaneous plastics include polycarbonate, polylactide, acrylic, acrylonitrile, styrene, fibreglass and nylon.

Of primary concern with No. 7 plastics is their risk for chemical leeching into food or drinks packaged in polycarbonate containers made using BPA (Bisphenol. A). BPA is a known xenoestrogen which disrupts the normal functioning of endocrine glands causing serious health problems.

No. 7 plastics are used for making baby bottles, sipy cups, water cooler bottles and automobile parts.

A bio-based compostable plastic, made from corn starch, called PLA (Polylactic Acid) is also included in No. 7 category. PLA coded plastics are not recyclable but obviously can be composted so need to be segregated from other plastics of this category for composting.

Table 6.2 shows the images of resin codes, polymer abbreviations, main uses and their recyclability or otherwise. A more comprehensive table is included as Appendix C.

Table 6.2: SPI/RIC Recycling Codes, Symbols, Abbreviations, Typical Uses and Recyclability of Plastics

Recycling number	Image	Alternate image # 1	Alternate image # 2	Abbreviation	Polymer name	Uses	Recyclability
1.	1	1 PETE	01 PET	PETE or PET	Polyethylene terephthalate	Polyester fibres (Polar Fleece), thermoformed sheet, strapping, soft drink bottles, tote bags, furniture, carpet, paneling and (occasionally) new containers.	Yes
2.	2	2 HDPE	02 PE-HD	HDPE or PE-HD	High-density polyethylene	Bottles, grocery bags, milk jugs, recycling bins, agricultural pipe, base cups, car stops, playground equipment, and plastic lumber	Yes
3.	3	3 V	03 PVC	PVC or V	Polyvinyl chloride	Pipe, window profile, siding, fencing, flooring, shower curtains, lawn chairs, non-food bottles, and children's toys.	Yes, but consult the recycler
4.	4	4 LDPE	04 PE-LD	LDPE or PE-LD	Low-density polyethylene, Linear low-density polyethylene	Plastic bags, six pack rings, various containers, dispensing bottles, wash bottles, tubing, and various molded laboratory equipment	Yes, but consult the recycler
5.	5	5 PP	05 PP	PP	Polypropylene	Auto parts, industrial fibres, food containers, and dishware	No

6		 PS	 PS	PS	Polystyrene	Desk accessories, cafeteria trays, plastic utensils, coffee cup lids, toys, video cassettes and cases, clamshell containers, packaging peanuts, and insulation board and other expanded polystyrene products (e.g., Styrofoam)	No
7		 OTHER	 O	OTHER or O	Other plastics, such as acrylic, nylon, polycarbonate, and polylactic acid (a bioplastic also known as PLA), and multilayer combinations of different plastics	Bottles, plastic lumber applications, headlight lenses, and safety shields/ glasses.	No

When number is omitted, the arrows arranged in a triangle form indicate the **Universal Recycling Symbol**, a generic indicator of recyclability. Subsequent revisions to the RIC have replaced the arrows with a solid triangle. This has been done to remove consumer confusion about the meaning of RIC and the fact that the presence of RIC symbol on an item does not necessarily indicate its recyclability.

In 2008, ASTM* International took over the administration of the RIC System and eventually issued ASTM *D7611-Standard Practice for Coding Plastic Manufactured Articles for Resin Indentification*. In 2013 this Standard was revised to change the graphic marking symbol of the RIC from the chasing arrows of the Recycling Symbol to a solid triangle instead.

Table 6.3 shows the RIC Symbols after their ASTM revision of 2013.

Table 6.3: Modified RIC Symbols for Polymer Recyclability vide ASTM Revision of 2013

Resin Identification Number	Resin	Resin Identification Code-Option A	Resin Identification Code-Option B
1.	Poly(ethylene terephthalate)	 PETE	 PET

* ASTM– American Society for Testing and Material

2.	High density polyethylene	2 HDPE	02 PE-HD
3.	Poly(vinyl chloride)	3 V	03 PVC
4.	Low density polyethylene	4 LDPE	04 PE-LD
5.	Polypropylene	5 PP	05 PP
6.	Polystyrene	6 PS	06 PS
7.	Other resins	7 OTHER	07 O

Modifications to the RIC are currently being discussed and developed by ASTM Sub-Committee D20.95 on recycled plastics.

6.3 Identification of Plastics

Although identification of plastics has been covered to some extent in chapter 3, under testing of plastics but it appears that much knowledge may not be adequate in the light of recent developments in the plastic industry so a more detailed information is needed.

For working with plastics it is often required to identify the specific plastic that has been used for making a given product. This is imperative to determine the cost and expected properties of the product, as also for replacement of broken/worn out parts, and *for recycling of plastics*. The identification of plastics is generally quite difficult due to the following :

1. The wide range of basic polymers available for use.
2. The wide range of additives which can be used to modify the properties of basic polymer.
3. The wide range of mixtures or compounds of polymers that can be produced to get the required properties.

Inspite of these constraints, there are some simple tests which can be conducted to get a basic idea of the possible base polymer used for the manufacture of any given item. These tests can be carried out without the need of special equipment. However, these tests are only the preliminary guess which need to be supplemented with more sophisticated method (s) if definite results are aimed at, but they need more elaborate and standard equipment to be conducted in a designated laboratory set up.

6.4 Identification Tests for Plastics

The tests to identify a plastic include the following :

1. Preliminary Examination
2. Elemental Analysis :
   (i) Lassaigne's Test
   (ii) Beilstein Test
3. Solubility Test
4. IR (Infrared Ray) Analysis (KBr Disc)
5. Specific Gravity Test
6. Flammability Test.

These tests are briefly described as follows.

6.4.1 Preliminary Examination

Based on a preliminary examination, a polymer can sometimes be classified as belonging to one of the following 4 groups.

1. Rubber
2. Flexible thermoplastic
3. Rigid thermoplastic
4. Thermoset.

6.4.1.1 Manufacturer's Labeling Code

The Composition of increasing number of plastics is identified using the SPI (Society for Plastic Industry) recycling code which is usually stamped on the bottom of the product. The number is enclosed in the chasing arrows triangle recycling symbol (N); where N varies from 1 to 7 depending upon the type of plastic as discussed earlier. The same are reproduced below for ease of access to the reader.

1. PET/PETE—Polyethylene Terephthalate, (1 PETE)
2. HDPE—High Density Polyethylene, (2 HDPE)
3. V—Vinyl/Polyvinyl Chloride
4. LDPE—Low Density Polyethylene
5. PP—Polypropylene
6. PS—Polystyrene
7. O—Others

6.4.1.2 End Use

In the case of a finished product, the intended use can very often indicate the nature of the polymer. For example :

- Elastic bands and tyres are rubbers (elastomers). Elastic bands are usually natural rubber (cis-1, 4-polyisoprene). Automobile tyres are usually a blend of styrene butadiene rubber (SBR) and natural rubber.
- Pot handles and cutlery handles are usually thermosets, e.g. phenolics. Dishes are likely to be melamine formaldehyde thermosets.
- Garden hoses, shower curtains and raincoats are usually PVC.
- Tents and screens are usually nylon.
- Floating rope is usually polypropylene.
- Foam upholstery cushions are usually polyurethanes.
- Bulletproof glass, sun glasses and reading glasses with plastic lenses are polycarbonate.

6.4.1.3 Feel

With practice one starts to get the feel for various plastics. For example, Teflon and polyethylene feel waxy. Thermosets feel hard and sound brittle when dropped or struck with a metallic object. Rubbers are flexible and sometimes stretchy.

The polyolefins have a very distinctive feel and one can often judge that. The presence of glass fibre or other reinforcement materials can alter the feel and the stiffness of the product and that can often be detected.

6.4.1.4 Colour

The colour of a plastic can give some information. Some plastics have restricted colour ranges, particularly the thermosets, for example phenol formaldehyde thermosets like Bakelite, are inherently dark coloured. Others tend to be more glossy in colour like polypropylene, whereas some are both glassy and glossy like acrylics.

In the absence of additives, pure amorphous polymers have high optical clarity (transparent), whereas crystalline polymers like Acetals, are translucent to opaque. The following are examples of transparent polymers : PET (pop bottles), PMMA (plexiglass), PC (bullet-proof glass), PVAc (blister packaging).

6.4.2 Elemental Analysis

Two tests are well established for elemental analysis of plastics :

(i) Lassaigne's Test.

(ii) Beilstein Test.

Lassaigne's Test or the **Sodium fusion test**, is used in elemental analysis for the qualitative determination of the presence of foreign elements, namely halogens (e.g. PVC - Polyvinylidene Chloride), nitrogen (e.g. Polyamides) and Sulphur (e.g. Vulcanized rubber and polysulfide) in an organic compound, i.e. a plastic.

The test involves heating the sample strongly with clean sodium metal fusing it with the sample. The fused sample is plunged into water, and the usual qualitative tests are performed to determine the presence of the halogens, sulfur and nitrogen.

Additives may interfer; e.g., some flame retardants contain chlorine (Cl) and Bromine (Br).

Sometimes **Beilstein Test** is also carried out to identify a polymer.

Beilstein Test : It is a simple qualitative chemical test for halides. It was developed by Friedrich Konrad Beilstein, hence its name.

In this test a copper wire is cleaned and heated in the flame of a Bunsen burner to form a coating of copper oxide. It is then dipped in the polymer test sample and again heated in the flame. A green flame will signify the formation of copper halide. This test does not detect fluorine/ fluorides. However, this test is no longer frequently used because of the possibility of generating highly toxic chloro-dioxins if the test material is a polychloroarene.

6.4.3 Solubility Test

Thermoplastics are soluble in appropriate solvents, whereas thermosets may swell but are insoluble. Plastics may require long exposure to a solvent to dissolve. Stirring significantly reduces dissolving time. The general rule of like-dissolves-like applies i.e.; hydrocarbon solvents (e.g., Kerosene, Benzene, White Spirit, Gasoline, etc.) tend to dissolve non-polar polymers like PP, PE, SEBS, PS and PTFE, etc. whereas polar solvents (e.g. water, acetic acid, methanol, ethanol, etc.) dissolve polymers containing polar functional groups* e.g., PA, PC, PMMA, and ABS. *Table 6.4* illustrates polymer solubility behaviour.

Table 6.4: Solubility Behaviour of Some Well Known Plastics

S.No.	Solvent → / Plastic ↓	Gasoline	Toluene	Acetone	Cyclohexanone
1.	ABS	SW	S	S	S
2.	LDPE, HDPE	I	I	I	I
3.	PA	I	I	I	I
4.	PET	I	I	I	I
5.	PMMA	I	SW	S	S
6.	PP	I	I	I	I
7.	PPO	I	SW	S	S
8.	PS	S	S	S	S
9.	PVC	I	SW	SW	S

I — Insoluble, SW — Swells, S — Soluble

Note : Solubility behaviour is shown at room temperature. Heating can solubilize thermoplastics such as PE and PP.

6.4.4 IR Analysis

Infrared Ray (IR) Spectroscopy deals with the infrared region of the electromagnetic spectrum, i.e. light having a longer wavelength and thus having a lower frequency than visible light.

The IR spectroscopy utilizes the concept that molecules tend to absorb specific frequencies of light that are characteristic of the structure and mass of the molecules involved. For instance the molecules can absorb the energy contained in the incident light and the result is a faster rotation or a more pronounced vibration.

---

* A polymer is considered *polar* if its charge can be measured as positive or negative while a *non-polar polymer* has no charge.

IR spectroscopy results can generally be analysed in three ways viz., by measuring reflection, transmission and absorption. The major use of IR spectroscopy is to determine the functional groups of both organic and inorganic compounds.

An IR spectrum is a graph plotted with the infrared light absorbed/transmitted on the Y-axis against frequency or *wavenumber* (cm^{-1}) on the X-axis, where wavenumber is the reciprocal of wavelength in cm^{-1}. Fig. 6.1 gives a conceptual representation of wavelengths for different regions of the e.m. spectrum.

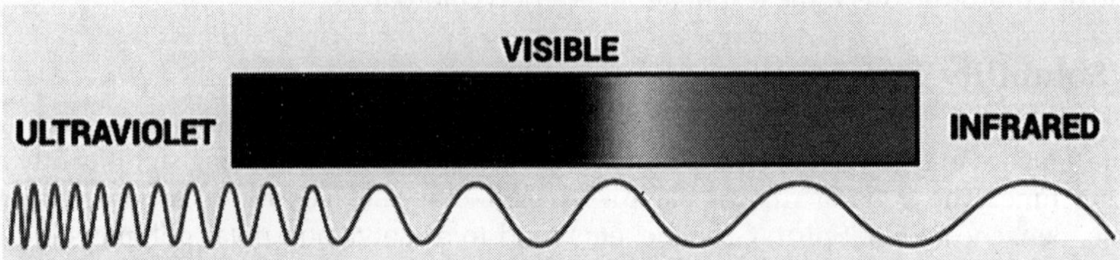

Fig 6.1: Three broad regions of electromagnetic spectrum and the corresponding wavelengths.

IR Spectroscopy detects frequencies of infrared light that are absorbed by a molecule. Molecules tend to absorb these specific frequencies of light since they correspond to the frequency of the vibration of bonds in the molecule.

The energy required to excite the bonds belonging to a molecule, and to make them vibrate with increased amplitude, occur in the infrared region. However, a bond will interact with electromagnetic infrared radiation only if it is polar i.e. having some charge on it. The intensity of the absorption depends on the polarity of the bond. Symmetrical non-polar bonds in N = N and O = O do not absorb radiation as they cannot interact with an electric field.

6.4.4.1 Ragions of the Infrared Spectrum

Most of the bands that indicate what functional group is present are found in the region from 4000 to 1300 wavenumbers. Their bands can be identified and used to determine the functional groups of an unknown compound.

Bands that are unique to each molecule similar to a finger print are found in the *fingerprint region* from 1300 to 400 wavenumbers. Only these bands are used to compare the spectrum of one compound with that of another. *Fig 6.2* shows graphically the Functional Group Region and the Fingerprint region as related to wavenumbers. The process of using wavenumbers from 4000 to 400 for Analysis is referred to as *Mid Infrared Spectroscopy.*

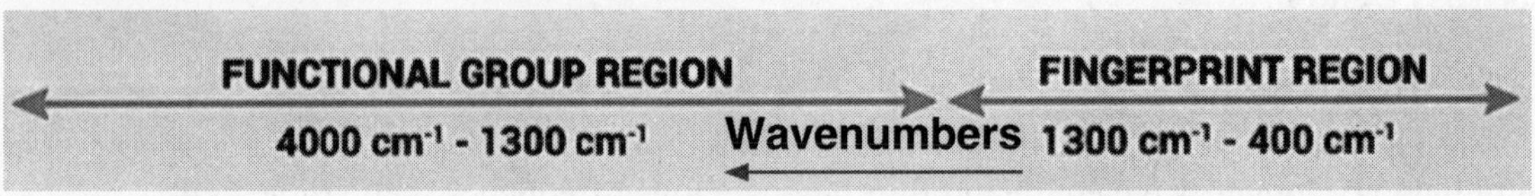

Fig 6.2: Functional Group Region and the Fingerprint Region as related to wavenumber.

The samples used for IR spectroscopy can be in any of the three physical states, viz., solid, liquid or gaseous.

6.4.4.2 IR Spectroscopy Instrumention

A schematic illustration of the instrumentation for IR spectroscopy is shown in *Fig. 6.3.*

First a beam of IR light from the source is split into two and passed through the *reference* and the *sample* respectively. Then, both these beams are reflected to pass through a detector. Finally the response data is printed out after the processor deciphers the data passed through the detector. Thus, IR spectroscopy involves the collection of absorption/transmission information and its analysis in the form of a spectrum.

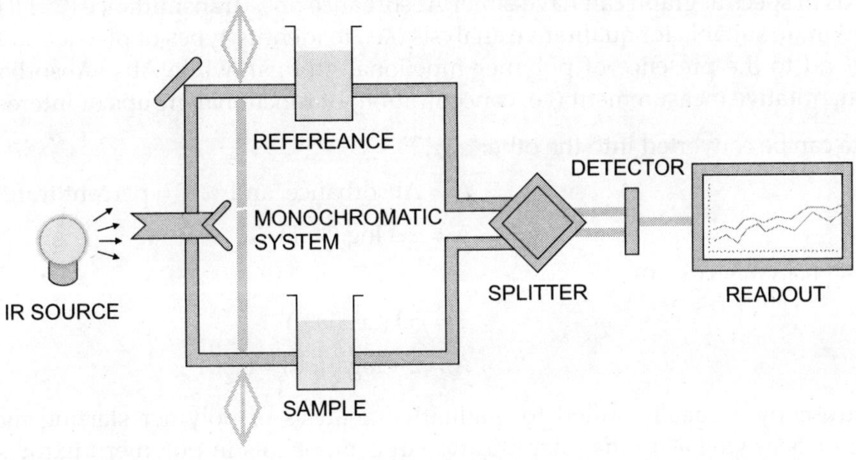

Fig 6.3: A Schematic of the instrumation for Infrared Spectroscopy.

6.4.4.3 Spectrum Analysis

The positions of absorption transmission bands in the spectrum give information about the presence or absence of specific functional groups in a molecule and as a whole the spectrum constitutes a *fingerprint* that can be used to determine the identity of the sample. A difference between two spectra indicates that the two samples are made up of different components.

Fig. 6.4 shows the IR spectra of 4 common plastics viz. polyethylene (PE), polypropylene (PP), polystyrene (PS) and polytetra fluoroethylene (PTFE). The clear differences between the spectra allow a ready discrimination between the materials by visual inspection.

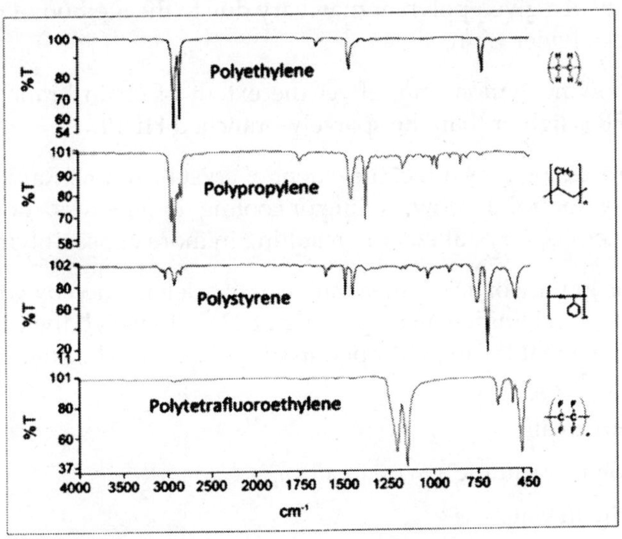

Fig 6.4: Mid-IR spectra of 4 common polymers.

Additional interpretation of the spectra can yield information about the structure. For example, looking at the C-H stretch region around wavenumber 2950 cm^{-1}, differences can be seen between PE and PP due to the differing ratios of CH_2 and CH_3 groups. Polystyrene has bands above 3000 wavenumbers, indicating the presence of aromatic group. PTFE has no C-H groups at all, so the very weak bands present must be due to impurities or surface contamination.

The Y-axis in spectral graph can have either Absorbance or % Transmittance (% T). Usually % T is considered more suitable for qualitative analysis (i.e., to identify types of plastics based on peaks that correspond to the presence of polymer functional groups), while Abs (Absorbance) is best suited for quantitative measurement (i.e. concentration) of functional groups of interest.

One data can be converted into the other e.g.,

if $\quad A$ = Absorbance, and % T = percent transmittance

then, $$A = 2 - \log(\%\ T) \qquad (6.1)$$

Where log is to the base 10.

So, if $\quad \%\ T = 50$ (i.e. 50%)

then, $$A = 2 - \log_{10}(50) = 0.301$$

IR Spectroscopy is ideally suited to qualitative analysis of polymer starting materials and finished products as well as to the quantification of components in polymer mixtures and to the analysis of in-process samples. IR Spectroscopy is reliable, fast and cost effective.

6.4.5 Specific Gravity Test

The specific gravity of a polymer is very helpful in determining its identity. Polyolefins, i.e., polyethylene, polypropylene and polyallomers (i.e., Block polymers of ethylene and propylene) have specific gravity less than 1.0 so float on water whereas almost all other (non-cellular) polymers sink. Specific gravity tests do not apply to plastics that are produced as expanded foams. Enclosed gas cells within such plastics will cause even dense polymers to float. Although the addition of fillers can change the relative density of polymers even so, this method narrows down the number of possible choices.

The specific gravity of a given polymer may vary due to the method of polymerization and the method by which the polymer is processed.

The method of polymerization may affect the extent of chain branching. For example, the highly branched LDPE is lighter than the sparsely-branched HDPE.

The method of processing may affect the extent of crystallization. Rapid cooling produces less dense amorphous polymers while slow cooling or cooling under stress, for example melt spinning or gel spinning often induces crystallization, resulting in more dense polymers.

Procedure : Specific gravity of a polymer can be easily determined by weighing a small sample of it in air and then in referal fluid; generally water at 23°C. To weigh the polymer sample in water a wire is attached to it so that it can be dipped in water held in a beaker. The specific gravity can then be calculated as follows :

Let, A = wt. of wire in air

B = wt. of polymer sample + wire in air

C = wt. of wire in water

D = wt. of sample + wire in water

Note : For C, immerse only that much of the wire as was immersed when weighing the polymer sample.

Then,

$$\text{Specific gravity} = \frac{\text{density of polymer}}{\text{density of water}}$$

$$= \frac{\text{wt. of sample in air}}{\text{wt. of water displaced by sample}}$$

$$= \frac{(\text{wt. of sample + wire n air})-(\text{wt. of wire in air})}{(\text{wt. of sample in air})-(\text{wt. of sample in water})}$$

$$= \frac{B - A}{(B - A) - (D - C)} \tag{6.2}$$

For plastics that float on water like olefins, secure the wire so that the sample is held under water otherwise hang a dead weight on the sample and carry out the test and incorporate the weight of wire and dead weight in the calculations as follows.

Let, A′ = wt. of wire + dead weight in air

B′ = wt. of polymer sample + wire + dead weight in air

C′ = wt. of wire + dead weight in air

D′ = wt. of sample + wire + dead weight in water

Then,

$$\text{Specific gravity} = \frac{B' - A'}{(B' - A') - (D' - C')} \tag{6.3}$$

The accuracy of calculated value of specific gravity can be checked by comparing that with the standard value given in *table 6.5* for plastics and other relevent materials like glass, wood and steel.

Table 6.5: Specific Gravities of Some Common Plastics and Other Referral Materials

Polymers (Alphabetized)	**S.G.**	**S.G. (Range)**	**Polymers (SG prioritized)**	**S.G.**	**S.G. (Range)**
acetal	1.42		oak	0.86	
acrylic	1.18		polyallomer	0.89	
ABS	1.04		polypropylene	0.89	
aluminum	2.50		polyethylene, L.D.	0.91	0.91-0.92
cellulose acetate	1.30		ionomer	0.94	
cellulose nitrate	1.25		polyethylene, H.D.	0.95	0.94-0.96
diallyl phthalate	1.50	1.34 - 1.78	ABS	1.04	
epoxy (cast)	1.11		poly (phenylene oxide)	1.06	
FEP	2.14		polystyrene	1.06	1.04 - 1.10
glass	2.50		polyester	1.10	1.01 - 1.20
ionomer	0.94		epoxy (cast)	1.11	
melamine formaldehyde	1.48		Nylon	1.15	1.13 - 1.20
Nylon	1.15	1.13 - 1.20	phenoxy	1.17	
oak	0.86		acrylic	1.18	
phenoxy	1.17		PMMA	1.18	

phenol formaldehyde	1.27	1.25 - 1.30	polyurethane	1.18	1.11 - 1.25
PMMA	1.18		polycarbonate	1.20	
poly (vinylidene chloride)	1.70		polysulfone	1.24	
polyallomer	0.89		cellulose nitrate	1.25	1.16 - 1.35
polycarbonate	1.20		poly (vinyl chloride)	1.25	
polyester	1.10	1.01 - 1.20	phenol formaldehyde	1.27	1.25 - 1.30
polyethylene, L.D.	0.91	0.91 - 0.92	cellulose acetate	1.30	
polyethylene, H.D.	0.95	0.94 - 0.96	acetal	1.42	
poly (phenylene oxide)	1.06		melamine formaldehyde	1.48	
polypropylene	0.89		diallyl phthalate	1.50	1.34 - 1.78
polystyrene	1.06	1.04 - 1.10	poly (vinylidene chloride)	1.70	
polysulfone	1.24		silicone	1.75	
polyurethane	1.18	1.11 - 1.25	FEP	2.14	
poly(vinyl chloride)	1.25	1.16 - 1.35	tetrafluoroethylene	2.16	2.14 - 2.20
silicone	1.75		aluminum	2.50	
steel	7.70		glass	2.50	
tetrafluoroethylene	2.16	2.14 - 2.20	steel	7.70	

6.4.6 Flame Test

This test is also known as *Burn Test* or *Melt Test.*

Flame test identifies the ease of ignition, rate of burning, colour of flame and soot as well as the nature of combustion products.

Flame test is a simplified guide to identify many common types of plastics using a flame from a cigarette lighter, torch flame, or a gas burner. However, the Burn Test should be used only for tentative identifications. Many complex compounds require a rigorous spectral or destructive chemical analysis for a definite identification.

To initially determine whether a material is thermoplastic or thermoset type, heat a metal or glass stirring rod until it grows red or orange (i.e., 260°C/500°F) and press it against the polymer sample. If the sample softens, the material is a thermoplastic, if it does not, it is possibly a thermoset.

Next, hold the test sample to the edge of a flame for 5 to 10 seconds. If the material burns, note the colour of the flame, the nature of the smoke, the presence of soot in the air and whether, while burning, the material drips. Hold the sample over an asbestos pad to catch the drips. Work near or in a fume hood and try to direct maximum fumes to the fume hood.

Many polymers have characteristic odours when burnt or melted. Avoid inhaling these fumes as most of these are harmful. Odours of melted polymers (e.g. in test tubes) are more characteristic than the odours of burning polymers. People in polymer business like polymer recycling identify a plastic by burning a test sample of a polymer and observing its flame and odour.

To identify the odour, samples of known plastics can be used for comparison against the known characteristics of each plastic listed in *Table 6.6*

Table 6.6: Flame Test Characteristics of Some Widely Used Plastics

Materials	**No Flame**	**Burns but Extinguishes on Removal of Flame Source**			**Continues to Burn after Removal of Flame Source**				
Thermo-plastics	**Odor**	**Odor**	**Color of Flame**	**Drips**	**Odor**	**Color of Flame**	**Drips**	**Speed of Bur-ning**	**Remarks**
ABS	–	Acrid	Yellow, blue edges	No	Acrid	Yellow, blue edges	Yes	Slow	Black smoke with soot in air
Acetals	–	–	–	–	Formal dehyde	Blue, no smoke	Yes	Slow	
Acrylics	–	–	–	–	Fruity	Blue, yellow tip	No (cast) Yes (molded)	Slow	Flame may spurt if rubber modified
Cellulosics Acetate	–	Vinegar	Yellow with sparks	No	Vinegar	Yellow	Yes	Slow	Flame may spark
Fluorocarbons FEP	Faint odor of burnt hair	–	–	–	–	–	–	–	Deforms; no combustion, but drips
PTFE	Faint odor of burnt hair	–	–	–	–	–	–	–	Deforms; does not drip
CTFE	Faint odor of acetic acid	–	–	–	–	–	–	–	Deforms; no combustion, but drips
PVF	Acidic	–	–	–	–	–	–	–	Deforms
Nylons Type 6 & 6/6	–	Burnt Wool	Blue, yellow tip	Yes	–	–	–	–	6/6 More rigid
Polycarbonates	–	Faint, sweet aromatic ester	Orange	Yes	–	–	–	–	Black smoke with soot in air
Polyethylenes	–	–	–	–	Paraffin	Blue, yellow tip	Yes	Slow	Floats in water
Polyimides	b	–	–	–	–	–	–	–	Chars; material rigid
Polypropylenes	–	Acrid	Yellow	No	Sweet	Blue, yellow tip	Yes	Slow	Floats in water; more difficult to scratch than polyethylene
Polystyrenes	–	–	–	–	Illuminating gas	Yellow	Yes	Rapid	Dense black smoke with soot in air
Polysulfones	–	b	Orange	Yes	–	–	–	–	Black Smoke
Polyurethanes	–	–	–	–	b	Yellow	No	Slow	Black Smoke

Vinyls Flexible-Rigid		Hydrochloric Acid	Yellow with green spurts	No	–	–	–	–	Chars; melts
Polyblends ABS/Carbonate	–	–	–	–	b	Yellow, blue edges	No	–	Black smoke with soot in air
ABS-PVC	–	Acrid	Yellow, blue edges	No	–	–	–	–	Black smoke with soot in air
PVC/Acrylic	–	Fruity	Blue, yellow tip	No	–	–	–	–	–
Thermosets									
Melamines	Formal-dehyde and fish	–	–	–	–	–	–	–	–
Phenolics	Formal-dehyde and phenol	Phenol and wood or paper	Yellow	No	–	–	–	–	May crack
Polyesters	–	Hydrochloric Acid	Yellow	No	b	Yellow, blue edges	No	Slow	Cracks and Breaks
Silicones	b	–	–	–	–	–	–	–	Deforms
Ureas	Formaldehyde	–	–	–	–	–	–	–	

a, Flame retardant grade. b, Non-descript. c, Inorganic fiber. d, Organic fiber ref. Materials Engineering, Penton/IPC, Cleveland, Ohio

N.B. Take care when smelling any burning plastic material. Make sure the flame is totally extinguished. Do not directly inhale the fumes from the burning sample. Hold the sample away from nose (about 20-30 cm) and *sniff* just enough to get the smell. Do not inhale the smell deeply. Take care of any dripping polymer it will be *hot* and can burn.

Once a tentative identification is made, it is usually desirable to repeat the flame test once or even twice to get a positive confirmation of the results of the original identification.

Remember that additives may affect the results, e.g., the use of flame retardants can considerably change the flame and smoke burning characteristics of the polymer under test.

6.4.6.1 Examples

A few examples of identification of plastics by flame test are included as follows .

(A) Thermoplastic Materials

First of all confirm that the plastics is a thermoplastic by heating a wire in a match stick flame and then touching the plastic sample with the hot wire. Melting of the plastic confirms that it is thermoplastic, otherwise it is a thermoset.

Drop Test : Drop the test sample onto a hard surface.

Result : Metallic ringing sound generally indicates the presence of styrene based polymer.

Styrene based Polymer

Absence of metallic ringing sound means it is generally not a styrene based polymer (unless it is foamed plastic in which case the foam will be evident or it may be High Impact Polystyrene, PS-HI, in which case it will have the polystyrene type feel to it).

Non-Styrene based Polymer

(a) Styrene based Polymers

Flame Testing : Hold the polymer sample in a match stick flame and smell.

Results :

(i) Characteristic smell of polystyrene.

The material is **Polystyrene.**

(ii) Characteristic smell of polystyrene + a bitter smell.

The material is **Styrene-acrylonitrite.**

(iii) Characteristic smell of polystyrene + smell of rubber.

The material is **Acrylonitrile-butadiene-Styrene Polymer.**

(b) Non-Styrene Based Polymers

1. Float Test

Put the polymer test sample in a bowl of water that has some detergent added to it to enhance the wettability of the test piece. If detergent is not added the test sample may not wet as surface tension will prevent the sample from sinking even if it is heavier than water. However, for Foam samples this test will not work.

Results

(i) Sample floats – the polymer is generally a **Polyolefin.**

(ii) Sample Sinks — the polymer is generally **not a Polyolefin.**

(c) Polyolefin Type Polymers

If the sample floats it generally indicates a polyolefin type of a polymer but two more tests are required to be conducted to check what type of polyolefin it is.

2. Scratch Test

Scratch the polymer sample with your fingernail.

3. Flame Test

Hold the test sample in a match-stick flame and smell.

Results

(i) Glossy surface, does not scratch, burns with the smell of paraffin wax.

It is **Polypropylene**

(ii) Glossy surface, slight scratching, burns and drips like wax.

It is **HDPE (High Density Polyethylene)**

(iii) Not very glossy surface, scratches easily and burns with the smell of paraffin wax. It is LDPE.

Low Density Polyethylene

(d) Non-Polyolefin Type Polymers (1)

Flame Testing

Difficult to ignite and self-extinguishing (note the colour of the flame and with what difficulty it starts to burn).

Results

(i) Sample burns with a green flame :

(a) Acrid i.e., strong bitter and unpleasant smell, material is soft and flexible.

It is **Plasticised Polyvinyl Chloride.**

(b) Acrid smell but the material is stiff and glossy.

It is **Unplasticised PVC.**

(ii) The sample burns with a yellow flame :

(a) Smell of formaldehyde

The material is **Polyacetal.**

(b) No smell, sample has slippery feel.

Then push a cold metal point (rod or wire) onto hot surface of the polymer sample and then withdraw.

If threads of plastic material are formed, it is Polyamide.

Polyamide (Nylon)

(c) No real flame formed, material forms a cellular structure and then decomposes.

The material is **Polycarbonate**

(B) Thermosetting Materials

Hold the polymer sample in a match-stick flame and smell.

Results :

(i) Test sample burns but is self-extinguishing. It gives Carbolic smell. The sample is usually black or brown.

It is **Phenol formaldehyde resin**

(ii) The test sample burns but is self-extinguishing. It gives sharp acrid odour. It is light in colour.

It is **Phenol formaldehyde epoxide resin.**

(iii) The test sample burns with fishy smell. The sample is usually of a bright colour or white.

It is either **Urea formaldehyde** or **Melamine formaldehyde resin**

To ascertain the true identity carry out **scuff Test.***

Results

(i) Scuffs with fingernail.

It is **Urea formaldehyde resin**

(ii) Does not scuff with fingernail.

It is Malamine formaldehyde resin

* **Scuff Test** is carried out to determine the scoop, scratch, or rub resistance of the material.

7 EPILOGUE*

7.0 Categorisation of Plastics

Plastics are a group of materials, either natural or synthetic which may be shaped when soft and then hardened to retain the given shape. Plastics are polymer based. A polymer is a substance made of many repeating units.

Naturally occurring polymers include tar, shellac, tortoise shell, animal horn, cellulose, amber**, and latex from tree sap. Synthetic polymers include polyethylene (used in plastic bags); polystyrene (used to make styrofoam cups), polypropylene (used for making fibres and bottles); polyvinylchloride (used for food wraps, bottles, garden hoses, etc.); and polytetra fluoro ethylene or Teflon (used for making non-stick kitchen wares).

Due to limited natural resources, today synthetic polymers in the form plastics are in wide use and the plastic industry is one of the fastest growing throughout the world.

Depending upon the service conditions a large variety of plastics has been developed which can be grouped into three categories.

1. General purpose or Commodity Plastics
2. Engineering Plastics
   (i) General purpose Engineering Plastics
   (ii) High Performance Engineering Plastics
3. Super Engineering Plastics

Brief discussion of these three categories of plastics follows :

7.1 Commodity or General Purpose Plastics

Commodity plastics are those plastics that are used in high volume and for wide range of applications, such as film for packaging, photographic and magnetic tapes, clothing, beverage and trash containers and a variety of household goods where mechanical properties and service conditions are not demanding or critical. Such plastics have relatively low mechanical properties and are of low cost. The range of products made from such plastics includes plates, cups, carrying trays, medical trays, containers, printed material, shopping carrybags and other disposable items.

Examples of commodity plastics are polyethylene (PE), polypropylene (PP), polyvinyl chloride (PVC), polymethyl-methacrylate (PMMA) and many more. They constitute nearly 90% of the total volume of plastics produced.

---

* The Concluding Remarks.

** Fossilized tree resin used in jewellary and even in medicines.

Fig 7.1: Some typical items made from Commodity Plastics.

7.2 Engineering Plastics

Engineering plastics are a group of plastics that have better mechanical and thermal properties than the most widely used Commodity Plastics.

Being more expensive Engineering Plastics are produced in lower quantities and are often used for smaller objects or low volume applications, e.g., mechanical components rather than for bulk and high volume use such as packaging and carry bags.

Engineering Plastics usually refer to thermoplastic materials rather than thermosets. Examples of Engineering Plastics includes acrylonitrile butadiene styrene (ABS), used for car bumpers, dashboard trims and logo bricks (toys); polycarbonate (PC), used in motorcycle helmets; and polyamide (PA or nylons), used for ski and ski boots.

Each Engineering Plastic has usually a unique combination of properties that may make it the material of choice for some specific application. For example, polycarbonates are highly resistant to impact, while polyamides (nylons) are highly resistant to abrasion. Other properties exhibited by different Engineering Plastics include heat resistance, mechanical strength, rigidity, chemical stability, self lubrication and the safety.

Some of the widely used Engineering Plastics include acrylonitrile butadiene styrene (ABS), Nylon 6, Nylon 66, polyamides (PA), polybutylene terephthalate (PBT), polycarbonate (PC), polyetherketone (PEK), polyether-etherketone (PEEK), polyethylene terephthalate (PET), polyimides (PI), polyoxymethylene (POM/Acetal), polyphenylene sulfide (PPS), polysulphone (PSU), polyphenylene oxide (PPO), polytetra fluoroethylene (PTFE), and more.

Sometimes PPS and PSU are separated to form a separate group of High Performing Engineering Plastics while the remaining are categorised as General Purpose Engineering Plastics.

Table 7.1 shows the engineering properties of some important Engineering Plastics and Fig. 7.2 shows some typical components made from such plastics.

Table 7.1: Mechanical Properties of Some Important Engineering Plastics

Polymer Type	Ultimate Tensile Strength (MPa)	Elongation (%)	Tensile Modulus (GPa)
ABS	40	30	2.3
ABS + 30% Glass Fiber	60	2	9
Acetal Copolymer	60	45	2.7
Acetal Copolymer + 30% Glass Fiber	110	3	9.5
Acrylic	70	5	3.2
Nylon 6	70	90	1.8
Polyamide-imide	110	6	4.5
Polycarbonate	70	100	2.6
Polyethylene, HDPE	15	500	0.8
Polyethylene Terephthalate (PET)	55	125	2.7
Polyimide	85	7	2.5
Polyimide + Glass Fiber	150	2	12
Polypropylene	40	100	1.9
Polystyrene	40	7	3

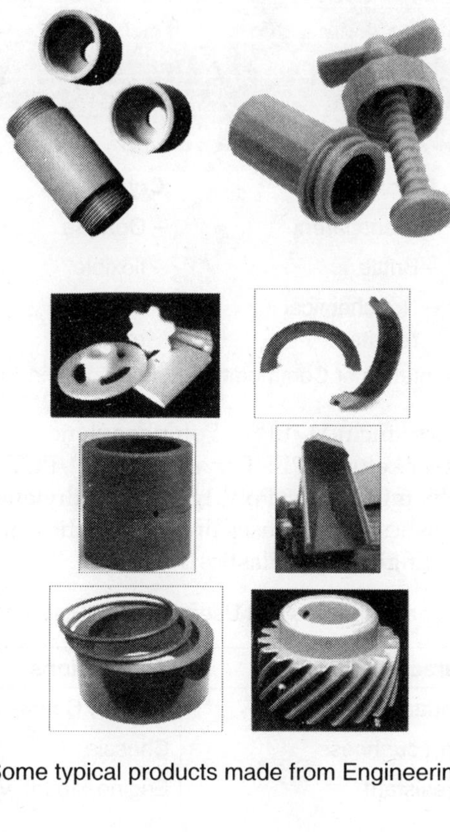

Fig 7.2: Some typical products made from Engineering Plastics.

7.3 Super Engineering Plastics

Super Engineering Plastics or Ultrahigh Performing Plastics are at the highest end of the spectrum of Engineering Plastics, Fig 7.3, with very high strength and thermal resistance. There are some important properties exhibited by the Super Engineering Plastics that includes resistance to impact, heat and abrasion, high mechanical strength, self lubrication, fire safety and chemical stability. These properties of the Super Engineering Plastics make them the material of choice in vast number of industries as well as domestic applications. The Super Engineering Plastics are generally used in high temperature, high stress, and harsh environmental conditions. Applications of the Super Engineering Plastics can be found in diverse industries including automotive and transportation, electrical and electronics, packaging, health care, textiles, and sports. The Super Engineering Plastics find their uses in manufacturing bumpers and dashboards of automobiles, and pipes for carrying drain waste. Being good insulators they are used in making handlebars and plug connectors of electronic devices.

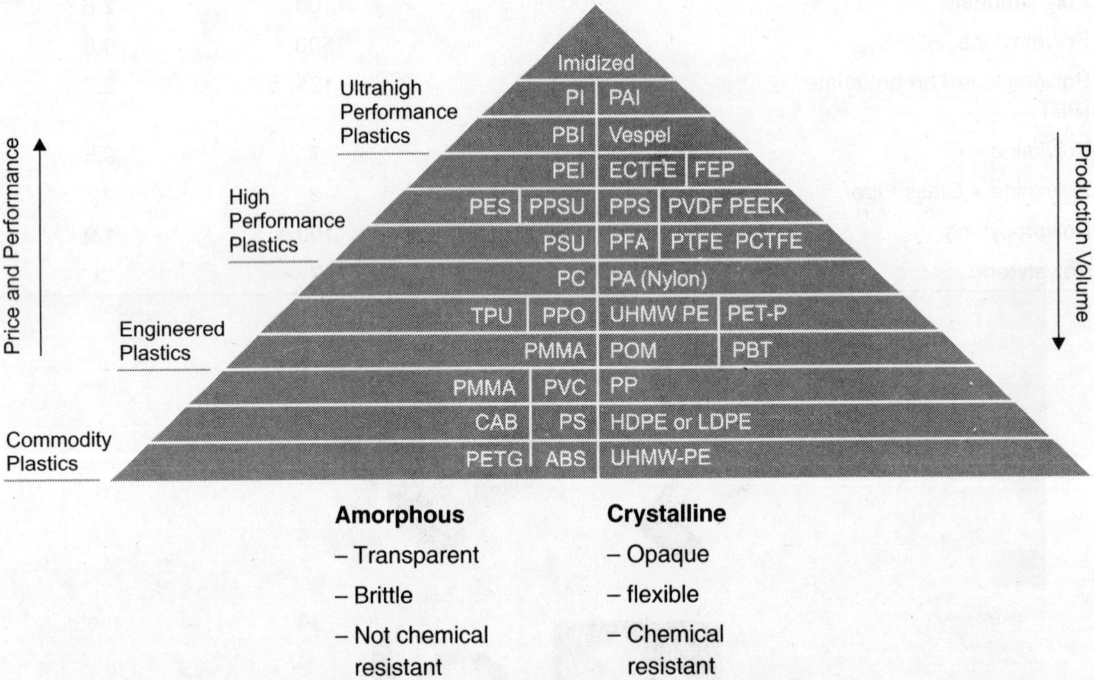

Fig 7.3: Pyramidial representation of Commodity, Engineering and Super Engineering Plastics.

Super Engineering Plastics include PPS (Polyphenylene Sulfide), LCP (Liquid Crystal Polymers), PEEK (Polyether ether ketone), PEI (Polyetherimide), PCT (Poly cyclohexane dimethanol terephthalate), and PA9T (Polyamide 9T or Poly 1,9 nonamethylene terephthal amide). Table 7.2 shows the important characteristics and the uses of these plastics; and Fig 7.4 shows some typical components made from Super Engineering Plastics.

Table 7.2: Typical Characteristics and Uses of Some Super Engineering Plastics

Plastic	Characteristics	Applications
PPS	Standard	Sensor, Connector
	High Toughness	Chassis
	Oil resistant	Engine Mount, valves

	Heat Resistant	Boiler Parts, Pump Parts
LCP	Heat Resistant Flame Retardant High Strength	IC Tray Connector, etc
PEEK	Standard Chemical Resistant Glass-fiber Reinforced	Automotive Engine Parts Oilwell Aerospace, etc
PEI	Heat Resistant Flame Retarded Chemical Resistant	Car Lens Reflector TFT Cassette Hard Disk Component, etc
PCT	Standard Chemical Resistance	Toy, Appliance Trays Displays Consumer Items
PA 9T	Standard High Heat Dimensional Stability	Surface Mounting Connector

(a) High strength high accuracy gears

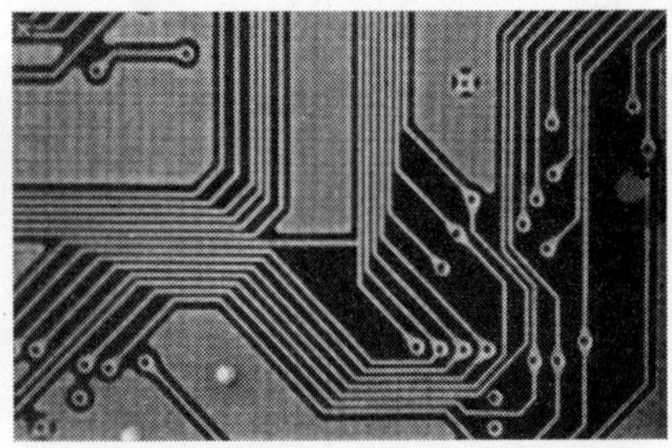

(b) Printed Circuit board

(c) Moulded Gear-Axle assembly

Fig 7.4: Typical components made from Super Engineering Plastics

All these categories of plastics are presented in a tabular form in Table 7.3 which also shows their classification by chemical structure.

Table 7.3: Classification of Plastics by Their Strength and Chemical Structure

				Classification by chemical structure
PLASTICS	Engineering Plastics	Super engineering plastics (Specialized engineering plastics)	Aromatic polyester (ARPES)	
			Polyetheretherketone (PEEK)	POLYESTER
			Polyethemitrile (ENP)	POLYETHER
			Polyetheimide (PEI)	
			Polyethersulfone (PES)	
			Fluoro Plastics (PTFE) (PFA) (FEP) (ETFE) (PVdF)	FLUORO PLASTICS
			Polyimide (PI)	POLYIMIDE
			Polyamideimide (PAI)	
		High engineering plastics	Polyphenylensulfide (PPS)	POLYSULFIDE
			Polysulfone (PSF, PSU)	
		General purpose engineering plastics	Polyacetal (POM)	POLYETHER
			Modified polyphenylene ether (Modified PPE)	
			Modified polyphenylene oxide (Modified PPO)	
			Polyethylene naphtherate (PEN)	POLYESTER
			Polyethylene terephthalate (PET)	
			Polybuthylene terephthalate (PBT)	
			Polycarbonate (PC)	
			Polyamide (PA)	POLYAMIDE
			Epoxy Resin (EP)	–
			Polyethylene (PE)	POLYOLEFIN
	General purpose plastics (Commodity Plastics)		Polypropylene (PP)	POLYOLEFIN
			Polyvinyl-chloride (PVC)	POLYVINYL
			Polystyrene (PS)	
			Acrylonitrile-butadiene-styrene Resin (ABS)	
			Polymethyl methacrylate (PMMA)	
			Polyurethane (PUR)	–
			Phenolic Resin(PF)	–
			Melamine Resin (MF)	–
			Urea Resin (UF)	–
			Silicone Resin (SI)	–

ELASTOMER
Fluoro oil (PFPE)
(PFAE)

It must be kept in mind that the dividing lines between different categories of plastics are not very sharp so there can be some overlap and thus one type of plastic may sometimes appear in two categories near the border line.

Super Engineering Plastics are going to make it easier to produce very light weight cars and even cars with transparent bodies. Fig. 7.5 shows two such conceptual cars of the future.

(a) Carbody made of transparent Plastic

(b) High speed car with body of Super Plastic

Fig 7.5: Cars of the future with bodies made of Super Engineering Plastics

Lastly it can be said that plastics have a vast field of applications and these are being produced in ever improving quality. This is possible because Plastic Engineering has become a large accumulation of knowledge contributed by different streams of Science and Technology as is depicted by Fig. 7.6.

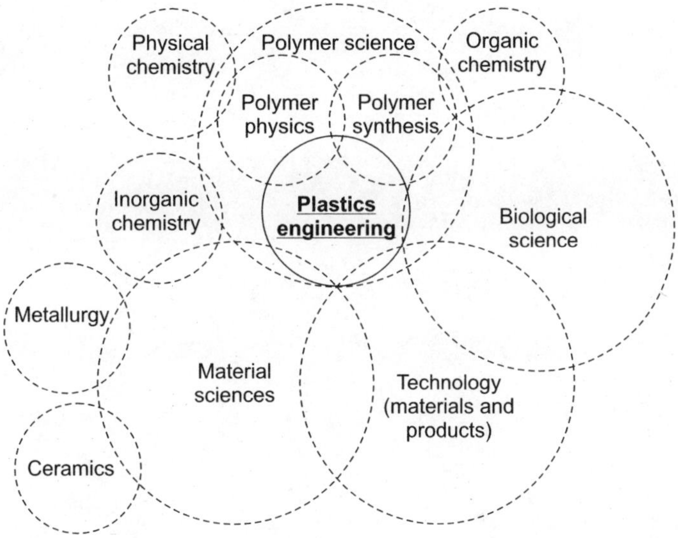

Fig 7.6: Plastics Engineering as related to other fields of Science and Technology.

It is evident from the above figure that it is essential to have background knowledge of a large number of subjects to have deeper understanding of the vast ever expanding field of Modern Plastics.

APPENDICES

APPENDIX : A

LIST OF FIGURES

CHAPTER-1 POLYMERS, PLASTICS AND ELASTOMERS (RUBBERS)

CHAPTER-2 POLYMERIC MATERIALS

CHAPTER-3 TESTING OF PLASTICS

CHAPTER-4 PROCESSING OF POLYMERS

CHAPTER-5 CASE STUDY

CHAPTER-6 RECYCLING OF PLASTICS

CHAPTER-7 (EPILOGUE)

APPENDIX : B

LIST OF TABLES

CHAPTER-1 POLYMERS, PLASTICS AND ELASTOMERS (RUBBERS)

CHAPTER-2 POLYMERIC MATERIALS

CHAPTER-3 TESTING OF PLASTICS

CHAPTER-4 PROCESSING OF POLYMERS

APPENDIX : C

ABBREVIATION FOR PLASTICS

ABBREVIATIONS FOR PLASTICS & RELATED MATERIALS

AA	Allyl alcohol
ABS	Acrylonitrile butadiene styrene
AGE	Allyl glycidyl ether
AMS	Alpha methyl styrene
AN	Acro nitrile
AR	Acrylic Rubber
BA	Butyl acrylate
BD	Butane diol
BOH	Hydroxyl terminated butadiene
BPA	Benzo phenone anhydride
BSA	Bisphenol A
CA	Cellulose acetate
CB	Cellulose butyrate
CEL	Cellulose
CELD	Cellulose derivatives
CEVE	Chloro ethyl vinyl ether
CHDM	Cyclo hexane di methanol
CMC	Carboxy methyl cellulose
CN	Cellulose nitrate
CP	Cellulose propionate
CR	Poly Chloroprene rubber
CS	Chloro Sulfonate
CSPE	Chloro Sulfonated polyethylene
CTFE	Chloro tri fluoro ethylene
DAIP	Di allyl iso phthalate
DAP	Di allyl phthalate
DADPS	Di amino di phenyl sulfone
DCDA	Di cyan di amide
DEG	Di ethylene glycol
DETA	Di ethylene tri amine
DHB	Di hydroxy bisphenol
DPE	Di phenyl ether
DPSU	Di phenyl sulfone
E	Ethylene
EA	Ethyl acrylate
EAR	Ethylene acrylate rubber
EC	Ethyl Cellulose
ECH	Epi chloro hydrin
ECHR	Epi chloro hydrin rubber
EDPSU	Ether di phenyl sulfone
EES	Ethylene ether sulfide
EG	Ethylene glycol
EPR	Ethylene propylene rubber
ES	Ethylene sulfide
EVA	Ethylene vinyl acetate copolymer
F	Formaldehyde
FUL	Furfural
FUN	Furan
G	Glycidyl group
HD	Hexadiene
HEC	Hydroxy ethyl cellulose
HEMA	Hydroxy ethyl methacrylate
HFI	Hexa fluoro propylene
HMDA	Hexa methylene di amine
HMDI	Hydrogenated MDI
IA	Isophthalic acid
M	Melamine
MA	Methyl acrylate
MAA	Methacrylic acid
MBAM	Methylene bis acryl amide
MC	Methyl cellulose
MDA	Methylene di aniline
MDI	Methylene di phenyl di isocynate
MLA	Maleic anhydride
MMA	Methyl methacrylate
MO	Methylene oxide
MOCA	Methylene bis ortho chloro aniline
MPA	m phenylene di amine
MPDMS	Modified PDMS
NBR	Nitrile rubber
P	Phenol
P	Phenyl
PAA	Poly acrylic acid

PAA	Poly amic acid
PAEK	Poly aryl ether ketone
PAI	Poly amide imide
PAM	Poly acryl amide
PAMS	Poly alpha methyl styrene
PAN	Poly acrylo nitrile
PA-n	Aliphatic poly amide
PA-(n1, n2)	Aliphatic poly amide
PA-R	Greneral poly amide
PA-(R1, R2)	Greneral polyamide
PAT	Poly alkylene terephthalate
PA (6, 3, T)	Poly amide
PA (MD, 12)	Poly amide
PA (MX, 6)	Poly amide
PB	Poly butadiene
PBA	Poly butyl acrylate
PBEG	Poly butylene ether glycol
PBSU	Poly bisphenol sulfone
PBT	Poly butylene terephthalate
PB 1	Poly butylene
PC	Poly carbonate
PCT	Poly cyclohexane di methanol terephthalate
PCTFE	Poly chloro tri fluoro ethylene
PDMS	Poly di methyl siloxane
PE	Poly ethylene
PEA	Poly ethyl acrylate
PEEG	Poly ethylene ether glycol
PEEK	Poly ether ether ketone
PEI	Poly ether imide
PEMA	Poly ethyl meth acrylate
PEO	Poly ethylene oxide
PESU	Poly ether sulfone
PET	Poly ethylene terephthalate
PFEP	Poly fluorinated ethylene propylene copolymer
PHBA	Poly p hydroxy benzol acid
PHFP	Poly hexa fluoro propylene
PHFI	Poly hexa fluoro isobutylene
PHFPO	Poly hexa fluoro propylene oxide
PIB	Poly isobutylene
PI	Poly imide
PMA	Poly methyl acrylate
PMAA	Poly meth acrylic acid
PMAN	Poly meth acrylo nitrile
PMMA	Poly methyl meth acrylate
PMO	Poly methylene oxide
PMP	Poly methyl pentene
PMPI	Poly m phenylene isophthal amide
PMS	Para methyl styrene
PNB	Poly norbornene
PNP	Poly nuclear phenol epoxy
PROR	Propylene oxide rubber
PP	Polypropylene
PPA	p′ phenylene diamine
PPEG	Poly propylene ether glycol
PPESU	Poly phenyl ether sulfone
PPFA	Poly per fluoro alkoxy
PPFBA	Poly per fluoro butyl acrylate
PPFVME	Poly per fluoro vinyl methyl ether
PPMS	Poly para methyl styrene
PPO	Poly phenylene oxide
PPRO	Poly propylene oxide
PPPT	Poly p phenylene terephthal amide
PPS	Poly phenylene sulfide
PPX	Parylene
PPZ	Poly organo phosphazene
PRO	Propylene oxide
PO	Phenylene oxide
PS	Poly styrene
PSU	Poly sulfone
PTA	Phthalic anhydride
PTFE	Poly tetra fluoro ethylene
PU	Poly urethane
PUR	Poly urethane General
PVAC	Poly vinyl acetate
PVACL	Poly vinyl acetal
PVAE	Poly vinyl alkyl ether
PVAH	Poly vinyl aldehydic
PVAL	Poly vinyl alcohol
PVB	Poly vinyl butyral
PVC	Poly vinyl chloride
PVCZ	Poly vinyl carbazole
PVDC	Poly vinylidene chloride

PVDF	Poly vinylidene fluoride
PVEE	Poly vinyl ethyl ether
PVF	Poly vinyl fluoride
PVFO	Poly vinyl formal
PVIE	Poly isobutyl ether
PVME	Poly vinyl methyl ether
PVPO	Poly vinyl pyrrolidone
P2 VP	Poly 2-vinyl pyridine
S	Styrene
SBR	Styrene butadiene rubber
SPPO	Substituted poly phenylene oxide
SR	Poly sulfide rubber
TA	Terephthalic acid
TAC	Tri allyl cyanurate
TDI	Toluene di isocyanate
TETA	Tri ethylene tetramine
TFE	Tetra fluoro ethylene
TFP	Tri fluoro propyl
TGMDA	Tetra glycidyl methylene di aniline
TMA	Tri mellitic acid anhydride
TMP	Tri methylol propane
TP-R	General thermoplastic polyester
U	Urea
UA	Urea linkage
UPPO	Unsubstituted poly phenylene oxide
UT	Urethane linkage
V	Vinyl
VA	Vinyl acetate
VDF	Vinylidene fluoride

APPENDIX : D

SPECIFIC USES OF PLASTICS

SPECIFIC USES OF SOME IMPORTANT POLYMERS

S.No.	Material	Symbol	Specific Uses
1.	Acrylic Rubber	AR	Oil seals for automobiles, gaskets and hoses.
2.	Acrylonitrile-Butadiene Rubber	NBR	Fuel and oil tubing and hose, gaskets and seals, conveyor belting, printing rolls and pads.
3.	Alkyds		Electrical and electronic insulators, arc chute, automotive coil tops, distributor caps and switch insulators.
4.	Cellular Polymers (Foams, sponges, etc.)		Insulation boards, packaging of frozen or perishable food like ice cream, fish, etc.; buoyancy material, automotive occupant protection, cushioning, bedding, acoustic insulation, dampening material and filtering material.
5.	Chlorosulfonated Polyethylene	CSPE	Hose products, footwear, roll covers, tank linings, wire and cable covers and building products. High cost limits its use for demanding conditions.
6.	Diallyl Isophthalate Polymer	DAIP	Spectacle lenses
7.	Diallyl Phthalate	DAP	Electrical components and insulation for use under adverse operating conditions, military communication gear requiring resistance to fungi.
8.	Ebonite		Car battery cases.
9.	Epichlorophydrin Rubber	ECHR	Fuel, lubricating fluid; air and vacuum hose for automobiles; seals, gaskets, diaphragms; wire and cable covers, and protective boots.
10.	Epoxy Systems	EP	Surface coatings and adhesives, floor and road surfacing, encapsulation of small electrical and electronic components like transistors, laminated printed circuits, large outdoor insulators, rocket motor casings; pipe, tank, drum and can linings.
11.	Ethylene-Propylene Rubber	EPR	Solid or Cellular (foam or sponge) weather strip for automobile wind screens, radiator and heater hose, and sheeting for outside use; wire and cable insulation, hose and belt products, and coated fabrics.
12.	Ethylene-Vinylacetate Copolymer	EVA	Medical tubing, squeeze bulbs, gaskets, food packaging film, coatings and hot melt adhesives.
13.	Furan Systems		Binder for sand in foundry work, used in cement for chemical plants, grinding wheels, etc.
14.	Ionomers	IO	Golf balls, bowling pin covers, wire coating, safety shield, food packaging film.
15.	Melamine formaldehyde	MF	Tableware like dishware safe moulded dinnerware; as adhesive for water-resistant plywood, electric boards, circuit breakers, switch gears, utensil handles and appliance knobs.

16.	Nylons (in general)		Surgical sutures (threads for stiching wounds and cuts), artificial hair and fur, tennis racket strings, tyre cords.
17.	Nylon 11 & 12		Film and tubing
18.	Nylon 610		Vending machine parts, tubing and food packaging.
19.	Nylons		Textiles, teeth in plastic zip fasteners. Castors for light furniture, carburettor linkages, small gears, brush bristles, buttons, fishlines, switch components in telephone exchanges.
20.	Phenol Formaldehyde	PF	Knobs, handles, instrument cases, bottle caps and closures, automobile distributor caps, switches, thrust washers, electrical components like connectors, sockets; as binder in grinding and cut-off wheels, chipboards, plywood fabrication, decorative laminates for counters and table tops; as fire-resistant rigid foams, etc.
21.	Polyamides (Transparent)	PA	Filter bowls, headlamp roflectors, rotameter parts.
22.	Polyamideimide	PAI	High performance parts in automative, aircraft and aerospace; high temperature insulants.
23.	Polyamide Thermoplastic Elastomers		Hoses, gaskets and seals and high performance athletic goods such as ski-boots and footballs.
24.	Polybutadiene	PB	Automobile tyres, production of HIPS (High Impact Polystrene), and ABS (Acrylonitrile Butadiene Styrene).
25.	Polybutylene	PB1	Pressure piping (Cold/hot water, gas, irrigation); aqueous mineral slurries e.g; gypsum.
26.	Polybutylene tetraphthalate	PBT	Push-button switch bodies, fuse boxes, motor housing and brush holders, cookware handles, sextant components, base for coffee-making machines, coil bobbins.
27.	Polycarbonate	PC	Safety glazing, windshields, bottles, partition shields, street light globes, hair dryer bodies, coffee makers. Automotive lenses and light assemblies, high voltage plugs and sockets, microwave oven wares, camera bodies capacitor films, safety helmets and goggles, gears, etc.
28.	Polychloroprene Rubbers	CR	Automotive products like fuel hoses, cable sheathing Protective boots, etc. They are also used for bridge pads, roofing membranes, fabric coatings, gloves, adhesives and corrosion resistant coatings.
29.	Polychlorotri Fluoro ethylene	PCTFE	Good resistance to oxygen – used for aerospace and medical applications; packaging medicines in strips and blister packs, gaskets, o-rings, seals, instrument windows, electronic flexible printed circuits, etc.
30.	Polyester Thermoplastic Elastomers		Relalively high cost elastomers used for automotive parts, gear wheels and footwear soles.
31.	Polyetheretherketone	PEEK	Wire and cable coverings for critical applications in aerospace and nuclear industry, pump impellers, valve parts, membrane, filters and conveyor belts in high temperature service.

32.	Polyethylene:		
(i)	Low Density Polyethylene	LDPE	Refrigerator dishes, domestic wares, plumbing, wire insulation, squeeze bottles, packaging film.
(ii)	High Density Polyethylene	HDPE	Dustbins, mudguards, toys, milk bottle crates, wrapping for wet food stuffs like meat and fish, tissue paper.
(iii)	Ultrahigh Molecular Weight Polyethylene	UHMWPE	Pen tips, gears, wear surfaces of body parts.
33.	Polyethylene terephthalate	PET	As a fibre; photographic and x-ray films, magnetic tapes, electrical insulation, food packaging films including boil-in bags, vials and bottles for alcoholic beverages and carbonated drinks under pressure.
34.	Polyisoprene/Natural Rubber	PI	Pharmaceutical or medical devices, tyres, gaskets, and shock absorption devices.
35.	Polymethylpentene	PMP	Spirits despensers, lighting elements, food packaging, medical wares like syringes; closures for cosmetic bottles, wire insulation, etc.
36.	Polymeric Fibres		Used as filters, cords and cables and also as fabrics.
37.	Polymeric Films		Innumerable uses including mechanical and barrier packaging, garbage or trash bags, packaging of sweetmeat boxes, food grain bags, silo covers, greenhouse covers, insulation tape, audio/video tape, as diffusion membrane, moisture-proof envelopes, high altitude balloons and sail material.
38.	Polymethylmethacrylate	PMMA	Has excellent transparency so used for hard contact lenses, automotive tail light lenses, safety and security glazing, skylights, illuminated signs. Also used as coatings e.g., floor waxes and emulsion (latex) paints.
39.	Polymethylene Oxide/ Acetals	PMO	Hot water plumbings, gears, bearings, cams, ratchets, carburettor and fuel pump parts, aerosol containers, zip fasteners, electric kettles and food mixer parts, etc.
40.	Polyolefin Thermoplastic Elastomers		Mainly in the automotive field for such parts as radiator, grills, headlight surroundings and rub strips.
41.	Polypropylene	PP	Moulded integral hinges, picnicware, carpet backing, sacks, ropes, textiles, outdoor furniture, lamination.
42.	Polyhenylene Oxide	PPO/ MPPO	Television yokes, tool handles, small appliance housings, business machine housings, sprinkler parts, decorative grills, etc.
43.	Polyphenylene Sulfide	PPS	Pump housing and impellers, valves, metering devices, etch tank coatings for chemical process industry, lamp sockets, coil bobbins, gears, bushings, carburettor parts, microwave compatible cookware, anti-stick coatings, adhesive and laminating resin.
44.	Polysulfide Rubbers	SR	Special applications that involve contact with fuels, paints and printing inks and outdoor exposure, i.e; as caulking material, mastics and putty.

45.	Polytetra fluoro ethylene	PTFE	Bridge bearing pads, arch chutes, human heart valves, cryogenic valves, expansion bellows, non-stick cookwares, laboratory thread tape sealant, wire and cable insulation for high frequency applications.
46.	Polyurethanes	PU	Pump liners and impellers, gears, sprocket wheels, bushings o-rings, seals and gaskets, solid tyres, roller conveyors for printing paper industry, sports boots for sking and hockey, conveyor belts, and chute liners.
47..	Polyurethane Thermoplastic Elastomers		Skiboots, strech fabrics and in fibrous form as foundation garments.
48.	Poly vinyl butyral	PVB	Adhesive layer for safety glass.
49.	Poly vinyl carbazole	PVCZ	High frequency dielectrics/xerography
50.	Poly vinyl flouride	PVF	Lighting panels, glazing for solar energy collection, electrical wrapping tape and parting layers for laminates.
51.	Poly vinylidene fluoride	PVDF	Pen tips, seals, gaskets, packagings diaphragms, valve parts, coil bobbins, wire insulation and jacketing, thermowelded liners, filters, pipes-tubing and ducting, etc.
52.	Poly vinylidene fluoride and Polytetra fluoro ethylene	PVDF/ PTFE	These are specialty polymers used in demanding applications like mechanical seals, packings, o-rings,gaskets, diaphragms, expansion joints and connectors, hose liners, roll covers, and wire & cable insulation.
53.	Poly vinyl pyrrolidone	PVPO	Artificial blood/blood plasma.
54.	Silicone Polymers/ Polydimethyl Siloxane	PDMS	Power cable insulation, high voltage leads and insulator boots, ignition cables, spark plug boots; o-rings, gaskets, seals for aircraft doors and windows, freezers, flapper valves, diaphragms, protective boots and bellows, printing rolls.
55.	Styrene-Butadiene Rubber	SBR	Mostly for automobile tyres, but also for footwear, wire and cable insulation, industrial rubber products, adhesives, paints, carpet backing and paper backing.
56.	Styrenic Thermoplastic Elastomers		Footwear soles and heels, elastic bands and adhesives.
57.	Unsaturated Polyester Systems	UPR	Automobile parts like car bonnet, front ends; window frames, chemical storage tanks, moulded chairs and containers, car bodies, electrical switch boxes, road lamp housings, television cabinets, and aircraft parts, etc.
58.	Urea formaldehyde	UF	Caps and closures for bottles and containers, electrical fittings, buttons and toilet seats; as wood adhesive in plywood and chipboard manufacture; as foam in wall insulation, and floral decoration.
59.	Vinylesters Resins	VER	As matrix for fibre-reinforced plastics (FRP), chemical process equipment like piping, tanks, coatings, etc.
60.	Cross – linked Polyethylene	XLPE/ PEX/XPE	Domastic water piping, and insulation for high voltage electrical cables, transportation of sewage and slurries; pipework in buildings; hydronic radiant heating and cooling systems. Also as substitute for PVC piping.

APPENDIX : E

SPECIFIC GRAVITY TABLE

COMPARATIVE VALUES OF SPECIFIC GRAVITY AND WEIGHT OF SOME INDUSTRIALLY IMPORTANT PLASTICS

S.No.	Material	Specific Gravity	Weight	
			lbs/in³	kg/m³
1.	Polypropylene	0.905	0.032	885.8
2.	Polyethylene: Low density	0.914	0.033	913.4
	High density	0.942	0.034	941.1
3.	Ethylene Vinyl Acetate	0.950	0.034	949.1
4.	Ionomer (Surlyn)	0.960	0.035	959.1
5.	Acrylonitrile Butadiene Styrene (ABS)	1.04	0.037	1024.2
6.	Polyethylene oxide	1.06	0.038	1051.8
7.	Polystyrene	1.06	0.038	1051.8
8.	Polystyrene Acrylonitrile (SAN)	1.07	0.038	1061.7
9.	Ethyl Cellulose	1.12	0.040	1107.2
10.	Nylon	1.14	0.041	1134.9
11.	Polycarbonate ABS alloy	1.14	0.041	1134.9
12.	Acrylic	1.18	0.043	1190.2
13.	Phenoxy	1.18	0.043	1190.2
14.	Cellulose Acetate Butyrate	1.19	0.043	1200.3
15.	Cellulose Propionate	1.20	0.044	1217.9
16.	Polycarbonate	1.20	0.044	1217.9
17.	ABS - PVC alloy	1.21	0.044	1228.0
18.	Polyurethane	1.24	0.045	1245.6
19.	Polysulfone	1.25	0.045	1255.6
20.	Cellulose Acetate	1.27	0.046	1273.3
21.	Polyvinyl Chloride (Rigid)	1.33	0.048	1328.6
22.	Acrylic PVC alloy	1.35	0.049	1356.3
23.	Chlorinated Polyether	1.40	0.050	1384.0
24.	Acetal	1.42	0.051	1411.7
25.	Polyvinyl Dichloride	1.50	0.054	1356.3
26.	Polyvinylidene Chloride	1.60	0.058	1605.4
27.	Kynar Vinylidene Fluoride	1.75	0.063	1743.8
28.	KEL-F trifluorochloroethylene	2.14	0.078	2159.0
29.	TFE (Teflon) fluorocarbon	2.15	0.078	2169.0
30.	FEP (Teflon) fluorocarbon	2.16	0.078	2179.2

COMPARATIVE VALUES OF SPECIFIC GRAVITY AND WEIGHT FOR SOME REINFORCED THERMOPLASTICS

S.No.	Material	Specific Gravity	Weight	
			lbs/in^3	kg/m^3
1.	Polypropylene - Glass filled 30%	1.13	0.041	1134.9
2.	Polyethylene - Glass filled 30%	1.19	0.043	1190.2
3.	ABS - Glass filled 20%	1.23	0.044	1217.9
4.	Polystyrene - Glass filled 30%	1.27	0.046	1273.3
5.	SAN - Glass filled 25%	1.31	0.047	1301.0
6.	Nylon 6 - Glass filled 30%	1.37	0.050	1384.0
7.	Nylon 66 - Glass filled 30%	1.37	0.050	1384.0
8.	Polycarbonate - Glass filled 20%	1.52	0.055	1522.4
9.	Acetal - Glass filled 20%	1.56	0.056	1550.1
10.	Polyurethane	1.57	0.056	1550.1

COMPARATIVE VALUES OF SPECIFIC GRAVITY AND WEIGHT OF THERMOSETTING PLASTICS

S.No.	Material	Specific Gravity	Weight	
			lbs/in^3	kg/m^3
1.	Phenolic - General purpose	1.35	0.049	1356.3
2.	Melamine Cellulose	1.50	0.054	1494.7
3.	Urea Cellulose	1.50	0.054	1494.7
4.	Diallyl Phathalate - General purpose	1.60	0.058	1605.4
5.	Epoxy Glass	1.80	0.065	1799.2
6.	Silicone Glass	1.88	0.068	1882.2
7.	Alkyd - General purpose	1.93	0.070	1937.6

APPENDIX : F

BURN TEST CHARACTERISTICS

PLASTIC MATERIAL BURN TEST CHARACTERISTICS

Materials	No Flame	Burns, but Extinguishes on Removal of Flame Source			Continues to Burn after removal of Flame Source				Remarks
	Odor	Odor	Color of Flame	Drips	Odor	Color of Flame	Drips	Speed of Burning	
THERMOPLASTICS									
ABS (Product/ ABS-Black)		Acrid	Yellow, blue edges	No	Acrid	Yellow, blue edges	Yes	Slow	Black smoke with soot in air
Acetals	-	-	-	-	Formalde-hyde	Blue, no smoke	Yes	Slow	
Acrylics	-	-	-	-	Fruity	Blue, yellow tip	No (cast) Yes (molded)	Slow	Flame may spurt if rubber modified
Cellulosics									
Acetate	-	Vinegar	Yellow with sparks	No	Vinegar	Yellow	Yes	Slow	Flame may spark
Acetate Butyrate	-	-	-	-	Rancid butter	Blue, yellow tip	Yes	Slow	Flame may spark
Ethyl Cellulose	-	-	-	-	Burnt sugar	Yellow, blue edges	Yes	Rapid	-
Nitrate	-	-	-	-	camphor	White	No	Rapid	-
Propionate	-	-	-	-	Burnt sugar	Blue, yellow tip	Yes	Rapid	-
Chlorinated Polyether	-		Green, yellow tip	No	-	-	-	-	Black smoke with soot in air
Fluorocarbons									
FEP	Faint odor of burnt hair	-	-	-	-	-	-	-	Deforms; no combustion, but drips
PTFE	Faint odor of burnt hair	-	-	-	-	-	-	-	Deforms; does not drip

CTFE	Faint ordor of acetic acid	-	-	-	-	-	-	-	Deforms; no combustion, but drips
PVDF	acidic	-	-	-	-	-	-	-	Deforms
Nylons									
Type 6	-	-	-	-	Burnt wool	Blue, yellow tip	Yes	Slow	-
Type 6/6	-	Burnt wool or hair	Blue, yellow tip	Yes	-	-	-	Slow	More rigid than Type 6 nylon
Phenoxies	-	Acrid[d]	Yellow[c]	No[c]	Acrid[d]	Yellow[d]	Yes[d]	Slow[d]	Black smoke with soot in air
Polycarbonates	-	Faint sweet aromatic ester	Orange	Yes	-	-	-	-	Black smoke with soot in air
Polyethylenes	-	-	-	-	Paraffin	Blue, yellow tip	Yes	Slow	Floats in water
Polyphenylene									
Oxides (PPO)	-	Phenol	Yellow-orange	No	-	-	-	-	Flame spurts; very difficult to ignite
Modified Grade	-	Phenol	Yellow-orange	No	-	-	-	-	Flame spurts; difficult to ignite, soot in air
Polyimides	[b]	-	-	-	-	-	-	-	Chars; material very rigid
Polypropylenes	-	Acrid[a]	Yellow[a]	Yellow[a]	Sweet	Blue, yellow tip	Yes	Slow	Floats in water; more difficult to scratch than polyethylene
Polystyrenes	-	-	-	-	Illuminating Gas	Yellow	Yes	Rapid	Dense black smoke with soot in air
Polysulfones	-	[b]	Orange	Orange	-	-	-	-	Black smoke
Polyurethanes	-	-	-	-	[b]	Yellow	No	Slow	Black smoke
Vinyls: Flexible	-	Hydrochlor-ich acid	Yellow with green spurts	No	-	-	-	-	Chars, melts

Rigid		Hydrochlor-ich acid	Yellow with green spurts	No	-	-	-	-	Chars, melts
Polyblends									
ABS/ Polycarbonate	-	-	-	-	b	Yellow, blue edges	No	-	Black smoke with soot in air
ABS/PVC	-	Acrid	Yellow, blue edges	No	-	-	-	-	Black smoke with soot in air
PVC/Acrylic	-	Fruity	Blue, yellow tip	No	-	-	-	-	
THERMOSETS									
Alkyds									
Diallyl Phthalates	-	-	-	-	Phenolic	Yellow	No	Slow	Black smoke, cracks
Diglycol Carbonate	-	-	-	-	Acrid	Yellow	No	Slow	Black smoke with soot
Epoxies	-	-	-	-	Phenol	Black smoke	No	Slow	Black smoke with soot in air
Melamines	Formalde-hyde and fish		-	-	-	-	-	-	-
Phenolics	Formalde-hyde and phenolic	Phenol and wood or paper[d]	Yellow[d]	No	-	-	-	-	May crack
Polyesters	-	Hydro-chlor ic acid	Yellow[a]	No	b	Yellow, blue edges	No	Slow	Cracks and breaks
Silicones	b		-	-	-	-	-	-	Deforms
Ureas	Formalde-hyde	-	-	-	-	-	-	-	-

a-Flame retardant, b-Nondescript, c-Inorganic Filler, d-organic Filler

APPENDIX : G

PRACTICAL/LAB EXERCISE ON BURN TEST

POLYMER IDENTIFICATION LAB EXERCISE

Procedure :

1. Perform a flame test and a float/sink test on all the types of polymers for which samples have been provided. There is a set of small beakers containing small polymer chips for these two tests.
   For the float/sink test, do not use a piece of plastic that has been burned. Use a fresh piece of plastic. Ensure that the surface of the plastic is completely wetted. Using tweezers, shake the piece under water to release any air bubbles adhering to the surface of the plastic. Adhering air bubbles may cause a dense plastic to float.
   Record observations for all the characteristics listed on the Polymer Identification Flame Test (Float/Sink Test) chart.
   Do not burn any other plastics without the instructor's permission and be sure to work in a fume hood over an asbestos pad or gypsum board to catch all drips. See the Flame Test instructions for details on performing the test in chapter 6.
   Each student must submit his/her own copy of the results.
2. Accurately determine the specific gravity (s.g.) of 3 different polymer samples supplied by the instructor. Compare the results with s.g. values listed in the tables of Specific Gravities in Appendix : E. Also consider other physical properties such as sink/float, feel, stiffness, clarity, color, etc. <u>Do not burn any of the S.G. samples</u>.
   Report the Sample Identification number (FT-1, FT-2, FT-3; SG-1, SG-2, SG-3, etc.), the calculated s.g., and the polymer identity you have determined. Calculations are not required in the report.

Polymer Identification Flame Test

Polymer	SE or P	Speed of Burn (slow, med., fast)	Chars (C) or Melts (M)	Drips	Flame Color	Smoke Color	Sinks (↓) or Floats (↑) in water		
PVC (Tygon tubing)									
Vulcanized black rubber									
PS									
Melamine (Melmac dish)									
PMMA (Plexiglass)									
Polycarbonate (Laxan)									
FRP Epoxy									
PP									
ABS (drain pipe)									
PE									

SE = Self Extinguishing, P = Propagating; TP = Thermoplastic, TS = Thermoset

APPENDIX : H

CONVERSION TABLE

CONVERSION FACTORS

To convert SI to A multiply by	A	SI	To convert A to SI multiply by
2.47096×10^{-4}	acre	m^2	4047
1×10^{10}	ångström (Å)	m	1×10^{-10}
0.0131578	atmosphere	cm of Hg	76.0
0.967741	atmosphere	m of H_2O	10.33
9.869232×10^{-4}	atmosphere (atm)	Nm^{-2}	1.01325×10^5
1	bar	dyn/cm^2	1
1×10^{-5}	bar	$N\ m^{-2}$	1×10^5
$9.478\ 17 \times 10^{-4}$	B ThU	J = N m	$1.055\ 06 \times 10^3$
$5.265\ 65 \times 10^{-4}$	B ThU/°F	$J\ K^{-1}$	$1.899\ 09 \times 10^3$
0.056 869	B ThU/min	$W = J\ s^{-1}$	17.584 26
$4.299\ 23 \times 10^{-4}$	B ThU/pound	$J\ kg^{-1}$	2.326×10^3
0.316 998	B ThU/Ft^2 h	$W\ m^{-2}$	3.154 59
0.5778	B ThU/ft^2 h for a temperature gradient of 1 °F/ft	$W\ m^{-1}\ K^{-1}$	1.730 73
0.238 846	calorie (cal)	J	4.1868
$2.388\ 46 \times 10^{-5}$	cal/cm^2 s	$W\ m^{-2}\ s^{-1}$	4.1868×10^4
$2.388\ 46 \times 10^{-4}$	cal/g	$J\ kg^{-1}$	4.1868×10^3
$2.388\ 46 \times 10^{-4}$	cal/g °C	$J\ kg^{-1}\ K^{-1}$	4.1868×10^3
2.3889×10^{-4}	kcal/g mol	$J\ mol^{-1}$	4.1868×10^3
35.3147	cubic foot	m^3	0.238 316 8
0.0353107	cubic ft/min	lit/min	28.32
6.1024×10^4	cubic inch	m^3	$1.638\ 71 \times 10^{-5}$
1×10^5	dyne (dyn)	N	1×10^{-5}
1×10^7	dyne cm	N.m	1×10^{-7}
1×10^3	dyn/cm	$N.m^{-1}$	1×10^{-3}
10	dyn/cm^2	$N.m^{-2}$	0.1
1×10^7	erg	J	1×10^{-7}
1×10^3	erg/cm^2	$J\ m^{-2}$	1×10^{-3}
1×10^7	erg/s	w	1×10^{-7}
3.280 84	foot (ft)	m	0.3048
0.019	ft. lb/in	J/m	53.35
0.183	ft.lb/in	kg·cm/cm	5.45
23.7304	foot poundal	J	0.042 14
0.737 56	foot pound force	J	1.355 82
$2.199\ 69 \times 10^2$	gallon (UK)	m^3	$4.546\ 09 \times 10^{-3}$

To convert SI to A multiply by	A	SI	To convert A to SI multiply by
$2.641\,72 \times 10^{2}$	gallon (US)	m^3	$3.785\,41 \times 10^{-3}$
1	gauss	lines/cm^2	1
$1.543\,24 \times 10^{4}$	grain	kg	6.4799×10^{-5}
1.0×10^{-3}	g/cm^3	kg m^{-3}	1.0×10^{3}
1.0	g mol (c.g.s.)	mol	1.0
1×10^{-7}	hectobar	N m^{-2}	1×10^{7}
0.1019991	horsepower (boiler)	kW	9.804
$1.341\,02 \times 10^{-3}$	horsepower	W	745.7
39.3701	inch	m	2.54×10^{-2}
2.36221	inch/min	mm s^{-1}	0.423 33
3.599712×10^{3}	Joule	watt-hr	2.778×10^{-4}
0.102	kg-cm/cm	J/m	9.803
0.101 972	kilogram force	N	9.806 65
$1.019\,72 \times 10^{-5}$	kilogram force/cm^2	N m^{-2}	$9.806\,65 \times 10^{4}$
0.101 972	kilogram force/mm^2	MN m^{-2}	9.806 65
3.224 62	kilogram force/$mm^{3/2}$	MN $m^{-3/2}$	9.310 11
0.145 04	ksi (thousand pounds force/ square inch	MN m^{-2}	6.894 76
0.910 05	ksi $\sqrt{\text{inch}}$	MN $m^{-3/2}$	1.098 84
$9.999\,72 \times 10^{2}$	litre	m^3	$1.000\,028 \times 10^{-3}$
0.4343	$\log_{10}$ N	$\log_e$ N I_n N	2.303
0.39 3701	mil (thou)	μm (micron)	25.4
0.621 371	mile	km	1.609 344
10	poise (dyn s cm^{-2})	N s m^{-2}	0.1
2.204 59	pound (lb)	kg	0.4536
7.233 01	poundal	N	0.138 255
0.204 809	pound force	N	4.448 22
$1.450\,38 \times 10^{-4}$	pound force/square inch	N m^{-2}	$6.894\,76 \times 10^{3}$
$3.612\,73 \times 10^{-5}$	pound/cubic inch	kg m^{-3}	$2.767\,99 \times 10^{4}$
1.55×10^{3}	square inch	m^2	6.4516×10^{-4}
10.7639	square foot	m^2	0.092 903
0.386 100 6	square mile	km^2	2.589 998
1×10^{4}	stoke, c.g.s. unit of diffusivity or thermal diffusivity (cm^2 s^{-1})	m^2 s^{-1}	1×10^{-4}
$9.842\,07 \times 10^{-4}$	ton (UK)	kg	$1.016\,05 \times 10^{3}$
$1.102\,31 \times 10^{-3}$	ton (short ton, US)	kg	907.185
0.100 361	UK ton force	kN	9.964 02
0.064 749	UK ton force/square inch	MN m^{-2}	15.4443
$7.500\,62 \times 10^{-3}$	torr (1 mm of mercury)	N m^{-2}	133.322

APPENDIX : I

US/INDIAN STANDARDS ON PLASTICS

I-1 ASTM PLASTICS STANDARDS
(Selected Few Relevent Standards)

D5630-13	Standard Test Method for Ash Content in Plastics
D6980-17	Standard Test Method for Determination of Moisture in Plastics by Loss in Weight
D7191-18	Standard Test Method for Determination of Moisture in Plastics by Relative Humidity Sensor
D1621-16	Standard Test Method for Compressive Properties of Rigid Cellular Plastics
D2842-19	Standard Test Method for Water Absorption of Rigid Cellular Plastics
D7710-14	Standard Test Method for Determination of Volume and Density of Rigid and Irregularly Shaped Molded Cellular Materials
D543-14	Standard Practices for Evaluating the Resistance of Plastics to Chemical Reagents
D1435-13	Standard Practice for Outdoor Weathering of Plastics
D3045-18	Standard Practice for Heat Aging of Plastics Without Load
D4674-19	Standard Practice for Accelerated Testing for Color Stability of Plastics Exposed to Indoor Office Environments
D6954-18	Standard Guide for Exposing and Testing Plastics that Degrade in the Environment by a Combination of Oxidation and Biodegradation
C1147-14	Standard Practice for Determining the Short Term Tensile Weld Strength of Chemical-Resistant Thermoplastics
D2103-15	Standard Specification for Polyethylene Film and Sheeting
D2463-15	Standard Test Method for Drop Impact Resistance of Blow-Molded Thermoplastic Containers
D3420-14	Standard Test Method for Pendulum Impact Resistance of Plastic Film
D5047-17	Standard Specification for Polyethylene Terephthalate Film and Sheeting
D6988-13	Standard Guide for Determination of Thickness of Plastic Film Test Specimens
D256-10 (2018)	Standard Test Methods for Determining the Izod Pendulum Impact Resistance of Plastics
D638-14	Standard Test Method for Tensile Properties of Plastics
D732-17	Standard Test Method for Shear Strength of Plastics by Punch Tool
D785-08 (2015)	Standard Test Method for Rockwell Hardness of Plastics and Electrical Insulating Materials
D1044-13	Standard Test Method for Resistance of Transparent Plastics to Surface Abrasion
D2990-17	Standard Test Methods for Tensile, Compressive, and Flexural Creep and Creep-Rupture of Plastics
D5023-15	Standard Test Method for Plastics: Dynamic Mechanical Properties: In Flexure (Three-Point Bending)

D5024-15	Standard Test Method for Plastics: Dynamic Mechanical Properties: In Compression
D5026-15	Standard Test Method for Plastics: Dynamic Mechanical Properties: In Tension
D5592-94 (2018)	Standard Guide for Material Properties Needed in Engineering Design Using Plastics
D6068-10 (2018)	Standard Test Method for Determining J-R Curves of Plastic Materials
D6110-18	Standard Test Method for Determining the Charpy Impact Resistance of Notched Specimens of Plastics
D7027-13	Standard Test Method for Evaluation of Scratch Resistance of Polymeric Coatings and Plastics Using an Instrumented Scratch Machine
D7774-17	Standard Test Method for Flexural Fatigue Properties of Plastics
D7791-17	Standard Test Method for Uniaxial Fatigue Properties of Plastics
D542-14	Standard Test Method for Index of Refraction of Transparent Organic Plastics
D1746-15	Standard Test Method for Transparency of Plastic Sheeting
D6290-18	Standard Test Method for Color Determination of Plastic Pellets
D4495-16	Standard Test Method for Impact Resistance of poly (Viny1 Chloride) (PVC) Rigid Profiles by Means of a Falling Weight
D4803-18	Standard Test Method for Predicting Heat Buildup in PVC Building Products
D6108-19	Standard Test Method for Compressive Properties of Plastic Lumber and Shapes
D5577-19	Standard Guide for Techniques to Separate and Identify Contaminants in Recycled Plastics
D6288-17	Standard Practice for Separation and Washing of Recycled Plastics Prior to Testing
D7611/D7611M-19	Standard Practice for Coding Plastic Manufactured Articles for Resin Identification
D2143-15	Standard Test Method for Cyclic Pressure Strength of Reinforced, Thermosetting Plastic Pipe
D2925-14	Standard Test Method for Beam Deflection of "Fiberglass" (Glass-Fiber-Reinforced Thermosetting Resin) Pipe Under Full Bore Flow
D6783-05a (2017)	Standard Specification for Polymer Concrete Pipe
D7745-11	Standard Practice for Testing Pultruded Composites
D6289-13 (2019)	Standard Test Method for Measuring Shrinkage from Mold Dimensions of Molded Thermosetting Plastics
D883-19a	Standard Terminology Relating to Plastics
D1600-18	Standard Terminology for Abbreviated Terms Relating to Plastics
D1525-17e1	Standard Test Method for Vicat Softening Temperature of Plastics
D1929-16	Standard Test Method for Determining Ignition Temperature of Plastics
D4804-19	Standard Test Method for Determining the Flammability Characteristics of Nonrigid Solid Plastics
D7309-13	Standard Test Method for Determining Flammability Characteristics of Plastics and Other Solid Materials Using Microscale Combustion Calorimetry

I-2 BIS : PLASTICS STANDARDS
(Selected Few Relevant Standards)

IS 2267 (pdf)	1995	Polystyrene Moulding and Extrusion Materials
IS 2267 (txt)		
IS 2508 (pdf)	1984	Low Density Polyethylene Films
IS 2508 (txt)		
IS 2543 (pdf)	1995	Cellulose Secondary Acetate materials for Moulding and Extrusion
IS 2543 (txt)		
IS 2561 (pdf)	1963	Rubber-based Adhesives for the Automobile Industry
IS 2562 (pdf)	1979	Rubber-based Adhesives for Tyres and Tubes, Curing
IS 2562 (txt)		
IS 3389 (pdf)	1994	Urea-formaldenhyde Moulding Materials
IS 3395 (pdf)	1997	Low Density Polyethylene (LDPE) and Linear Low Density Polyethylene (LLDPE) Materials for Moulding and Extrusions
IS 3395 (txt)		
IS 3669 (pdf)	1994	Melamine-Formaldehyde Moulding Materials
IS 3693 (pdf)	1966	Plastic Spectacle frames
IS 3693 (txt)		
IS 3730 (pdf)	1988	High Density Polyethylene Bucket
IS 3730 (txt)		
IS 5210 (pdf)	1995	High Impact Polystyrene Sheets
IS 6307 (pdf)	1985	Rigid PVC Sheets
IS 7328 (pdf)	1992	High Density Polyethylene Materials for Moulding and Extrusion
IS 9220 (pdf)	1979	Tableware Made of Melamine Plastics
IS 9220 (txt)		
IS 9848 (pdf)	1987	Briefcases, Plastic Body
IS 9848 (txt)		
IS 10889 (pdf)	2004	High density polyethylene films
IS 10889 (txt)		
IS 10951 (pdf)	2002	Polypropylene Materials for Moulding and Extrusion
IS 10951 (txt)		
IS 11273 (pdf)	1992	Woven roving fabrics of 'E' glass fibre
IS 11273 (txt)		
IS 11320 (pdf)	1997	Glass fibre rovings for the reinforcement of polyester and of epoxide resin system
IS 13463 (pdf)	1992	Polyamide (nylon 6) material for moulding and extrusion
IS 13463 (txt)		
IS 13464 (pdf)	1992	Polyamide (nylon 66) material for moulding and extrusion
IS 14425 (pdf)	1997	Guide for manufacture of glass reinforced polyester resin mouldings
IS 14425 (txt)		
IS 14434 (pdf)	1998	Polycarbonate Moulding and Extrusion Materials
IS 14434 (txt)		

IS 14443 (pdf)	1997	Polycarbonate sheets
IS 14443 (txt)		
IS 14500 (pdf)	1998	Linear low density polyethylene (LLDPE) films
IS 14500 (txt)		
IS 14534 (pdf)	1998	Guidelines for recycling of plastics
IS 14534 (txt)		
IS 14635 (pdf)	1999	Polytetrafluoroethylene (PTFE) Materials for Moulding and Extrusion
IS 14635 (txt)		
IS 14753 (pdf)	1999	Polymethyl Methacrylate (PMMA) (Acrylic) Sheets
IS 14753 (txt)		
IS 14995 (pdf)	2001	Stretch Cling Films
IS 14995 (txt)		
IS 15225 (pdf)	2002	Chlorinated Polyvinyl Chloride Compounds Used for Pipes and
IS 15225 (txt)		Fittings
IS 15226 (pdf)	2002	Rigid Polyvinyl Chloride (PVC) Compounds
IS 15226 (txt)		
IS 15621 (pdf)	2005	Terabromobisphenol-A
IS 15621 (txt)		
IS 16864 2018	2018	Polycarbonate Protective Shield Specification
IS 17077 (Part I)	2019	Plastics Acrylonitrile-Butadiene-Styrene ABS
IS/ISO 17088 (pdf)	2008	Specifications for compostable plastics
IS/ISO 17088 (txt)		
ISO 19062-1	2015	Moulding and Extrusion Materials Part-1 Designation System and Basis for Specifications

APPENDIX : J

PLASTICS INSTITUTES IN INDIA

POLYMER SCIENCE/ENGINEERING/TECHNOLOGY

S.No.	Institute/University	Place	State	Course
(A) DEGREE INSTITUTES/UNIVERSITIES				
1.	Adikavi Nannaya University	Rajah Rajah Narendhra Nagar Rajamundhry	Andhra Pradesh	M.Sc. (Polymer Science)
2.	Panjab University (PU)	Sector-14, Chandigarh (U.T)	Chandigarh (U.T)	M.Tech. (Polymer)
3.	Bhaskaracharya College of Applied Sciences	Sector-2, Dwarka, New Delhi-10075	Delhi	B.Sc. (Hons) (Polymer Science)
4.	Indian Institute of Technology (IIT), Department of Materials Science & Engineering	Hauzkhas, New Delhi-110016	Delhi	B.Tech., M.Tech., Ph.D (Polymer Technology)
5.	University of Delhi, (DU)	University Road, Delhi-110007	Delhi	B.Sc. (Hons) (Polymer Science), B.Tech. (Polymer Science & Chemical Technology), Ph.D. (Polymer Science & Technology)
6.	Delhi Technological University (DTU)	Shahbad Daulatpur, Bawana Road	Delhi	B.Tech (Polymer Science and Chemical Technology) M.Tech. (Polymer Technology)
7.	Institute of Science & Technology for Advanced Study & Research (ISTAR)	Vallabh Vidyanagar	Gujarat	M.Sc. (Polymer Science)
8.	Birla Institute of Technology (BIT)	Mesra, Ranchi	Jharkhand	M.E. (Polymer Technology)
9.	JSS Science and Technology University	Mysore	Karnataka	B.Tech. (Polymer Science and Chemical Technology)
10.	Jawaharlal Nehru Centre for Advanced Scientific Research (JNCASR)	Jakkur, Bangaluru	Karantaka	B.Tech. (Polymer Engineering)
11.	Visvesvaray Technological University, (VTU)	Jnana Sangma, Machhe, Belagavi, Belgaum	Karnataka	B.E. (Polymer Engineering), B.Tech. (Polymer Technology)
12.	Cochin University of Science & Technology (CUSAT) Cochin Univ. P.O.	Kochi	Kerala	B.Tech. (Polymer Technology), Ph.D. (Polymers)

13.	Mahatma Gandhi University	Priyadarshani Hills, Kottayam	Kerala	B.Tech. (Polymer Engineering)
14.	Dr. Baba Saheb Ambedkar Marathwada University (BAMU)	Jaisinghpura, Aurangabad	Maharashtra	B.Tech (Polymer Technology)
15.	MIT-World Peace University (MIT-WPU)	Kothrud, Pune	Maharashtra	B.Tech. (Polymer Engineering), M.Tech (Polymer Science & Technology)
16.	North Maharashtra University (NMU)	Umavi Nagar, Jalgaon	Maharashtra	M.Tech. (Polymer Technology)
17.	University of Mumbai	M.G.Road, Fort, Mumbai	Maharashtra	M.Tech. (Polymer Technology), Ph.D. (Polymer Science & Technology)
18.	Institute of Chemical Technology., Opp. Khalsa College	Matunga, Mumbai	Maharashtra	B.Tech. (Plastic and Polymer Engg.), M.Tech. (Polymer Technology)
19.	Savitribai Phule Pune University, (SPPU)	Ganeshkhind, Pune	Maharashtra	B.E. (Polymer Engineering), M.E. (Polymer Technology), M.Sc. (Polymer Science)
20.	Sant Longowal Institute of Engineering & Technology (SLIET)	Longowal, Distt. Sangrur	Punjab	B.Tech. (Polymer Science & Chemical Technology), B.E. (Polymer Science & Chemical Technology)
21.	Mohanlal Sukhadia University	Udaipur	Rajasthan	M.Sc. (Polymer Science)
22.	Anna University of Technology	Palayamkottai Tirunelveli	Tamil Nadu	B.E. (Polymer Engineering), M.Tech (Polymer Technology)
23.	B.S. Abdur Rahman Crescent Engineering College	Seethakathi Estate, GST Road, Vandalur, Chennai	Tamil Nadu	B.Tech. (Polymer Engineering)
24.	Kamraj College of Engineering and Technology (KCET)	Virudhunagar, Vellakulam	Tamil Nadu	B.E. (Polymer Engineering), M.Tech. (Polymer Science)
25.	Hindustan Institute of Technology and Science (HITS)	Padur, Chennai	Tamil Nadu	B.Tech. (Polymer Engineering), Ph.D. (Polymer Science & Technology)
26.	Indian Institute of Technology (IIT)	Roorkee	Uttrakhand	B.Tech. (Polymer Science & Chemical Technology)
27.	University of Calcutta	College Street, Kolkata	West Bengal	B.Tech. (Polymer Engineering), M.Tech. (Polymer Science & Technology)
28.	Central Institute of Plastic Engineering & Technology	Vijyawada Guwahati Hajipur	Andhra Pradesh Assam Bihar	B.Tech. (Plastic Technology), M.Tech. (Plastic Engg) M.Tech. (Polymer Nano Technology), M.Sc. (Polymer Science)

		Raipur Bhopal, Gwalior Bhubneshwar, Balasore Amritsar Jaipur Hyderabad Lucknow	Chhatisgarh Madhya Pradesh Odisha Punjab Rajasthan Telengana Uttra Pradesh	
(B) POLYMER ENGINEERING DIPLOMA INSTITUTES				
1.	Guru Nanak Dev Institute of Technology (GNDIT)	Sector-15, Institutional Area, Rohini	Delhi	Diploma in Polymer Technology
2.	Govt. Polytechnic College	Nattakam, Kottayam	Kerala	Diploma in Polymer Technology
3.	Karnataka Govt. Polytechnic	Kadri Hills, Mangalore	Karnataka	Diploma in Polymer Technology
4.	Mar Gregorius College of Engineering & Technology	Near Press Club, Kottayam	Kerala	Diploma in Polymer Engineering
5.	Central Institute of Plastic Engineering & Technology, (CIPET)	Vijyawada Guwahati Hajipur Raipur New Delhi Ahmedabad, Valsad Murthal, Baddi, Gurgaon Ranchi Bengaluru, Mysore Kochi Bhopal, Gwalior Aurangabad Chandarpur Imphal Bhubneshwar, Balasore Amritsar Jaipur Chennai Madurai Hyderabad Agartala Dehradun Lucknow Haldia	Andhra Pradesh Assam Bihar Chhatisgarh Delhi Gujarat Haryana Jharkhand Karnataka Kerala Madhya Pradesh Maharashtra Manipur Odisha Punjab Rajasthan Tamil Nadu Telengana Tripura Uttrakhand Uttra Pradesh West Bengal	Diploma in Polymer Engineering

BIBLIOGRAPHY

1. Akira Kobayashi: Machining of Plastics, Mc Graw Hill Book Co., New York, 1967.
2. Amstead, B.H., Phillips F. Ostwald and Begeman, Myron L.: Manufacturing Processes, John Wiley and Sons, New York, U.S.A. 1969.
3. Anon: Identification of Polymer, 2019.
4. Arthur W. Birley and Martyn J. Scott: Chapman and Hall, New York, 1982.
5. A. T. Mc Pherson and Alexander Klemin: Engineering Uses of Rubber; Reinhold Publishing Cropn., New York, 1956.
6. Bureau of Indian Standards: PCD12 Plastics, 2019.
7. G. Nikolaev and N. Olshansky: Advanced Welding Process MIR Publishers, MOSCOW, 1977.
8. G. R. Moore and D. E. Kline: Properties and Processing of Polymers for Engineers; Prentice–Hall, Englewood Cliffs, N. J., USA, 1984.
9. Jean-Michel Charrier: Polymeric Materials and Processing; Hanser Publishers, New York, 1991.
10. J. M. Margolis, Editor: Engineering Thermoplastics; M. Dekker, New York, USA, 1985.
11. K. J. Saunders: Organic Polymer Chemistry; Chapman and Hall, London, 1988.
12. Lawrence E. Doyle: Manufacturing Processes and Materials for Engineers, Prentice–Hall Inc., New Jersey, 1969.
13. M.H.A. Kempster: Materials for Engineers, Hodder and Stoughton, Kent, U.K:1984
14. N. G. McCrum, C. P. Buckley and C. B. Bucknall: Principals of Polymer Engineering; Oxford University Press, Oxford, U. K., 1988.
15. P. C. Powell: Engineering with Polymers; Chapman and Hall-Methuen, New York, USA, 1983.
16. Perkin Elmer, Inc., Polymer Identification Using Mid Infra-red Spectroscopy, Waltham, MA, 2011.
17. QLP Quality logo Products: Types of Plastics and their Recycling Codes, 2019.
18. R. J. Crawford: Plastics and Rubbers Engineering Design and Applications; Pergamon Press, Oxford, U.K., 1985.
19. R. J. Crawford: Plastics Engineering; Pergamon Press, Oxford, 1981.
20. R. L. E. Brown: Design and Manufacture of Plastic Parts; Wiley-Inter science, New York, 1980.
21. Ronalds D. Beck: Plastic Product Design; Van Nostrand Reinhold Company, New York, 1970
22. R. T. Fenner: Principles of Polymer Processing; Chem. Publ. Co., New York, USA, 1980.
23. S. S. Schwartz and S. H. Goodman: Plastic Materials and Processes; Van Nostrand-Rein hold, New York, 1982.
24. T. A. Richardson: Industrial Plastics: Theory and Applications; Bobbs-Merrill, Indianapolis, USA, 1983.
25. Victor James Noik: FTIR Peaks: Absorbance or % Transmittance? Univ. of Malaysia, Sabah, 2014.
26. Welding Institute: How to .... Identify Unknown Plastics? Cambridge, U.K., 2019.
27. W. J. Paton: Modern Manufacturing: Prentice–Hall Inc., New Jersey, U.S.A., 1970.
28. Z. H. Stachurski: Engineering Science of Polymer Materials; Royal Australian Chemical Institute, Victoria, Australia, 1987.

INDEX

R

S

PLASTICS & PLASTICS TECHNOLOGY

R.S. PARMAR

FEEDBACK FORM

1. Which topics from your examination point of view are inadequately or not covered in this book?

___

___

2. Is there any misprints/ mistakes/ factual inaccuracy in this book? If so, please specify.

___

___

3. What is your assessment of this book as regards the presentation of the subject-matter, expression, precision and price in relation to the other recommended books available on this subject?

___

___

4. Which competing books you regard as better than this? Please mention their author and publishers.

___

___

5. Any further suggestions/comments you would like to make for the improvement of the book?

___

___

PS: If needed, please attach a separate sheet.

Name : ______

College : ______

Address : ______

Email : ______

Mobile : ______

Please mail the feedback form at:

KHANNA BOOK PUBLISHING CO. (P) LTD.

4C/4344, Ansari Road, Daryaganj, New Delhi-110002

You can also email this feedback form at contact@khannabooks.com or Whatsapp at +91 – 99109 09320

BEST FEEDBACK WILL BE REWARDED SUITABLY